American Agriculturist Handbook for 1909

by Orange Judd Company

with an introduction by Jackson Chambers

This work contains material that was originally published in 1909.

Introduction

I am pleased to present yet another title on Agriculture.

The work is in the Public Domain and is re-printed here in accordance with Federal Laws.

As with all reprinted books of this age that are intended to perfectly reproduce the original edition, considerable pains and effort had to be undertaken to correct fading and sometimes outright damage to existing proofs of this title. At times, this task is quite monumental, requiring an almost total "rebuilding" of some pages from digital proofs of multiple copies. Despite this, imperfections still sometimes exist in the final proof and may detract from the visual appearance of the text.

I hope you enjoy reading this book as much as I enjoyed making it available to readers again.

Jackson Chambers

Kellerstrass Farm
Arthur Oscar Schilling
1907

1

FOREWORD

WE present to our friends a new idea in reference books. This American Agriculturist Hand Book for 1909 makes no pretense of covering the whole field of human knowledge. We do claim that it contains more reliable information upon the subjects that it is most important for the farmer to know about than can be found together elsewhere. It contains the largest directory of national and state officers and officers of agricultural and kindred societies ever gathered in one volume. There is valuable information for the women and young people of the household; there are portraits and instructive pictures, and a wealth of other information useful for live people in every walk of life.

While we have compiled much matter of general interest, so that the Hand Book will be useful to the business and professional man of village and city, we have made this book for the farmer and his good wife. Our purpose has been to get into one book an attractive resume of all the most important facts pertaining to soils, crops, farm animals, marketing, farm mechanics, domestic economy and the other things that will interest the progressive farmer in 1909.

Study of this book will doubtless stimulate your desire to know more about soils, more about live stock, more about horticulture, more about many other specialties here treated briefly. The broader knowledge will be found in the always authoritative books published by the Orange Judd Company, a partial list of which appears in the latter part of the Hand Book. Make this Hand Book a daily companion. Study it. When a question arises, refer to it for the answer. Let it supplement the habitual reading of the seasonable, progressive and newsy American Agriculturist weeklies.

Here is a great combination that spells success: The American Agriculturist weeklies, for agricultural and world news, for discussion and advice, for efficiency and happiness in the household; a carefully selected farm library, for thorough study of agricultural and kindred subjects; and always at hand the American Agriculturist Hand Book, for ready reference.

Do well your work, whatever other men may do. Be not content to do things as they have been done, unless that Way is surely Best. Avoid the Ruts the Average Men have made. Be keen to find the Way that's Best. That is the Way to Learn, the Way in which to Work and Live. Study the Wisdom of the men who know best How to Do the things you want to Do. Get Now the Latest Word from High Authority. Strive for Excellence. Be Kind to all, especially to those within your home and those who have least sunshine in their lives. Forgetting not to Sympathize with those in sorrow and in sin, Live in Life's Joys; be Happy in the Wealth of Nature and of Mind and Soul that is and may be yours—not yours to Keep, but yours to Share. Avoid the Waste of time misspent and pleasures that cost more than they are worth. Seek always all that's Best and make your little sphere a better one than it would be had you not lived.

Commercial Side of Agriculture—
Facts for Business-like Farmers

Agriculturists of the United States, who Conduct the Nation's Greatest Industry, Should Be Equipped for Driving a Bargain —How to Use this Hand Book

"You can lead a horse to water, but you can't make him drink," is an old saw, homely yet full of suggestiveness. That is exactly what ails a lot of farmer folk, as well as sober and sane people in every walk of life. Half the trouble with a good many farmers in conducting their business is the fact that they are not business men. This is an Irish bull, but no matter. Now the American Agriculturist Hand Book for 1909 is like the well-loaded table on a New Year's or Thanksgiving day or the well-filled watering trough to which allusion has been made. A mighty good bill of fare is presented; but it is directly up to you, Mr Farmer, to appropriate some of these good things. You will recall the homely expression of the good farmer's wife who prepared a splendid dinner for the company and then simply announced at the beginning of the meal: "Now jes' reach to and help yourselves."

Things Worth Knowing

So it is in this Hand Book, which sets before every farmer and every farmer's boy who turns over its pages, a big array of things worth while. It is up to you. If you listen to the smooth argument of that onion buyer from the city who next August will try to get your crop for 30 cents a bushel, and tell you that onions have never sold higher than that, you being a beginner in the business, it is up to you to possess yourself of the facts, fling them in his teeth and toss him out in the road. If John Henderson, living in the next town, and buying hogs for Swarmour of Porkopolis, tells you in April that the market is always "off" in May and June, urging you to sell to him at a ridiculously low price, you have a chance to give him the "G B," because you have first fortified yourself with something on the hog movement, and the course of prices one year with another, and probabilities as you glean this opinion from these pages. Don't be a clam. Use your eyes, keep posted on what is going on in the world's crops and markets, and thus "get next."

The commercial agriculture department of the American Agriculturist Hand Book for 1909 presents valuable information of a general and also a specific character. This relates to the world's crops and markets, and is worth while for better understanding the crop and market conditions, not only at home, but the world around. In no other publication is there so simple and helpful an array of figures relating to the crops grown in the United States, together with hints about the world's requirements.

Many Helpful Tables

Helpful tables will be found showing the magnitude and movement of domestic crops and something of the foreign situation. Study them. They will afford direct testimony regarding drafts made upon surplus countries for breadstuffs. A study of the figures relating to the wheat crop will show the world's aggregate output by countries; will accentuate the very prominent part taken by the United States in this, but it will also hint at the increasing competition which our farmers have to face from Argentina and other parts of the world.

Making the Most of the Opportunity

First study the general arrangement of this book and then the direct application which certain parts of it carry to you. It is well to know something of the surplus of a given farm crop in the last year or the last ten years. This may be determined by glancing at some of the tables printed in these pages. If it be a cereal, study the world outlet for the surplus of your own and a million other farmers. If you are obliged to compete in the world's markets (and this is generally true in the case of wheat, corn, oats, etc), you should know something of what you are obliged to face. Widen your horizon. Know the facts in the city distributing markets as well as in your own street, at home and abroad. An illustration of this general statement will be found in the matter of wheat. For example, what are the influences which affect the price of every bushel of wheat grown, whether in Pennsylvania, Texas or North Dakota? Study the situation as outlined in the Hand Book and amplified by figures of world production, movement and price. This will include knowledge of the world's crops in all surplus countries, and the world's requirements in the buying countries. England and Belgium, as an illustration, are always large consumers of foreign breadstuffs, and will buy in the cheapest world's markets, whether it be in the United States, or Russia or Argentina. The figures will afford an adequate idea of the increase in the area under wheat in the United States during a series of years, and also the competition our farmers must face in the increasing attention given to this cereal in Argentina, that crop going to Europe and competing directly with our own. A proper understanding of these facts will afford something of a guide to the progressive farmer in the rotation of his crops one season with another. All such practical questions of the business farmer will be made more intelligible by studying the range of prices by months and years as applied to crop production during such period, together with the export movement. Similar lessons may be learned from scanning the situation relating to other products of the farm.

Methods Followed in Marketing Produce

It would be well to present right here a few facts about the requirements of some of the big city markets in handling perishable produce. Always remember that the ins and outs of marketing fruits, vegetables, etc, must be thoroughly understood in order to secure best results. In catering to the family trade, you must remember that whatever is offered to customers direct from the farmer's wagon should be first-class in every respect, attractive in appearance and sound in quality.

Another method of disposing of the farm produce is to sell to the local retail grocer or market man. This has some advantages, but can be carried on only in a limited way. Not infrequently the local retailer with an established family trade can handle the produce of the small farmer and gardener to mutual advantage. The farmer who produces on a larger scale can sell his grain, live stock, wool, poultry, eggs, fruits, vegetables, etc, direct to a local buyer, who in turn has his established outlet at distant large points of consumption and distribution,

and after accumulating a carload or more from various farmers, ships as market conditions warrant. If a local buyer takes your grain, live hogs, poultry or potatoes, it is reasonable to assume that he is getting a fairly large slice of the middleman's profits. Shipping to customers at distant points, previously secured, perhaps through working up a trade from small beginnings, is another method. In many instances this is successful, especially where the farmer has taken pains to establish a reputation for fine quality goods and honest pack and count.

How to Use Printed Quotations

A word right here about printed quotations. Usually those appearing in the so-called market columns of the daily newspaper are woefully deficient as to accuracy. The reporter on the big city daily is usually smart enough, but is utterly lacking in the sympathetic touch with agriculture. He prepares so-called market reports in a perfunctory way and may be a mile off in a given quotation. It has been well said that a market report which is not accurate is worse than useless—it is absolutely misleading. Another reprehensible market report is such as we sometimes find in the way of a so-called market letter sent out by some commission merchant who purposely withholds the facts. This statement is not an assault upon commission merchants as a class; there are good, bad and indifferent. Fortunately there are a great many good commission merchants, but unfortunately there are some with easy consciences who make a bid for the farmer's business and then are not slow to "do him." The commission merchant, assuming that he is both capable and intelligent, may be a very useful member of the business world. He stands as the representative of the farmer or the producer who is shipping the produce of the farm to the big consuming and distributing centers. The true commission merchant should therefore be the agent of the producer, getting the best returns on the product consigned to his care, taking his legitimate toll in the way of brokerage and making returns to his principal in the country.

Quality Accounts for Price Ranges

It is at times impossible for even this trustworthy commission merchant to give perfect satisfaction. This is notably true with regard to produce which must be handled quickly, and where price changes are liable to be rapid and violent. Country shippers are often misled through failure to fully understand true values and market quotations. While these may and should properly portray the situation when published, conditions may be entirely changed at the time the prospective shipper reads them. Note also quality and attractiveness. The farmer who studies the market report finds peaches quoted at 50 to 75 cents a basket, or dressed fowls 8 to 10 cents a pound, or choice beef steers $6 to $7 a 100 pounds live weight. He at once assumes his property ranks with the best and expects that returns will show outside quotations secured. He fails to realize that his own produce must take its place with an enormous quantity from all sections and that what might appear strictly choice at home would grade far below that when placed by the side of the best consignments on the market on a given day, and naturally he should not expect that his goods will always command outside quotations. Requisite allowance must be made for quality, realizing that even for sound stock there are several prices obtained.

Get Acquainted with the Market

Where possible, the farmer who intends to ship direct to market should first go to that market, get acquainted with commission merchants or other receiving houses, and get posted. Ask your commission merchant to give you references. This is perfectly fair and proper. It is not enough that he should have a sizable bank account and that his checks are always good. It is up to him to show you that he will tote fair in the matter of selling your goods and making returns. As to commission charges, they vary greatly. Uniformity in different markets is quite lacking, although goods which quickly deteriorate in keeping quality are obliged to pay heavier charges than others. Such perishable produce as fresh fruits, vegetables, poultry, butter, eggs, etc, must be handled expeditiously, and must stand greater charges in proportion for selling than grain, live stock, cotton, wool, etc. From the time the goods leave the country shipping station until returns are made and check received, several charges must be borne. On goods sent to a distant town or city, to be sold on commission, the railroad or vessel freight is always charged up, next comes cartage from wharf or depot to salesroom or warehouse. Goods not to

be sold immediately, but held for a time in order to catch a better market, must, as a rule, pay storage and sometimes insurance.

Rates of Commission

Rates of commission are usually 5 to 10% on perishable produce. Where potatoes are sold by the carload at a 5% rate, the commission per car is usually understood to be not less than $10, otherwise 10% on amount of sale. The commission on apples is 5 to 15% in carloads; in some instances a fixed rate of about 15 cents per barrel is made and when jobbed from the car, an additional charge of 3 to 5 cents per barrel for cartage is usually made to country consignor. Eggs pay 35 to 40 cents per case of 30 dozen when sent to cold storage for the season of six months, or 10 cents per case per month for short storage. In warm weather a charge is frequently made of 10 cents per case for candling, work which must necessarily be repeated in the course of a few weeks, unless the eggs are sold in the meantime. During the season of candling, the loss through bad eggs runs from one-half dozen, which would be called first-class, to about two dozen per case of 30 dozen, or occasionally higher. Goods sent to cold storage usually pay a fixed rate per month or for the entire season.

Principal Farm Crops by Years

[In round millions]

Crop of	Cotton, bales	Wheat, bushels	Corn, bushels	Oats, bushels	Rye, bushels	Barley, bushels	Potatoes, bushels	Hay, tons.
1908	12.9	675	2610	757	33	171	241	61
1907	12.5	603	2558	659	33	150	271	52
1906	12.5	776	2963	930	30	148	283	53
1905	11.3	720	2703	1003	31	144	251	58
1904	13.5	554	2574	973	30	144	289	58
1903	10.1	703	2346	823	32	139	255	58
1902	10.7	760	2556	1028	34	138	272	60
1901	10.7	752	1419	700	30	110	183	51
1900	10.4	510	2188	832	24	59	255	52
1899	9.1	565	2207	869	24	73	243	59
1898	11.1	715	1868	799	26	56	204	68
1897	10.0	589	1823	814	—		174	67
1896	8.5	470	2269	714	24	70	245	59
1895	7.2	460	2272	904	27	87	286	48
1894	9.5	460	1213	662	27	61	171	55
1893	7.5	396	1619	639	27	70	183	66
1892	6.7	516	1628	661	*	*	*	*
1891	9.0	612	2060	738	*	*	*	*
1890	8.7	399	1490	524	29	*	*	*
1889	7.5	491	2113	751	28	*	*	*
1888	6.9	416	1987	701	28	64	202	47
1887	7.0	456	1456	660	21	57	134	41
1886	6.3	457	1665	624	24	59	168	42
1885	6.6	357	1936	629	27	58	175	45
1884	5.7	513	1795	584	29	61	191	48
1883	5.7	421	1551	571	28	50	208	47
1880	5.7	499	1717	418	25	45	168	32
1875	4.6	292	1321	354	18	37	167	28
1870	3.0	236	1094	247	15	26	115	25
1865	0.3	149	704	225	20	11	101	24
1860	4.9	173	839	173	21	16	111	19

*No estimate for year indicated by asterisk.

Marketing the Nation's Wheat Crop, Source of Fourth of World's Bread

United States Yearly Finds Competition with Argentina for the European Trade Growing Sharper—Canada as a Producer

Wheat has been the chief cereal crop for so many centuries that it is necessary to go back to the pyramids of Egypt to learn about its beginnings; and then one is no better off. Figures do not mean much when it comes to talking about crop yields. Yet it may interest some schoolboy who reads these pages to know that the world's crop of wheat exceeds 3 billion bushels, of which our own country turns off, in a good year, a quarter of the grand total.

While the Christmas bells are ringing, and while the New Year is being ushered in, the wheat harvest is under way in such parts of the southern hemisphere as Argentina, which, by the way, has grown to an enormous producer. Thenceforward every month hears the whir of the reaper on some portion of the globe, even though our own wheat crops in North America and Canada are pretty well out of the way by late August. Rye pushes wheat in popularity in northern Europe, notably Russia and Germany, as a breadstuff, and of course rice maintains, as for a thousand years, the lead in that direction in eastern and southern Asia.

But, after all, wheat is the dominating factor in the world's markets for cereals and the standard around which has been fought many a battle in the speculative markets.

Canada as a Competitor

The coming giant in wheat culture is northwest Canada. This rather loosely worded phrase, as it appears so frequently in the public prints, really applies to the remarkable stretch of territory from Manitoba running westward to the Pacific ocean and lying directly north of Montana and our own Pacific northwest. How far extends toward the north the area of wheat-growing possibilities? Ask the stars. Certainly when our grandfathers were youngsters and studied geography, all of that great country was regarded as a territory for fur-bearing animals and little else. But in the past few years this newer Canada has been given

SECRETARY JAMES WILSON

The Honorable Secretary of Agriculture of the United States is Secretary Wilson of Iowa. For the past 12 years Secretary Wilson has given the country remarkable service, bringing the great department of agriculture conspicuously forward until it has won the attention and admiration of every farmer throughout the United States.

over to some extent to the plow, and wheat production up there increased from zero to, we will say, 150 million bushels in a favorable season. What of it? Well, our cousins across the line, and for that matter some of our own boys too, are growing untold millions of wheat which must find a market outlet in Europe. Chiefly in England, we will say, because the mother country will favor the youngster. But all in all, it means present and future competition for wheat growers in the United States worth the reckoning.

Our Customers for Breadstuffs

In studying this wheat proposition it should be remembered that the world may be roughly divided into two classes, one having a surplus and the other showing a deficit in this valuable breadstuff, which is to be made up from the more fortunate countries. For all practical purposes it may be said that the surplus countries are limited to a few—the United States, Canada, Argentina, Russia and to some extent Australia. Others cut little

The Wheat Crop of 1908

Winter:	Acres	Per acre	Bushels
N. Y.	450,000	18	8,160,000
Pa.	1,650,000	18	29,700,000
Tex.	1,015,000	11	11,165,000
Ark.	160,000	10	1,600,000
Tenn.	820,000	10	8,200,000
W. Va.	365,000	12	4,380,000
Ky.	755,000	11	8,305,000
O.	2,100,000	15	31,500,000
Mich.	890,000	17	15,130,000
Ind.	2,650,000	16	42,400,000
Ill.	2,309,000	18	41,562,000
Wis.	75,000	20	1,500,000
Minn.	90,000	18	1,620,000
Ia.	70,000	22	1,540,000
Mo.	2,250,000	11	24,750,000
Kan.	6,128,000	13	79,664,000
Neb.	2,175,000	18	39,150,000
Cal.	1,100,000	13	14,300,000
Ore.	375,000	24	9,000,000
Wash.	415,000	25	10,375,000
Okla.	1,300,000	12	15,600,000
Other	3,340,000	12	40,080,000
Total	30,482,000	14.4	439,601,000
Spring:			
N. E.	10,000	18	180,000
Mich.	30,000	17	510,000
Ill.	110,000	18	1,980,000
Wis.	400,000	17	6,800,000
Minn.	5,200,000	13	67,600,000
Ia.	750,000	15	11,250,000
Kans.	100,000	5	500,000
Neb.	350,000	14	4,900,000
N. D.	5,400,000	12	64,800,000
S. D.	3,150,000	13	40,950,000
Cal.	38,000	10	380,000
Ore.	500,000	20	10,000,000
Wash.	1,030,000	14	14,420,000
Other	753,000	15	11,295,000
Total	17,821,000	13.2	235,565,000
	Acres	Per acre	Bushels
Winter	30,482,000	14.4	439,601,000
Spring	17,821,000	13.2	235,565,000
	48,303,000	14.0	675,166,000

figure either way. France, an enormous producer, is occasionally a surplus country, but sometimes shows a slight deficit. The United Kingdom is always hungry

Half Century of Wheat Crops

[In millions of bushels]

1908	675	1891	612
1907	603	1890	399
1906	776	1889	491
1905	720	1888	416
1904	554	1887	456
1903	703	1886	457
1902	760	1885	357
1901	752	1884	513
1900	510	1883	421
1899	565	1882	504
1898	715	1881	383
1897	589	1880	499
1896	470	1877	364
1895	460	1875	292
1894	400	1873	281
1893	396	1870	236
1892	516	1865	149
		1860	173

for foreign wheat, and so of Holland and Belgium. All importing countries require every year upward of 400 million bushels of wheat more than they produce at home.

Thus it is that western Europe is the world's great market place for wheat. And, furthermore, western Europe buys where it can get the goods the cheapest. Very largely such great distributing centers as Liverpool and London go far toward controlling the price of every bushel of wheat grown by the farmer in America.

Last year's crop, that of 1908, was considerably short of a bumper, yet made a good total somewhere around 675 million bushels. In a full year our threshing machines run off a hundred millions more

Wheat Prices at Chicago

[No. 2, Cents per bushel]

	Jan.	May	July
1908	91@103	98@111	84@ 92
1907	71@ 76	79@100	100@107
1906	85@ 90	86@ 95	72@ 85
1905	116@121	87@111	87@105
1904	82@ 94	100@106	95@112
1903	70@ 79	70@ 84	64@ 75
1902	74@ 80	72@ 76	71@ 79
1901	71@ 76	70@ 75	68@ 71
1900	61@ 67	63@ 67	74@ 81
1899	66@ 76	68@ 79	68@ 75
1898	89@110	117@185	65@ 88
1897	71@ 94	68@ 98	68@ 80
1896	55@ 69	57@ 68	54@ 62
1895	48@ 55	60@ 85	61@ 75
1894	59@ 64	53@ 60	50@ 60
1893	72@ 78	68@ 76	54@ 66
1892	84@ 91	80@ 86	76@ 80
1891	87@ 96	99@108	84@ 95
1890	74@ 78	89@100	85@ 94
1889	92@102	77@ 87	76@ 85
1888	75@ 79	80@ 90	79@ 86
1887	77@ 80	80@ 89	67@ 71
1886	77@ 85	72@ 79	73@ 79
1885	76@ 82	86@ 91	85@ 91
1884	88@ 96	85@ 95	79@ 85
1883	93@104	107@114	96@103
1882	125@135	123@129	126@136
1881	95@100	101@113	108@122
1880	114@133	112@119	86@ 97
1879	81@ 87	90@103	88@105
1875	88@ 91	89@107	99@129
1873	119@126	122@134	114@146

According to records kept by the Chicago Trade Bulletin, wheat touched $1.61 in August, 1872; $2.47 in August, 1869; $2.85 in May, 1867, and sold at range of 80 cents to $1.15 in 1863.

than that. The wheat crop is made up, roughly speaking, in a proportion of about four-sevenths autumn-sown and three-sevenths spring. Accompanying tables afford a good idea of wheat, recent crops and prices.

"Protect the American hen," says Uncle Sam, and places a duty of 5 cents a dozen on foreign eggs imported into this country. Years ago, eggs came in free of duty, as many as 16 million dozens in a year. In 1907 the imports of eggs were 231,000 dozen, in 1908, 231,000 dozen.

From 5,000 to 6,000 horses are brought into this country every year. These imports are largely for breeding purposes, and the pure-bred animals of recognized breeds enter free of duty.

King Corn is Ruler Both East and West

About 80 per cent of World's Supply of This Grain Grown in United States

Corn makes cattle and cattle make money. Nothing new in this time-honored phrase, yet each year seems to more fully emphasize the merits of this magnificent cereal. If corn is high, as it has been for many months, it still may be profitably utilized in the surplus states in the making of beef and pork. If corn is low, it is almost sure to be due to the fact of a magnificent crop; this in turn is of the greatest moment. From whatever standpoint, corn is one of the most profitable of all the cereals. No matter what the bulk of the crop is, most of it is used within the borders of the counties where grown. This is also true of our older middle and eastern states and the south, where home requirements always exceed the home supply; it is true in the corn belt, where stock feeding and dairying form the principal business.

Some of these days the corn crop will amount to 3 billion bushels from a round hundred million acres. Neither figure has yet been quite reached, although 3 years ago, owing to a good rate of yield of 31 bushels to the acre, the crop was practically the amount named. Last year, 1908, only a moderate crop was secured, slightly better than 2,600,000,000 bushels. The average was

large enough, as shown in accompanying table, but the weather vicissitudes cut down the rate of yield.

This country turns off perhaps 80% of all grown on the globe. Argentina and Austria-Hungary, both exporting countries, produce fairly liberal crops; considerable amounts are also produced

Prices of No 2 Cash Corn at Chicago

[In cents per bushel]

Year	Jan.	May	July	Sept.
1908	57-60	68-82	70-78	78-82
1907	40-44	49-56	52-55	60-64
1906	41-43	47-50	49-53	47-50
1905	42-43	48-64	54-59	51-55
1904	42-48	47-50	47-50	51-55
1903	44-48	44-46	49-53	45-53
1902	56-64	59-65	56-88	57-62
1901	36-38	43-58	43-58	54-60
1900	30-32	36-40	38-45	39-43
1899	35-38	32-35	31-35	31-35
1898	26-28	33-37	32-35	29-31
1897	21-23	23-26	24-29	27-32
1896	25-28	27-30	24-28	19-22
1895	40-46	46-55	41-47	31-36
1894	34-36	36-39	40-46	48-58
1893	40-45	39-45	35-42	37-43
1892	37-39	40-100	47-52	43-49
1891	47-50	55-70	57-66	48-68
1890	28-30	32-35	33-47	44-50
1889	33-36	33-36	34-37	30-34
1888	47-50	54-60	45-51	40-46
1887	35-38	37-39	34-38	40-44
1886	36-37	34-37	34-45	36-41
1885	34-40	44-49	45-48	40-45
1880	36-41	36-38	33-38	39-41
1875	64-70	60-76	67-77	54-62
1874	49-61	55-66	58-80	66-86

in Italy, Egypt, etc, but the latter do not affect the world's markets. The short seasons in our northern latitudes and in Canada, with cold nights, are against the best growth and maturity of dent corn, and in New England, for example, flint corn takes the lead. Yet the little state of Connecticut produced in 1908 the prize

Exports Principal Farm Crops From the United States

[In round millions]

Year ended June 30	Flour, bbls.	Wheat, bushels	Corn, bushels	Oats, bushels	Rye, bushels	Barley, bushels	Clover seed, lbs.	Timothy seed, lbs.	Cotton, bales	Apples, barrels*	Hay, tons*	Hops, lbs.	Oil cake and meal, lbs.	Tobacco leaf, lbs.
1908	14	100	52	1	2	4	4	26	7	1049	77	23	1691	323
1907	16	77	83	4	†	8	4	19	9	1539	59	17	2064	331
1906	14	35	118	46	1	18	2	11	.7	1209	70	13	1918	302
1905	9	4	89	5	†	11	11	16	8	1500	66	15	1895	328
1904	17	44	56	1	1	11	6	13	6	2018	60	11	1503	305
1903	20	114	75	5	5	8	16	18	7	1656	50	8	1671	357
1902	18	155	27	10	3	9	7	6	7	460	153	11	1633	291
1901	19	132	178	37	2	6	12	8	7	884	89	15	1714	307
1900	19	102	209	41	2	24	32	15	6	526	72	13	1627	345
1899	18	139	174	30	10	2	19	16	7	380	65	21	1567	284
1898	15	148	209	69	16	11	31	10	8	605	82	17	1356	263
1897	15	80	177	35	9	19	13	17	6	1495	62	11	1056	315
1896	15	61	100	13	1	8	6	12	5	360	59	17	798	288
1895	15	74	28	1	—	2	23	5	7	819	47	18	734	294
1894	17	88	65	6	—	5	45	10	5	79	54	17	745	269
1893	17	117	46	2	1	3	8	7	4	408	33	11	802	248
1892	15	157	76	9	12	3	20	10	6	930	35	13	826	241
1891	11	55	31	1	—	1	21	9	6	135	28	9	633	237
1890	12	54	102	14	2	1	27	11	5	545	36	7	712	242
1889	9	46	70	1	—	1	34	10	5	942	22	13	588	214
1888	12	66	24	—	—	1	13	2	5	490	18	8	563	249
1887	12	102	40	—	—	1	8	7	4	592	14	—	622	294

*In thousands.
†Less than a million.

acre of dent corn, according to the awards at the national corn show, held at Omaha in December.

One of the important developments in maize growing in recent years is the rapid spread of the silo and silage corn. This is particularly true of dairy sections, both east and west. The foreign demand for American corn is considerable when the price is low, and that normally means

A Billion Bushel Oats Crop is Bumper Mark

Yield in Recent Years Has Been Under the Normal Possibilities of the Country

One billion bushels of oats may be harvested in the United States under favorable conditions. This is about one-third

Corn Crop of 1908, by States

[Acres and bushels in round thousands]

	Acres	Per acre	Bushels
New York	675	35	23,625
Pennsylvania	1,490	34	50,660
Texas	6,569	21	137,949
Arkansas	2,991	22	65,802
Tennessee	3,287	24	78,888
West Virginia	795	28	22,260
Kentucky	3,150	25	78,750
Ohio	3,710	35	129,850
Michigan	1,409	30	42,270
Indiana	4,728	29	137,112
Illinois	9,596	30	287,880
Wisconsin	1,668	30	50,040
Minnesota	1,554	30	46,620
Iowa	8,961	34	304,674
Missouri	5,919	26	153,894
Kansas	7,462	23	171,626
Nebraska	7,972	29	231,288
North Dakota	62	25	1,550
South Dakota	1,912	29	55,448
California	43	32	1,376
Oregon	21	24	504
Washington	14	25	350
Oklahoma	2,829	28	79,212
Other	20,870	22	459,140
Total	97,687	26.8	2,610,768

The Oats Crop for 1908

	Acres	Per acre	Bushels
New York	1,340,000	31	41,540,000
Pennsylvania	1,237,000	28	34,636,000
Texas	726,000	30	21,780,000
Arkansas	242,000	25	6,050,000
Tennessee	183,000	20	3,660,000
West Virginia	87,000	18	1,566,000
Kentucky	233,000	17	3,961,000
Ohio	1,250,000	24	30,000,000
Michigan	1,010,000	26	26,260,000
Indiana	1,228,000	21	25,788,000
Illinois	4,200,000	20	84,000,000
Wisconsin	2,362,000	32	75,584,000
Minnesota	2,387,000	21	50,127,000
Iowa	4,500,000	25	112,500,000
Missouri	709,000	19	13,471,000
Kansas	1,100,000	21	23,100,000
Nebraska	2,134,000	25	53,350,000
North Dakota	1,250,000	24	30,000,000
South Dakota	904,000	26	23,504,000
California	121,000	29	3,409,000
Oregon	305,000	33	10,065,000
Washington	149,000	35	5,215,000
Oklahoma	420,000	27	11,340,000
Other	2,636,000	25	65,900,000
Total	30,713,000	24.6	756,806,000

when our home crop is large. But the situation is about as broad as it is long; when our crop is medium or small, we need it all at home, and can get along very well without the export trade. Figures covering the corn crops of a good many years are presented here.

of the world's crop. For two years in succession the domestic crop has been poor; prices are high. In fact, they would rule higher but for the very general use of barley to mix with oats for feed purposes. In the production of oats the United States leads all other countries, but Russia is a close second. Next in im-

Imports Grain, Cotton and Other Crops into United States

[In thousands]

Year ended June 30	Flour, bbls.	Wheat, bushels	Corn, bushels	Oats, bushels	Rye, bushels	Barley, bushels	Cotton†	Flax tow, tons	Hemp tow, tons	Hay, tons	Flaxseed, bushels	Raw tobacco, pounds
1908	40	342	20	364	*	199	71	10	6	10	57	32,056
1907	48	375	11	75	*	38	105	9	9	61	90	39,540
1906	45	58	10	23	*	18	71	9	5	69	52	37,355
1905	41	3,102	15	39	20	81	60	8	4	46	296	33,288
1904	47	7	17	171	33	91	49	10	6	114	213	31,162
1903	*	1,077	40	137	*	56	75	8	5	293	129	34,016
1902	*	119	18	25	*	57	99	9	6	48	477	29,429
1901	*	600	5	21	*	171	47	7	4	143	1,632	26,851
1900	*	317	2	41	*	190	67	7	3	144	67	20,619
1899	*	1,871	4	12	*	110	50	6	4	20	82	14,035
1898	2	2,047	3	9	33	125	53	6	4	4	138	10,477
1897	2	1,534	2	46	*	1,272	52	9	5	120	105	13,805
1896	1	2,110	4	48	*	837	55	8	8	303	755	32,883
1895	2	1,430	17	308	13	2,117	49	7	7	202	4,166	26,668
1894	*	1,181	2	8	*	791	28	5	2	87	593	19,663
1893	*	966	2	21	9	1,970	43	7	5	104	112	28,110
1892	*	2,460	15	20	84	3,146	29	8	5	80	285	21,989
1891	8	546	2	10	141	5,079	21	6	11	58	1,516	23,061
1890	1	157	2	21	198	11,333	8	8	37	125	2,391	28,721
1889	1	131	2	22	*	11,368	8	8	56	105	3,253	20,107
1888	3	583	37	68	*	10,831	5	6	48	100	1,534	18,600

*Less than 1,000.
†In millions of pounds.

portance follow Germany, France, Austria, England and Canada. The crop is little grown in the southern hemisphere. On our side of the equator it does well in far northern latitudes.

In accompanying tables are valuable figures showing the oats crop of 1908 by states, prices covering a long series of years, the foreign trade, etc. Iowa leads in the production of oats, followed by Illinois, Wisconsin, Nebraska, New York

Price of No 2 Cash Oats at Chicago

[In cents per bushel]

Year	Jan.	May	July	Sept.
1908	48–52	53–57	51–61	48–51
1907	33–37	44–49	41–46	51–57
1906	29–32	32–35	30–39	30–34
1905	29–31	29–32	27–35	25–30
1904	36–42	39–45	38–45	29–34
1903	31–34	33–38	33–45	35–38
1902	38–46	41–49	30–56	26–27
1901	23–24	28–31	27–39	33–36
1900	22–23	21–24	21–24	21–22
1899	26–28	24–28	20–25	21–23
1898	21–24	26–32	21–26	20–22
1897	16–17	17–19	17–18	19–21
1896	17–19	18–20	15–19	14–17
1895	27–29	25–31	22–25	18–21
1894	26–29	32–36	29–41	27–31
1893	30–32	29–32	22–30	23–29
1892	28–30	28–34	30–34	31–34
1891	41–44	45–54	27–45	26–30
1890	20–21	24–30	27–35	44–51
1889	24–25	21–24	22–23	19–20
1888	30–32	32–38	28–33	23–25
1885	25–29	31–36	26–33	24–27
1880	32–36	29–34	23–26	27–35
1875	52–53	57–65	48–56	34–40
1873	24–26	30–34	27–30	26–31
1869	45–50	56–63	57–71	42–46

and Ohio in about the order named. The area under oats does not change very much from year to year, being close to 30 million acres. In recent seasons, with the crop somewhat indifferent, with prices high and with an enormous demand, it has been urged that farmers generally should go more largely into the production of the oats crop. Good years show 36 to 40 bushels to the acre in favored states, but the average rate is much less than that.

Rye and Barley, Two of Nation's Interesting Minor Cereal Crops

Among the minor cereals, rye and barley have long held a prominent position in a few states. They are not popular crops when compared with oats or corn, yet the acreage, particularly in barley, is somewhat larger than a decade of years ago. Eastern Europe is the country above all others where barley takes precedence; notably Russia, which ships heavily of its surplus to Germany, where this cereal is also very liberally grown. The crop is popular in Scandinavia. France and the United Kingdom. In our own country. Minneapolis, Wisconsin, Iowa and the Dakotas are heavy producers, and

California turns off a liberal quantity, some of the latter being exported. The domestic crop of barley in 1908 was one of the largest ever grown, aggregating 170 million bushels, from a liberal acreage and a fairly good rate of yield of about 26 bushels to the acre. Throughout all of last year low grades of barley sold at very good figures, the grain being taken to mix with oats for feed purposes.

Rye at a Standstill

Rye has made no gain in the favor of American farmers. The area given over to this crop is nearly stationary around 2 million acres, this yielding each year 30 to 35 million bushels of rye. States of leading production are Pennsylvania, Michigan, Wisconsin and New York in about the order named. Bread eaters in America have little use for rye. On the other hand, it is consumed in enormous quantities in eastern Europe, particularly Russia and Germany. Russia produces about half the rye grown in the world, and Germany, Austria and France are all important producers. Our own requirements are small, some rye going to distillers, some to millers and some is used on the farm for feeding purposes.

Rice Crop of the United States Is Valued at 17 Million Dollars

Cultivation of rice in the United States continues on a large scale, but the yield is still insufficient to meet home demands. Introduced in this country in 1647, its development into a great industry is within the memory of some of the present growers. The crop now forms one of the important items so far as value is concerned, in the great column of assets of the American farmer. While Georgia and the Carolinas have in years past grown large quantities of rice, the somewhat warmer climate of the bottom lands of Louisiana and Texas, particularly that swampy region along the gulf coast from Algiers. La, to Beaumont, Tex. has proved especially adapted to this grain.

The growth of rice culture is probably best told in figures of crop production. In 1906 the crop amounted to 17,854,000 bushels. grown on 575,000 acres. Last year the crop was 21,890,000 bushels, valued at about $17,000,000 and grown on 627,300 acres. Louisiana is the greatest producer of rice of any of the states. its crop last year amounting to 11,826,000 bushels. Texas came second with 9,741,-000 bushels. In South Carolina, while the industry has declined, 491,000 bushels were harvested, and in Arkansas, where rice planting is increasing, the yield was 406,000 bushels.

In spite of the increased production, imports increase. In 1908 the imports of whole rice, rice flour, rice meal and broken rice were 212,803,392 pounds, compared with 209,603,180 pounds in 1907, and 166,547,957 pounds in 1906. Before it is milled, rice weighs 45 pounds to the bushel.

Flax, Valuable for Its Seed, Has Possibilities not yet Developed

Cultivation of flax, one of the pioneer crops of the country, shows no important change in recent years, so far as production is concerned. The crop of 1908 was estimated at 25,717,000 bushels. Gradually the cultivation of this crop has worked westward until it seems to have taken a rather firm stand in Wisconsin, Minnesota, Iowa, Missouri, North and South Dakota, Montana, Nebraska and Kansas. Oklahoma grows a little, while the banner state of the Union for flaxseed in 1908 was North Dakota, with South Dakota a close second.

While the seed is the money producing item of the flax plant, great possibilities lie in the production of flax fiber, when manufacturers shall solve the problem of converting it into as fine linen as is manufactured abroad. Development of machinery is probably necessary to accomplish this. An outlet for the tow is found in the manufacture of stuffed furniture.

Commercial Side of Grass Seed Told in Tables of Quotations

Grass seed crops form an important item in the commerce of a good-sized group of states in the middle west, or, to be more specific, in Indiana, Michigan and Ohio. While the products of these states are consigned chiefly to Chicago and Toledo, the distribution is country-wide, and large quantities are also exported.

Timothy, which at one time was the chief grass seed of the country, continues to be a standard, but nevertheless has rivals which are multiplying. Among these might be mentioned alfalfa, which was imported from Europe and found such favor that now it is necessary to import seed to meet the demands of this country.

Along with timothy, though not so important, ranks clover and German millet; Hungarian might also be mentioned. The trend of the grass seed industry may be observed by a perusal of the accompanying tables, which set forth a comparison of prices this year and previous years:

Clover Prices Compared

Quotations for clover, prime quality per 100 pounds in Chicago, quarterly last year, with comparison with previous years, follow:

Year	Jan. 1.	Mar. 1.	Oct. 1.
1908	$16.75	$19.50	$9.00
1907	13.75	13.00	12.75
1906	13.25	13.00	12.75
1905	12.75	12.60	12.75
1904	11.00	10.90	11.75
1899	7.00	6.10	8.50
1894	10.75	8.85	8.50
1890	5.66	5.33	6.83

Cost of Timothy Grass Seed

The range of prices for timothy in Chicago are compared in the following table, quotations being for prime seed per 100 pounds:

Year	Jan. 1.	Mar. 1.	Oct. 1
1908	$4.40	$4.85	$3.50
1907	4.35	4.60	4.30
1906	2.90	2.70	3.85
1905	2.75	2.95	3.35
1904	2.90	3.15	2.80
1899	2.25	2.40	2.35
1894	4.35	4.10	5.20
1890	2.64	2.71	2.95

Bulk of the World's Peppermint Supply Grown in United States

Peppermint is raised in England, Japan, Germany and some other countries on a small scale, but its most extensive culture is in the United States. Michigan, northern Indiana and Wayne county, N Y, are the most important peppermint raising sections. Japan is now important, also.

In Michigan the peppermint is raised on black muck land. Most of the fields are drained swamp lands and marshes. After fall plowing the land, it is harrowed in early spring and provided with furrows about 3 feet apart in which the slender roots are planted.

In August or early September the crop is in full bloom and is then mowed with a scythe, dried until only enough moisture remains to prevent falling of the leaves and hauled to the distillery. After the oil is removed the hay is used as fodder for stock or allowed to rot for fertilizer. Peppermint oil has sold as low as 75 cents a pound, and as high as $4 or better. In recent years it has been around $1.75 to $2. An acre with a really good crop of mint should yield 40 to 50 pounds of oil.

Milk Making and Marketing—Many Problems in Supplying the Big Cities

Dairy Industry Offers the Greatest Opportunity of All the Agricultural Pursuits for Improvement Along the Line of Co-operative Effort—Statistics Worth a Dairyman's Time

In no branch of agriculture is co-operative effort more needed than in the dairy industry. Especially is this true in the matter of market milk. Changed conditions and improved methods of transportation whereby city dealers can bring milk from distant territory have so widened the sources of milk supplies that prices to the farmer are not commensurate with the cost of production. The most certain relief for these milk producers is co-operative effort, and where properly applied it has proved of great assistance in bringing living prices for milk.

The work of the Farmers' union in New England, now known as the Boston co-operative milk producers' company, is well known. For a dozen years this association has been hammering away at the middlemen and insisting upon a square deal. So successful has the company been that the price of milk has advanced from 32 cents per 8½-quart can in 1898 to 46 cents in 1908. This is for milk delivered in Boston. Massachusetts, New Hampshire, Vermont, Connecticut and Maine all contribute to the Boston milk supply, and the farmers organized their association by electing one to three directors from each state. These directors hire a manager or clerk and meet Boston dealers twice a year to negotiate prices and virtually sell the milk for the farmers. While there are many things the company has not yet accomplished, there can be no question about the great good it has done.

Other sections, such as Rhode Island, Pennsylvania, New York and Chicago, have producers' associations more or less efficient. Hundreds of letters reach editors of Orange Judd company each year, asking about how milk producers can start an association. This indicates that farmers are alive to the possibilities of associated effort and perhaps a little data on starting such an association will prove of great aid in this connection. One very good thing about co-operation is that results can be accomplished whether there be a half-dozen farmers in a certain neighborhood, or 5,000 in a larger section. The underlying principles are substantially the same.

Milk Prices at New York

[In cents per quart]

Year	April	June	Aug.	Oct.	Dec.
1908	3¼	2½	3	3¾	4
1907	3½	2½	3½	4	4
1906	2⅞	2¼	2⅞	3½	3¾
1905	2¾	2	2½	3	3½
1904	2¾	2	2¼	2⅞	3½
1903	3½	2⅜	2½	3	3½
1902	2⅝	2¼	2⅜	3	3¾
1900	2½	2¼	2⅝	3	3¼
1899	2¼	2	2¼	2⅞	3¾
1898	2¼	1¾	2¼	2½	3
1896	2⅝	1⅝	2⅜	2⅝	2¾
1895	2¼	2	2½	3	3
1893	2¾	2	2½	3	3
1892	2½	2	2¼	3	3¼
1891	2½	2	2¼	3	3½
1889	2⅝	2	2	2½	3⅝
1888	3	2	2⅞	3	3½
1881	3	2	2½	4	4½
1878	3	2	2¼	2½	3¼
1876	3	2½	3	3½	4
1874	3¼	2¾	3	3¾	4¼
1870	4	3	4	6	6
1868	4½	3	4	5	7

If you wish to form a milk producers' association in your neighborhood, talk the matter over with some of your friends and neighbors, stir up a little enthusiasm in connection therewith. It would be a good thing to discuss the matter in the grange. You should also write for all the information obtainable on the subject. Then call a meeting and

Freight Zones and Rates for New York Milk

	ZONES			
	A	B	C	D
Miles from market	40	41 to 100	101 to 190	Over 190
Freight rate per can of 40 quarts	$0.23	$0.26	$0.29	$0.32
Add for ferriage on milk delivered on west side of the Hudson, per can	.05	.05	.05	.05
Total cost of getting milk to market from west side of river	.28	.31	.34	.37
Suppose exchange price at New York is, per can	1.31	1.31	1.31	1.31
Deducting freight and ferriage leaves presumable net price at farmers' stations, per can	.98	1.00	.97	.94
Equal to cents per quart	2.45c	2.5c	2.425c	2.35c

name a chairman. The next order of business would be to appoint a committee on by-laws and perhaps a canvassing committee, and information committee. After a little further discussion adjournment can be taken for a week or 10 days, being sure to place the date near enough by, so interest will not fall off.

At the second meeting the organization can be completed by electing officers and hearing reports of committees. It may be advisable to have some outside speaker at this meeting to keep up enthusiasm, and perhaps offer counsel. There is no great secret about any of this work, and you and your neighbors can just as well enjoy the profits that come from co-operation, whether it be in obtaining better prices for milk, vegetables or poultry supplies. Remember it is results that count in this world, and results will not come unless a start is made.

***Milk Prices at Boston**

Season	Summer, April 1 to October 1	Winter, October 1 to April 1	Yearly average—Per can, 8½ quarts	Yearly average—Per quart
1908-9	—c	46c		
1907-8	39	42½	40¾c	4.79c
1906-7	35½	38½	37	4.35
1905-6	35½	37½	36½	4.3
1904-5	35½	37½	36½	4.3
1903-4	37½	37½	37½	4.4
1899-0	31	33	32	3.8
1890-1	32	36	33	3.9
1887-8	30	36	33	3.9
1882-3	35	43	38	4.5

*Prices are Boston basis, subject to zone rates which are high or low according to distance from Boston. In the middle zone the rate is 9 cents per can.

What is Legal Milk and Cream; Standards in the Various States

Standards for milk and cream throughout the United States vary in a marked way. Some of the commonwealths require no standard whatever, while others have a rather stringent one. The editor of the Hand Book has been in communication with the proper authorities in all of the states with a view of obtaining the latest changes, if any. The accompanying table gives corrected list. Arizona, Arkansas, Kentucky, Nevada and Oklahoma have no standards whatever. In Alaska and Colorado the matter is left with local towns and cities. West Virginia has no regular state standard other than regulations concerning adulteration. It will be noted that in the table one column gives total per cent of solids in milk, another column solids not fat, another column per cent of fat, and the last column gives the per cent of fat required in cream.

"The New York state statute provides that milk containing more than 88% of water or less than 12% solids or less than 3% butter fat shall be declared adulterated." Thus milk may "contain 87% of water and 13% of solids," including 4½% butter fat and yet not be below the standard, even though there would be only 8½% solids not fat.

Milk Business of New York City; How Enormous Supply is Handled

The various tributaries of the milk supply are under supervision by the New York city board of health, which sends its inspectors throughout the districts in which the milk is produced. These inspectors may legally charge 50 cents a cow for each animal inspected. Inspectors sent out by the state department of agriculture, however, may not charge for their services, as the state pays all expenses. Whenever inspection is desired, therefore, the farmer should apply direct to the commissioners of agriculture at Albany, N Y.

As an aid to improve milk handling, both on the farms and at every point on the route to the final consumer, whether in New York city or in other cities of the state, the state department of health at Albany inaugurated a campaign of education during the autumn of 1908. Efforts are being directed

Monthly Philadelphia Milk Prices, per Quart

Year	Jan.	Feb.	Mar.	Apr.	May	June	July	Aug.	Sept.	Oct.	Nov.	Dec.	Total	Mo. av.
1907†	4½	4½	4½	4½	4	4	4	‡3¾	4	4½	4½	5	51¾	4.31
1906	4½	4	4	4	4	3½	4	3½	3½	4½	4½	4½	48	4.00
1905	4	4	4	4	4	3½	4	3½	3½	4½	4½	4½	47½	3.95
1904	4	4	4	4	4	3½	3½	3½	3½	4	4½	4½	47	3.91
1903	4	4	4	6	4	3½	4	3½	3½	4½	4½	4½	48	4.00
1902	4	4	4	4	4	4	4	3½	3½	4	4½	4½	48	4.00
1901	3½	3½	3½	3½	3½	3	3	3	4	4	4	4	42½	3.54
1900	4	3½	3½	3½	3½	3½	3½	3½	3½	3½	4	4	44½	3.70
1899	3½	3½	3½	3½	2½	2½	3	3	3½	3½	4	4	40	3.33

*Freight is included in these prices and averages about ½ cent per quart.
†January 1 to June 30, dry quarts; July 1 to end of year, liquid quarts.
‡August 1 to 15, price was 4 cents; August 15 to 31, 3½ cents.

more especially toward the consuming end through local health boards. Circulars on various topics are being issued from time to time and may be secured free of charge from the department.

How the Milk is Handled

Milk for the New York city market is shipped in cans which hold 40 quarts. These are generally furnished by the shipper. There is much complaint of lost cans. Instead of the uniform freight rate of 32 cents a can that had prevailed for

PROF W. M. HAYS

The Assistant Secretary of the United States Department of Agriculture is W. M. Hays of Minnesota, a man long identified with agricultural education. It was Prof Hays who gave impetus to the plant improvement, and it was he who lent the force, energy and right direction to the movement of teaching agriculture in the schools of the country.

years (except for very short hauls at 25 cents a can), the interstate commerce commission decreed in March, 1897, a 23-cent rate for all stations in the zone within 40 miles of the terminal at New York city, Jersey City, etc; between 40 and 100 miles, 26 cents; between 100 and 190 miles, 29 cents; beyond 190 miles, 32 cents. The exchange price is based on zone B; thus $1.31 a can, less freight 26 cents and ferriage 5 cents, nets the producers $1 a can, or 2½ cents a quart.

Hence, $1.31 a can is said to make the exchange price 2½ cents a quart.

Instead of shipping direct to New York and other city dealers, many farmers sell to creameries operated by city dealers at the local railroad station. They often agree in such cases to take ¼ to ½ cent less than the exchange quart rate.

Milk Men's Organizations

The New York city milk exchange, composed of a few of the larger dealers and some producers, "fixes the price of milk shipped to the New York market" monthly or oftener. The officers of the exchange are: President, W. A. Wright of Brooklyn, N Y; secretary, Joseph Laemmle, 6 Harrison street, N Y. On

Milk and Cream Standards

State	% Total Solids	% Solids not Fat	% Fat	% Fat in Cream
California	11.5	8.5	3.00	18
Connecticut	11.75	8.5	3.25	—
District of Col.	12.5	9.00	3.5	20
Florida	11.75	8.5	3.25	—
Georgia	11.75	8.5	3.25	18
Hawaii	11.5	—	2.5	—
Idaho	11.00	8.00	3.00	18
Illinois	11.5	8.5	3.00	18
Iowa	12.5	9.5	3.00	15
Kansas	11.75	8.5	3.25	18
Maine	12.00	9.00	3.00	18
Massachusetts	12.15	—	3.35	15
Michigan	12.5	—	3.00	—
Minnesota	13.00	9.5	3.5	20
Missouri	11.75	8.5	3.25	18
Montana	12.00	9.00	3.00	15
Nebraska	—	—	3.00	18
New Jersey	12.00	9.00	3.00	16
New Hampshire	*13.00	†9.5	‡3.5	—
New York	12.00	9.00	3.00	—
Ohio	12.00	—	3.00	—
Pennsylvania	12.5	—	—	—
Porto Rico	12.00	9.00	3.00	—
Rhode Island	12.00	—	2.5	—
South Dakota	12.5	9.5	3.00	18
Texas	12.5	9.00	3.00	—
Utah	12.00	8.80	3.2	18
Vermont	12.5	9.25	3.25	—
Wisconsin	—	8.5	3.00	18
Wyoming	12.00	9.6	2.4	—

*April to September 12 per cent.
†April to September 9 per cent.
‡April to September 3 per cent.
—No regulation.

milk from west of the Hudson river, 5 cents a can is deducted from this price as an allowance for ferriage or transfer.

The Five States Milk Producers' Association, of which Ira L. Enell of Kenwood, N Y, is president, and H. T. Coon of Homer, N Y, is secretary, has a New York city representative, James L. Bennett of 220 Broadway.

The officers of the Dairymen's League, organized and incorporated under the laws of New Jersey, are: President, J. Y. Gerow of Washingtonville, N Y; secretary, Henry Youngs of Goshen, N Y. The New Jersey representative is Leroy H. Morris of Newton, N J. Officers of the Five States Co-operative Creameries

Association are: President, D. C. Markham of Port Leyden, N Y, and secretary, William C. Hunt of Great Bend, Pa.

Milk Prices at Philadelphia and Officers of the Shippers' Union

The Philadelphia Milk Shippers' Union is the farmers' organization, with headquarters at Odd Fellows' Temple, corner of Broad and Cherry streets, Philadelphia, Pa. The officers are: President, Charles S. Atkinson of New Hope, Pa; secretary, A. B. Huey of Lenape, Pa.

The dealers' organization is the Philadelphia Milk Exchange. President, Louis D. Sloan; secretary, Joseph H. Miller of 1009 Columbia avenue. For monthly prices see table on page 15.

Milk Prices in Chicago Increase; Daily Supply is 240,000 Gallons

Chicago paid more for its milk in 1908 than in any previous year, the average being 3.8 cents a quart. This was paid to the dairymen of Illinois, Indiana, Michigan and Wisconsin for milk delivered at the depots in Chicago. Two factors play permanent parts in determining the price, which producers complain is insufficient to repay them for their investments and labor.

The large quantity of milk consumed by the western metropolis, more than 240,000 gallons a day, makes it necessary to bring at least a part of the supply a long distance. Second is the restrictive policy of the health authorities. Federal,

Chicago Milk

Month	1908	1907	1906	1905	1904	1895
Jan.	$1.45	$1.20	$1.10	$1.15	$1.15	$0.80
Feb.	1.45	1.15	1.10	1.10	1.15	.80
March	1.30	1.10	1.05	1.10	1.15	.75
Apr.	1.25	1.10	1.05	1.05	1.05	.70
May	.95	.90	.80	.80	.75	.60
June	.90	.85	.80	.80	.75	.60
July	1.00	1.00	.90	.85	.85	.65
Aug.	1.10	1.05	.90	.85	.85	.65
Sept.	1.10	1.10	1.00	.95	.95	.70
Oct.	1.15	1.10	1.00	.95	.95	.70
Nov.	1.35	1.45	1.20	1.15	1.15	.85
Dec.	1.35	1.45	1.20	1.15	1.15	.87
Av. per can	1.20	1.12	1.00	.99	.99	.72
Av. per qt.	.038	.035	.031	.030	.030	.022

state and city laws and regulations are constantly making production and distribution more expensive. Those supplying milk to Chicago are not organized. Many problems of transportation and price await solution by some really earnest association of producers, which has been frequently agitated. Since 1905 the price of milk has advanced 1.6 cents a quart at Chicago.

Butter and Cheese Divorced from Farm

Production of These Two Commodities is Becoming More of a Manufacturing Industry

One of the side issues of farming in the old days was buttermaking. In the recent decades of specialization, this important adjunct of farm industry has been passing out of the realm of the farm and into the province of the everyday manufacturing business. Cheese perhaps led the way from the farm to the factory. But now these two dairy commodities are related to agriculture chiefly through the fact that they are made of milk. Farmers have found that by cooperation they can establish creameries which will turn out uniformly better butter than was ever made in a small way on the farm. To be sure, butter-

Butter Imports and Exports

Exports of butter were cut nearly in half last year while imports were about doubled as shown by the following table giving statistics for the years ending June 30:

Year	Exports, pounds, millions and tenths	Average value, cents	Imports, pounds	Average value, cents	Rate of duty, cents
1908	6.4	21.8	780,608	23.4	6
1907	12.5	19.5	441,755	26.7	6
1906	27.3	18.0	196,642	29.0	6
1905	10.0	16.3	593,104	20.9	6
1904	10.7	16.5	153,536	22.6	6
1899	20.2	16.1	23,700	16.7	6
1894	11.8	17.5	144,000	16.2	6
1889	15.5	16.6	179,000	13.7	4
1885	21.7	16.8	187,000	18.7	4

Cheese Prices Compared

In the following table are shown the range of prices of cheese in New York and Chicago for last year and previous years, quotations being in cents per pound for full cream cheese:

Year	Chicago	New York
1908	*10 @15½	10½@16
1907	11 @15	11½@15½
1906	9 @14	9½@14½
1905	9½@14½	9 @14½
1904	7 @12½	6½@12½
1903	10 @14½	9¾@15½
1902	9 @13	9 @13½
1901	9 @12	8 @12½
1900	7 @12½	9 @13½
1899	8 @13	7½@13
1898	8 @11	7 @11
1897	7 @11	7 @12
1896	6 @10	6 @11
1895	7 @11	6 @12
1894	8 @13	9 @12
1893	7 @12	9 @12
1892	8 @12	9 @13
1891	7 @12	†10 @11
1890	7 @11	†9 @10
1889	†9	†10 @11
1888	†9	†10 @11
1887	†9	†11 @12
1886	†8	†9 @11

*January 1 to December 1.
†Average price for the year.

making is by no means passing entirely from the farm, but the country at large depends more largely upon the creamery, either co-operative or corporation owned.

The development of the co-operative creamery has perhaps reached its greatest efficiency in Minnesota, one of the great dairying states, together with Wisconsin, New York, Iowa, Illinois, Vermont and Pennsylvania. These states produce upward of 65% of the total butter, cheese and condensed milk manufactured in the United States.

Last year saw the further curtailment of available milk for cheese factories and creameries, a large part of the supply being absorbed by the great cities. Chicago, St Louis, Pittsburg, Philadelphia,

Range of Butter Prices

The range of butter prices at Chicago and New York for the years indicated are shown in the following table, quotations being for western extra creamery butter:

Year	Chicago	New York
1908........	*20 @ 34	21 @ 35
1907........	22 @ 33	23 @ 34
1906........	17 @ 31	19 @ 32
1905........	18 @ 34	19½ @ 35
1904........	17 @ 28	17½ @ 28
1903........	18½ @ 28½	19 @ 29
1902........	19 @ 28	20 @ 29
1901........	18 @ 24	19 @ 25
1900........	18 @ 26	16 @ 27
1899........	17 @ 26	18 @ 27
1898........	16 @ 22	17 @ 23
1897........	14 @ 23	14 @ 24
1896........	15 @ 24	15 @ 26
1895........	16 @ 25	17 @ 28
1894........	15 @ 27	17 @ 28
1893........	19 @ 33	20 @ 35
1892........	17 @ 31	17 @ 32
1891........	17 @ 35	†22 @ 26
1890........	14 @ 29	†20 @ 23
1889........	‡16	†21 @ 23
1888........	‡18	†24 @ 26
1887........	‡15	†23 @ 25
1886........	‡15½	†25 @ 27
1885........	16 @ 40	†20 @ 24
1884........	‡18	†20 @ 25
1883........	‡18	†20 @ 26
1882........	‡19	†28 @ 32
1881........	‡19	†24 @ 27
1880........	18 @ 37	

*January 1 to December 1.
†Average for western extra creameries.
‡Represents average price for all.

New York and Boston are gradually extending the territory from which they gather their milk supply. As these cities grow, and as they are able, by improved facilities to bring milk even a greater distance than they are now doing, the amount thus consumed will increase.

In addition to the consumption of milk by the large cities, last year was severe on the cheese factories and creameries because of extended drouth, which reduced pasturage, cutting down the milk production. The prevailing prices in New York and Chicago for butter and cheese reached a higher level in 1908 than for several years. The limited supply of cheese was noted both in the domestic

markets and the export trade. The exports were cut nearly in half compared with those of 1907 and 1906. The average value of the exports was the highest in the history of that branch of the trade, reaching 12.9 cents per pound. Imports of cheese were 32,500,000 pounds compared with 33,800,000 for 1907. The exports were 8,400,000 compared with 17,-300,000 in 1907.

Poultry Raising and Winter Egg Problem in Centers of Population

Raising poultry has been part of the farm work for time beyond memory, but in recent years this important adjunct of the farm, the poultry business, has been specialized until now it has developed to a high plane of scientific farming. Poultry are profitable for the eggs, for meat, and for the feathers. Improved machinery for hatching and caring for the young poultry has increased production. Science in care and feeding has lent aid to the hen, encouraging her to lay more eggs. Better commercial facilities have also helped the poultryman, and now cold storage warehouses at shipping points, refrigerator cars, and again cold storage plants in the large cities, make possible the utilization in the crowded cities of larger quantities of eggs and dressed poultry than in the old days. In spite of the improved facilities there is a scarcity every winter of desirable fresh eggs.

Egg Prices in Large Cities

Prices of eggs in the large distributing centers at the three important times in the year follow:

(All prices in cents)

Year	New York Apr. 1	New York Sept. 1	New York Dec. 1	Chicago Apr. 1	Chicago Sept. 1	Chicago Dec. 1	Boston Apr. 1	Boston Sept. 1	Boston Dec. 1
1908...17		30	52	16	21	29	20	31	48
1907...17½		22	38	16	19	28	18	23	40
1906...21		25	40	18	20	30	22	26	42
1905...19		26	39	16	21	27	19	27	39
1904...19		24	38	18	21	27	19	25	38
1899...14½		21	24	13	16½	20	13½	17	25
1894...12		19	27	10	15	21	13	22	30
1889...14½		16½	29	10½	14½	24	15	22	32

Foreign Trade in Eggs

Imports and exports of eggs in thousand dozens and the average value per dozen are shown in the following table:

Year	Exports Dozens	Exports Avg. val. cents	Imports Dozens	Imports Avg. val. cents	Duty cents
1908....	7,591	20.3	232	11.1	5
1907....	6,968	22.0	231	11.4	5
1906....	4,952	20.9	241	8.8	5
1905....	2,475	21.9	352	10.9	5
1904....	1,776	22.3	496	12.4	5
1898....	2,754	15.9	166	——	5
1893....	143	23.1	3,318	11.8	5
1890....	380	15.4	15,062	13.7	free
1883....	360	20.8	15,279	17.4	free

How to Handle Perishable Produce at a Profit, Reducing the First Cost

One of the Problems Involved is the Elimination of Some of the Middleman's Profit—Much is Being Done Along This Line by Farmers' Co-operative Associations

At a certain hearing of Pres Roosevelt's commission on country life, this story was told: A farmer living in northern Maine grew potatoes for market in the season of 1908, sold them at his loading station at 37 cents a bushel, they were shipped to southern Connecticut and the consumer there paid $1.03. Here was a "leak" of 66 cents. Who got it? Well, the railroads had about 15 cents and four different dealers between producer and consumer each took a nip. It is the old question, as true in Kansas as in Connecticut, how to eliminate some of the middleman's toll.

Cut Middleman's Tolls

No single rule will insure profit to the farmer or truck grower, yet much may be gained by wise packing, shipping and handling of perishable produce. Much is being accomplished by the co-operative associations; but, after all, untold numbers of producers must handle individually their crops, and helpful hints are always in order.

In catering to a family trade through house to house canvass, it is strictly up to the man on the wagon. In shipping to a distant point, many things must be studied, understood and observed. Fruits or vegetables, the produce should be available for the seller at the earliest moment after it is picked and shipped. Often several hours and sometimes a day may be saved by changing the time of shipment. Products that must be held over from one afternoon to the next day are liable to sell at low figures and may often appear to poor advantage.

Appearance is an Asset

The most successful fruit and produce growers now recognize that attractive appearance is a valuable asset. Berries should be placed in fresh, clean boxes, the fruit well assorted, packages being free from damaged or overripe specimens and in all cases uniform as to size and quality. Good measure will also bring good profits. Buyers appreciate honest count. Be sure your packages are of liberal, standard size and above all

make them uniform. In the state legislatures, the tendency just now is toward uniformity in size and capacity of shipping packages, such as the barrel and bushel box.

Early peaches are usually packed in crates, each containing a number of small baskets. This package is quite a favor-

DR SEAMAN A. KNAPP

Dr Knapp is the special agent in charge of Farmers' Co-operative Demonstrative Work for the U S Department of Agriculture. The fruit of Dr Knapp's labors is now being gathered in. This consists of a new style of farming throughout the southern states. Based upon good tillage, deep plowing, thorough cultivation and good seed, it has been proved that ordinary southern crops can be doubled or trebled.

ite and justly so, as fruit thus prepared is attractive to the eye, merchantable as to bulk and attracts buyers. Strawberries require careful handling and should not be picked immediately after a heavy rain. Grapes should be marketed quickly and handled as little as possible, although or course through the co-operative associations the cold storage warehouses nowadays come into play. In packing grapes, place the stems downward, package well filled, so that when the cover is removed, an even, stemless surface is presented. In handling asparagus only the best stalks should be cut for shipment, neatly tied in bunches of equal size and quality, and these packed in boxes holding a dozen or more bunches, well protected by loose soft paper or moss. Ordinarily cauliflower is profitably grown, meeting good sale when the heads are of reasonable size, bright and creamlike in appearance; wilted or badly stained specimens sell poorly. In establishing a trade, it often pays the producer to first visit the distributing market and study the conditions.

Apple Crops by States

[In round thousands of barrels]

	1908 bbls.	1907 bbls.	1906 bbls.	1905 bbls.	1904 bbls.
New England:					
Me.	625	1,700	900	630	1,425
N. H	350	800	710	500	940
Vt.	300	750	600	350	700
Mass.	450	600	700	525	895
R. I.	100	100	100	100	140
Conn.	200	400	275	450	670
Total	1,950	4,350	3,285	2,555	4,770
Middle:					
N. Y.	3,700	3,800	5,200	3,330	7,200
N. J.	300	400	650	750	1,250
Penn.	1,500	3,000	3,750	2,100	4,150
Del.	200	150	180	150	195
Ohio	1,250	1,800	2,750	950	3,275
Mich.	1,200	1,900	3,500	1,800	3,515
Wis.	250	200	250	100	375
Total	8,400	11,250	16,280	9,150	19,960
Central West:					
Ind.	500	600	1,000	510	730
Ill.	475	850	2,100	525	632
Mo.	350	400	2,275	400	600
Kan.	375	125	450	360	540
Neb.	150	140	375	200	345
Iowa.	425	350	650	365	1,550
Ark.	250	300	900	550	1,100
Total	2,525	2,865	7,750	2,910	5,397
Far West.	2,975	2,675	2,565	2,195	2,449
South.	4,100	4,450	4,900	5,500	8,735
All other	3,000	3,200	3,500	2,000	3,950
U S Crop.	22,950	28,690	38,280	24,310	45,360

Foreign cheese finds its way into this country to the amount of 35 million pounds a year, paying a duty of 6 cents a pound. Actual imports in 1908 were 32,500,000 pounds. Our own cheese industry has not yet fully recovered from the woeful mistake made years ago when some conscienceless manufacturers foisted great quantities of filled cheese onto foreign consumers. That cooked our goose. Going back as far as 1890, official figures show that more than 95 million pounds of cheese were exported. Within a few years after that, the business had dropped to 10 million pounds in a year.

Red-Cheeked Apple Everybody Wants It

Wide Varieties in Quality and Attractiveness Count in the Summing Up

One day, not so very long ago, an editor of the American Agriculturist Annual paid 25 cents for an apple. Luxuries of this character do not come our way very often, and the national "apple day" would make slow progress were each loyal member of it obliged to get needed supplies at such prohibitive figures. This apple was a beautiful fellow, showing his radiant and rosy face in front of a fruit stand on 42d street, New York city, after having traveled 3,000 miles from the state of Washington. It was neither better nor worse than millions of its family grown each year in our Pacific coast states; and some hard-headed Yankee friends, in fact, might insist that it is no better than the Maine Baldwin or a Hudson River Spy. It commanded, along with its companions, this out-of-reach price because it was superb in appearance and practically perfect in condition, owing to the methods employed in producing it. Limited quantities of such fruit will always sell at figures which are prohibitive to most of us. The point is, quality counts, and the fact that in a full year domestic orchards can turn off 50 million barrels does not stand in the way of producers securing top prices for top quality.

Fancy Apples Always Scarce

Each season, as it rolls around, finds a big aggregate bulk of apples on the trees, but a comparatively small percentage can be classed as commercially perfect. Therefore, prices are bound to cover a very wide range. Producers in favored orchard sections may receive as high as $2.50 to $3 and even better for perfect stock at the orchard, providing it is sound and with good keeping qualities. At the other extreme, untold quantities of apples net the producer as little as 50 cents a barrel or even worse. Queer, isn't it, that with all the new orchards set every season and other new orchards coming into bearing each year, the business is not overdone! Some are so cautious as to really fear too many farmers are going into apple growing.

But the general fact remains that there is practically never an oversupply of sound, merchantable apples. Another criticism is more logical; that there are altogether too many varieties. These number literally into the hundreds. Some of the dealers who make a specialty of this fruit are willing to go on record that one might almost count on the fingers of both hands all the varieties of apples that are needed to make up the commercial supply, and drop the remainder into the bottom of the sea and forget they ever existed. But, of course, there is a happy medium, and the up-to-date farmer delights in growing a number of varieties, even if only in a small way for home use.

Commercial Orcharding

When it comes to counting on one's fingers, certainly a show of the two hands would point to the very few sections where apple orcharding is carried on in a large way, and successfully, too, catering to domestic and foreign markets. New York leads in importance, followed closely by Michigan, Maine, the Pacific coast and growing orchard sections in some of the states of the west and southwest, such as Colorado, Kansas, Missouri and Arkansas. Other sections, of course, produce in the aggregate a good many apples, but those named furnish the bulk of the winter supply. The favorite commercial varieties in the eastern states and Canada include, as for many years past, Baldwin, Greening, Northern Spy, Spitzenberg, with perhaps more attention than formerly given to the Ben Davis, so popular in the west and southwest.

The market end of the apple business might readily fill a volume. Suffice it here to touch upon only a few points which stand out prominently. The barrel is still the favorite package. Coupled with that, is the unfortunate fact that prices on barrel stock have been advanced for several years and now the cost of packages cut sharply into the profits. The bushel box is slowly gaining favor, although commission merchants as a rule are against it. In many ways it is the logical and ideal package and ought to appeal to a great many families who would not consider buying a barrel of apples.

It is the old story with this fruit as with many other kinds of farm produce—there is altogether too much toll in the handling; the price which the consumer pays is altogether out of reason with the figure the producer actually gets. Co-operative effort in distribution is doing something toward whittling down this middleman's profit; there is still much more to be done.

Foreign Trade in Apples

The export trade in apples is encouraging, but develops rather slowly. Europe grows a good many apples, mostly summer and autumn varieties, and these are well out of the way by the time American and Canadian winter fruit is ready. But the price to the English, French and German consumer is such as to make the fruit really a luxury. So far as that is concerned, there is still something woefully lacking right here at home in getting the producers' fruit into distributive channels at a rate which will insure heavy consumption. Occasionally exports from the United States and Canada will closely crowd 4 million barrels, of course, the best class of stock. During the last few seasons the exports have approximated annually 2,500,000 barrels. The season runs from September 1 to June 1. The heavy trade is recorded in dried apples now largely the evaporated stock, and this shows reasonable profits when well managed. Exports of dried apples from the United States have increased wonderfully in recent years, and now run something like 50 million pounds annually at an average value of 7 cents.

Story of the American Cranberry as Viewed from the Business Side

The cranberry, that toothsome delicacy when made into sauce, reaches perfection on American soil. A little red, rather unpalatable berry, bearing the same name, is grown in the peat bogs of Europe, but

Cranberry Production and Prices

In the following table showing the production of cranberries in the United States three naughts (000) are omitted from quantities given:

Crop of	N. E. & N. Y.	New Jersey	West	Total	Boston price per bushel Oct.	Jan.
1908	700	300	90	1,090	*$7.00	$6.00
1907	650	340	80	1,070	2.25	2.75
1906	710	325	135	1,170	2.15	2.50
1905	415	275	125	815	2.25	4.50
1904	775	250	110	1,135	2.00	2.50
1903	425	475	100	1,000	2.25	2.50
1902	410	135	130	675	2.00	3.00
1901	540	300	110	950	2.00	2.00
1900	475	250	75	800	1.75	3.00
1899	600	240	120	960	1.50	2.00
1894	185	200	25	410	2.50	3.00
1889	350	200	70	620	2.00	3.00
1884	130	124	24	280	3.00	4.75
1880	251	129	113	493	2.00	2.00

*Price Dec. 1, 1908.

the wholesome, attractive small fruit as found in the United States is nowhere else equaled. Famed throughout the country for its cranberries is the Cape Cod region, Massachusetts. New Jersey is an important rival and Wisconsin comes third.

The crop of 1908, as reported in the American Agriculturist weeklies, which give considerable attention to the cranberry industry, was 1,090,000 bushels, compared with 1,280,000 bushels in 1907. The crop is packed both in barrels and crates, barrels containing 96 quarts and crates about one-third that amount.

The shortage of the 1908 crop began to be reflected in the market price early in the season, and within the closing weeks of the year 1908 record prices were established in the large distributing centers.

Potatoes, Raised by Every Farmer, Second in Importance to Wheat

Indigenous to the American soil, attracting attention first as a fodder plant, and later adopted by philanthropic societies of Europe as food for the poor in time of famine, the white potato, sometimes called the Irish potato, has steadily grown in importance until it is now one of the principal articles of food in every civilized country on the globe. The combined crop weighs more than the world's crop of wheat and in some countries takes the place of wheat to a considerable extent.

Used to Make Starch and Alcohol

In addition to being a universal article of food, it enters into the manufacture of starch and denatured alcohol. Every farmer recognizes the importance of the potato and knows how to grow it. So much attention has been given to this vegetable that the ravages of insects are no longer feared, the farmer having

learned that the judicious use of paris green kills the Colorado bug, the chief source of annoyance. Agricultural scientists have yet to discover how to entirely prevent fungus in seasons of excessive dampness. A mixture for spraying has been discovered, the application of which in July and August prevents rot, provided rains are not so heavy as to wash the mixture off the plants.

World's Potato Crop

The world's crop of potatoes is about 4,000,000,000 bushels; that of the United States has been in the neighborhood of 300,000,000 bushels for the past 9 years. The banner year was 1904, when 302,500,000 bushels were produced in the United States. In 1908 the crop was 293,600,000, compared with 299,000,000 in 1907. Looking over a period of years a gradual increase in production is noted, with occasional reversals. As far back as 1884 there were gathered in the United States 222,100,000 bushels. In recent years, outside of New York, Michigan has been the banner potato state. Wisconsin was a close second, Iowa came third. New York produced a crop of 30,176,000 bushels last year. In 1906 it had a banner crop of 36,000,000 bushels. Last year the three great western potato states yielded the palm to Maine, so that the order of importance was New York, Maine second, with 17,420,000 bushels, Michigan 15,405,000, Wisconsin 15,120,-000 and Iowa 13,485,000. In former years Michigan has raised a crop of more than 23,000,000 bushels. Maine equaled its

Great American Potato Crop

Acreage, given in thousands, is shown by states in the following table; the yield to the acre and total production is given in full in bushels:

	Acres grown				Av. Yield per acre, bush.				Total production, bush.			
	1908	1907	1906	1905	1908	1907	1906	1905	1908	1907	1906	1905
Me..............	82	80	74	67	210	140	210	236	17,420	11,200	17,500	12,730
N. H............	19	20	20	20	110	120	112	120	2,090	2,400	2,240	2,400
Vt.............	20	21	21	22	100	100	90	100	2,000	2,100	1,890	2,200
Mass...........	24	25	24	24	90	110	110	105	2,160	2,750	2,640	2,520
R. I...........	5	5	5	5	105	100	110	100	525	500	550	500
Ct............	19	20	20	20	92	95	90	95	1,748	1,900	1,800	1,900
N. Y...........	368	355	360	367	82	92	100	77	30,000	32,660	36,000	28,259
N. J...........	45	46	46	45	87	90	95	90	3,915	4,140	4,370	4,050
Pa............	228	212	210	212	80	85	97	75	18,240	18,020	20,370	15,900
O.............	154	160	168	171	73	75	90	75	11,242	12,000	15,120	12,825
Mich..........	237	250	235	255	65	95	94	86	15,405	23,750	22,210	20,400
Ind...........	79	83	86	90	60	98	80	85	4,740	8,134	6,880	7,650
Ill...........	133	144	152	154	62	75	87	84	8,246	10,800	13,224	12,936
Wis.	216	220	240	230	70	82	77	70	15,120	18,040	18,480	16,100
Ia............	155	160	165	160	87	97	100	90	13,485	15,520	16,500	14,400
Minn..........	123	135	145	148	76	94	90	60	9,348	12,690	13,050	8,800
Mo............	82	84	86	87	80	89	92	90	6,560	7,476	7,912	7,830
Kan...........	85	84	87	88	64	58	83	91	5,440	4,872	7,221	8,008
Neb...........	93	91	91	92	88	78	88	100	8,184	7,098	7,735	9,426
S. D..........	45	41	44	42	80	81	90	90	3,600	3,321	3,960	3,780
N. D..........	23	26	24	22	70	86	87	85	1,610	2,236	2,088	1,870
Col...........	46	45	42	45	140	150	135	142	6,440	6,750	5,900	6,390
Cal...........	38	40	41	40	125	148	148	150	4,750	5,920	6,068	6,000
Ore...........	34	37	36	33	140	150	125	140	4,760	5,550	4,500	4,620
Wash..........	31	33	27	28	150	175	140	130	4,650	5,775	3,780	3,640
Other..........	552	540	550	540	70	73	75	70	38,640	39,420	41,250	37,800
Total..........	1286	2957	2999	3002	82	91	95	84	101,535	268,022	283,238	252,984

1908 crop in 1906. Another important state is Illinois, which usually raises about 12,000,000 bushels, but last year suffered a reverse to 8,264,000 bushels, owing to drouth.

Onions One of the Specialized Farm Products in This Country

Varying with climate and soil, onions are grown in nearly all parts of the United States, and in many foreign lands. In a few states onion growing has become a specialized industry and there most of the crop is produced. The chief states are New York, Massachusetts, Ohio, Michigan and Indiana. Connecticut might also be mentioned, but in that state interest is flagging. In 1908 the total crop of onions in the United

Onion Prices, Crops and Movement

In the following table showing the production of onions in the United States, also imports and exports, the quantities are given in thousands of bushels:

		Price per bu. at New York		Exports bu.	Imports bu.
Crop of	Bushels	Oct.	*Jan.		
1908	4,089	.50@ .80	——	——	174	1,275
1907	4,067	.70@ .85	.65@ .85	——	
1906	3,753	.65@ .90	.70@1.25	257	1,126
1905	3,588	.60@1.00	.65@ .90	205	872
1904	3,341	.75@1.00	1.00@1.25	234	856
1899	4,615	.40@ .50	.40@ .70	171	546
1894	1,944	.60@ .70	.50@ .80	53	
1891	3,200	.40@ .70	.80@1.00	——	

*Quotations for January of the year following date given on side of table.

States was 4,089,000, compared with 3,626,000 bushels in 1907, and the following totals for the preceding years: 1906, 3,756,000 bushels; 1905, 3,588,000 bushels; 1904, 3,341,000 bushels; 1903, 3,090,000 bushels; 1902, 3,794,000 bushels.

In addition to the large home crop there are annually imported into the United States hundreds of thousands of bushels of onions from the Bermudas and occasionally from Egypt and Europe. In 1908 1,275.273 bushels were imported. In 1907 the imports were 1,000,000 bushels.

Hay, Its Importance; Methods of Handling

Much Land Devoted to This Important Feed Crop—Many Varieties in Use

In the northern part of the United States east of the Mississippi grasses and clovers are the principal hay crops. In the southern states cowpeas, soy beans, Japan and crimson clovers are mostly used. In parts of the Rocky mountain regions and Pacific coast states alfalfa is grown almost exclusively for hay. Where dairying is an important industry, many of the grain crops are cut for hay before the seed is matured. Varying amounts of millets and Hungarian grasses are grown in all of the states. These are especially valuable as late sown crops, because of the short season they require for maturing. In some sections, particularly the southern states, kafir corn and sorghum are grown and used as a hay.

Area Devoted to the Crop

About 15% of the improved land is devoted to hay. The highest per cent is in the North Atlantic states, where hay is raised on about one-third of the farm land. In the South Atlantic states only about one-twentieth of the land is used for hay crops.

The average value per acre of hay throughout the United States is about $8. This is only 3% less than the average value of the grain crops. The average yield per acre of all kinds of hay has varied but little during the past 25 years. American Agriculturist estimated the yield for 1908 as 1.47 tons per acre; total 61,383,000 tons. This is slightly higher than the average for the past five years, but during the past 25 years the yield has not fallen below 1.1 tons per acre.

Hay Yields and Prices

	Yield				Price timothy per ton					
					New York			Chicago		
Year	Acres, millions	Per acre tons	Total millions tons	Value, millions	March 1	August 1	December 1	March 1	August 1	December 1
1908	41	1.49	61	$714	$16.00	$20.00	$17.00	$18.00	$13.50	$14.00
1907	37	1.40	52	740	22.00	23.00	22.00	18.00	20.00	17.00
1906	38	1.36	52	592	16.00	20.00	21.00	12.50	16.00	16.50
1905	40	1.45	58	515	17.00	16.50	16.50	13.00	13.50	12.50
1900	39	1.3	50	446	18.00	19.00	19.00	11.50	12.50	14.00
1890	—	1.2	50	387	17.00	16.00	14.00	9.00	11.00	11.00
1880	26	1.2	32	372	17.00	22.00	24.00	13.00	16.00	16.00

A careful study of this table will show one the changes that have occurred in the hay industry during the past few years. The lower prices quoted at Chicago are due to the fact that the area of cheapest production is closer to that center. The prices given are for the market standard, choice timothy.

Yield of Various Varieties

Wild hay gives the greatest yield in the South Atlantic states and the smallest rate yield in the western states. Of the crops used for hay, wild grasses give the smallest average yield, varying from three-fourths ton to 1 1-5 tons per acre. Millet and Hungarian grass give the highest average yield in the North Atlantic and north central states. These grasses usually yield more than 1½ tons per acre. Alfalfa gives the highest yield of all the hay crops; average production for the United States is above 2½ tons per acre. The largest yields are reported from the western states. In irrigated districts four or five cuttings and in some cases even more are made in one season.

The western states also obtain the largest yields of clovers, the southern states being the least successful with this crop. The yield varies from 1 to over 2 tons per acre. More than one-half of all the grains cut green for hay is harvested in the western states. The yield is from 1 ton to 1½ tons per acre.

Marketing the Crop

The market for the hay crop has become so important that a national association and many district associations have been formed to control and regulate grades and prices. Choice timothy hay is the standard grade. This must be properly cured, of a bright, natural color, sound, well baled and not mixed with over one-twentieth other grasses. There are three other grades of timothy hay varying slightly in value as their quality varies.

No 1 clover hay is medium clover, properly cured, sound, well baled and not mixed with over one-twentieth other grasses. That not good enough to grade as No 1 is sold as No 2. Choice prairie is the standard grade of wild hay. As to prices, Boston usually pays the highest. The Chicago market is generally $3 to $4 above the Kansas City market. Minneapolis is little above the Kansas City price.

Broom Corn now Almost Entirely an Industry of the Western States

Many years ago the culture of broom corn was confined to New England and eastern New York, but now practically all of the crop is raised in western states. For a time Kansas and Illinois were the greatest producers, but within the last few years Oklahoma has increased its acreage rapidly, now turning off almost as much as these two states put together. The larger part of the dwarf variety is produced in Kansas and Oklahoma. Illinois grows the standard variety almost entirely.

Dwarf broom corn grows only 3 or 4 feet high and about half of this is brush It is used mostly for making small brooms. The greatest complaint of dealers is that brush raised in Oklahoma is carelessly cured. The essential part of curing is to keep the color from changing. After the brush is cut from the stalks it must be dried for some time. If this is done out of doors. there is great danger of the color being damaged. The difference in prices between green and badly colored is much more than enough to pay the difference in cost of curing in a shed.

The Seed as a Feed for Stock

Broom corn seed is used as a feed for pigs and milch cows. Chickens eat it for a while as a variety, but are not fond of it as a regular feed. It weighs 40 to 50 pounds per bushel. It does not pay to raise broom corn for the seed alone. In the large broom corn fields of the west the cattle are turned in after the harvest, and they are allowed to feed upon the leaves. It is generally agreed that the fodder of broom corn is worth about half as much as that of Indian corn.

The Culture Required for Success

Land that will produce a good crop of Indian corn will produce broom corn. The land must be as free as possible from weeds, because the young broom corn plant is so small and delicate that it cannot stand a very hard struggle for existence. Because of this it is often raised year after year on the same land, for land once cleared of weeds is much easier to keep clear. Those who use it as a rotation crop get good results from planting it on clover or timothy sod.

Before shipping, the seeds must be removed and the brush pressed into bales. The proper size of the bale is 3 feet 10 inches long, 24 inches wide and 30 inches deep. This is bound by four or five wires. Some place a stout lath at each corner to protect the brush from being cut by the wires. Bales of this size, if properly packed, generally contain about 300 pounds. It is much more profitable to bale the crooked and inferior stock by itself than to mix it in with the better grades and varieties.

The principal markets for broom corn are New York, Chicago, Philadelphia and Baltimore. In these centers there are commission merchants and dealers who devote their entire time to handling this commodity. Those who have suitable land and give the crop proper attention, find broom corn as profitable as any of the staple crops. It must not be understood, however, that anyone can take to raising broom corn and get rich in a short time. Like many other of the special crops, it requires careful study to insure success. The broom corn crop of 1908 was estimated at 50,000 tons.

The Oil Producing Castor Bean, Easy to Raise—Profits are Small

The castor bean is cultivated commercially in the states of Oklahoma, Illinois, Missouri and Kansas. Oklahoma produces about half the total product. The consumption of the United States is not entirely supplied by the home production. India furnishes a large amount of the castor oil used in this country.

The beans are planted from the middle of April to the first of May and the fruit ripens in July. The pods are so constructed as to throw the seeds to a considerable distance when the wall of the pod dries and breaks. This makes it necessary to collect the entire fruit clusters as soon as they turn a dark brown. These clusters are put into a tight bin, where they are allowed to dry and discharge the beans. If frosted beans are mixed with good ones, the value of the whole will be greatly reduced, hence this crop requires careful handling at harvest.

Insects and Disease Unknown

The castor oil plant has no serious pests among either fungi or insects. The oil extracted from the castor bean is used largely in the dyeing of cotton goods. The substance made from this oil is called turkey red oil. It is also used in the manufacture of sticky fly paper and soap, and in various other ways. The yield is only 10 to 12 bushels to the acre, and the beans command about $1 a bushel. Foreign beans pay 35 cents a bushel duty.

Field Beans, One of the Special Crops of the Northern Territory

Many fields of beans are raised on soils that will not produce corn or potatoes. The best results, however, are obtained on limestone soils. Most of the commercial crop of field beans is raised in northern United States, southern Canada and California. There are a few fields in the southern states.

Used in a Rotation

Where beans are used in a rotation they do best on clover sod and are usually given that place. Clover, beans and wheat is the most common rotation. When raised on sod, the work of cultivation is quite small and is practically all done by machinery. The harvesting is accomplished with a special machine, which cuts off the vines just below the ground. The vines are then raked into piles and allowed to cure for a short time, after which they are threshed. Where beans are grown on a large scale, the threshing is done with a power machine. Some growers, however, still use the flail.

Even though the field bean is a very easy crop to grow, still there are large quantities imported each year for domestic use. The United States raises 5,000,-000 to 7,000,000 bushels annually. The bulk is grown in Michigan, New York and California. Maine, Wisconsin, Ohio, Pennsylvania and Iowa also produce considerable quantities. Florida is the most important state in the south.

The Yields Obtained

An ordinary yield is from 14 to 20 bushels per acre, but yields of as high as 30 to 35 bushels are often reported. From $16 to $18 is about the average return from an acre.

The standard varieties of beans throughout the United States are few. The pea bean, commonly called navy bean in the west, is the most popular of all; 60% of all the beans handled in New York are this variety. The next in popularity is the red kidney. The medium and yellow eye varieties are also popular. While California produces successfully all of the different varieties, the lima bean is the most popular there. All white beans are sold on a 60-pound basis in New York city, but red beans are reckoned at 58 pounds to the bushel.

The price of beans varies considerably from year to year. During the last few years the price has held around $2 per bushel, although they have sold within the last 10 years as low as 70 cents.

Ten years ago the imports of sheep were about 400,000 head annually. In 1907 the imports were 225,000, and in 1908 21,900.

Leaf Tobacco Culture a Thriving Industry from Florida to Vermont

Yield of Cigar Leaf in 1908 was 458,000 Cases — Manufactures Reduced by Financial Depression — Growers Bending Their Efforts Toward Fine Sumatra

Of all the products of the soil probably none is so burdened with taxation as tobacco. Discovered in this country by Columbus, according to history and tradition, its cultivation has spread to other lands, but in its native soil it continues to thrive and forms one of the great commercial products of the country. Used for smoking and chewing chiefly, the plant is divided into two distinct types. These are known in the trade as heavy leaf and cigar leaf. The various classes of each type are familiar to all tobacconists and are designated as burley, dark, bright, broadleaf, Sumatra, Havana seed and numerous other nomenclatures.

Closely Watched Commodity

From the planting season until the crop is harvested, through the curing period, and also into the markets, the course of each year's crop is closely followed in the American Agriculturist weeklies. Systematic canvasses are made from time to time to obtain facts concerning the acreage planted, condition of the crop, quantity and quality of the yield, and the trend of the markets. These facts are promptly laid before the readers, and the astute tobacco grower finds them profitable.

Cigar Leaf Territory

Cigar leaf tobacco is grown mainly in Pennsylvania, New York, Wisconsin and the Miami valley of Ohio, and in the Connecticut and Housatonic valleys of New England, and in more recent years in Florida, Texas and Georgia. The latter states are becoming more and more important in the growth of high grade wrapper. Sumatra, which has become a synonym for the finest quality of tobacco for the outside covering or wrapper of a cigar, is now grown in yearly increasing quantities in Florida and Georgia. Extensive farms equipped with canvas covers produce a very fine grade of "shade grown" tobacco, giving promise that

DR A. C. TRUE

For many years Dr True has been Director of the Office of Experiment Stations at Washington, D C. In this trying position, Dr True has performed his work with great exactness, wonderful tact and with satisfaction to all with whom he works. His broad sympathy has given encouragement to hundreds of young teachers and investigators, and his clear thinking has helped in solving many perplexing difficulties connected with scientific development in agriculture.

eventually the American farmer may produce enough Sumatra wrapper to make the importation of the six or seven million pounds which now come annually from the Dutch East Indies unnecessary.

Heavy Leaf States

Heavy leaf tobacco is one of the chief products of Kentucky and Tennessee, and it is grown in large quantities in southern Ohio, Indiana, Maryland, Virginia, West Virginia, North and South Carolina. It is used in the manufacture of plug chewing tobacco, various forms of smoking tobacco, and in snuff. One of the most interesting features of the year just closed was the progress made in co-operative marketing by the growers of heavy leaf tobacco. By pooling their interests, millions of pounds of heavy leaf tobacco, some of which has been held back for two seasons, were marketed at prices greatly in advance of bids originally made by the large tobacco manufacturing corporations. Cigar leaf growers are only beginning to profit by this co-operative method of marketing their products. In Wisconsin some progress has been made along this line, and Pennsylvania, New York and New England have co-operative societies.

Effect of the Panic

Information gathered by American Agriculturist in the year 1908 shows that the tobacco industry was curtailed along with other enterprises by the financial disturbances. At the same time, crops were reduced by the unusual drouth. In place of a yield of 476,000 cases of 350 pounds each, as in 1907, the cigar leaf crop of 1908 was only 457,700 cases. On the manufacturing and distributing end there was a shrinkage caused by the stress of finances, as shown in the report of the commissioner of internal revenue, which places the amount of tobacco manufactured at 386,000,000 pounds, compared with 392,000,000 pounds the previous year. The number of cigars manufactured was also reduced more than 3½ millions. Cigarettes alone showed an increase. The total revenues to the United States from manufacturers of tobacco were reduced $2,000,000.

Crops of Cigar Leaf Tobacco

Cigar leaf tobacco harvested in the United States in 1908 and in previous years is shown in the following table compiled by American Agriculturist. The numbers are for cases of 350 pounds each, 000 omitted.

Crop of	1908	1907	1906	1905	1904	1889
Ohio	103	116	131	124	122	107
Wisconsin	115	129	138	122	125	55
Penn	94	94	102	84	83	82
N. England	88	86	91	86	76	34
New York	24	24	24	19	16	27
Southern	33	27	22	16	15	2
Total	458	476	508	451	437	307

Growth of Tobacco Industry

In the following table, compiled from the report of the Commissioners of internal revenue, may be seen the growth of the tobacco industry in the United States. In the figures given six naughts (000,000) are omitted.

	1908	1907	1906	1905	1904
Manuf'd tobacco lbs.	386	392	377	356	348
Cigars, millions	7914	8642	8071	7588	7404
Cigarettes, millions	5402	5167	3793	3376	3226
Internal revenue	$50	$52	$48	$45	$44

Hops Form $9,000,000 Item in Commerce

United States Second in Point of Production —Crop of 1908 was 240,000 Bales

Hops are one of the special lines to which American farmers devote considerable attention. Its importance may be measured by the fact that the crop is an item of about $9,000,000 in commerce of the country. They are grown chiefly on the Pacific coast and in New York. California, Oregon and Washington raise the bulk of the crop of the United States. Wisconsin, which at one time bid fair to be an important hop territory, has

Hop Prices Quarterly

The quarterly range of prices of Pacific Coast hops in New York City is shown in the following table, the figures being for cents per pound:

	March 1	July 1	Oct. 1	Dec. 1
1908	8@9	8 @ 9	10 @11	10@11
1907	16@18	11 @12	12 @14	11@12
1906	13@14	14 @15	20 @24	20@24
1905	31@32	25 @26	19 @20	15@17
1904	32@34	29 @30	29 @31	35@36
1903	28@30	22 @23	29 @ 30	25@27
1902	17@18	21 @22	25 @29	30@32
1901	19@20	17½@18½	14 @15	14@15½
1900	13@14	13½@14½	18 @19	19@21
1899	19@20	16 @18	13½@14½	14@15

turned its attention in other directions. In New York, hop growing is an old, established industry which has shown no material change in recent years. The past year was probably one of the poorest in the history of the industry, because of the prevailing low prices. The spread of prohibition under local option laws has been a disturbing factor to the hop growers.

Useful to Hop Growers

By their careful and systematic canvass of the hop growers, and their reports from the distributing markets, the American Agriculturist weeklies have long made themselves useful to the hop growers of the country. In the year just closed their investigations of the acreage planted and finally of the yield to the acre, showed that the crop amounted to 240,000 bales, compared with 268,500 bales last year. Since the fact of a small crop has been established, there has been an improvement in prices, but the average for the year is 3½ to 4 cents below that of 1907. The world's hop crop in 1907 was 1,139,000 bales of 180 pounds; 1906, 1,032,000 bales; 1905, 1,430,000 bales; 1904, 935,000 bales. Exports have ranged from 7,794,000 pounds in 1903, the lowest in many years, to 16,809,000 pounds in

1907. The banner export year was 1899, when 21,146,000 pounds were exported.

The leading producer of hops is Germany. The United States is a close second, the United Kingdom comes third, Austria and Bohemia following. The surplus from the United States is shipped to the United Kingdom. In the past year there has been temperance agitation in England, which, together with a large crop from Germany in 1907, lessened the demand for American grown hops, demoralizing the home markets.

Cotton Planters Struggle with Cultivation and Market Problems

Struggling against the boll weevil during the season of cultivation and against the speculative system which is blamed for low prices obtained in the markets, the cotton grower continues to produce a crop well above the 10,000,000-bale mark established in 1900. The bumper crop was in the fall of 1904, amounting to 13,557,-000 bales. According to an estimate of the department of agriculture the 1908-9 crop is 12,920,000 bales. Growers are discussing in their local societies and in interstate gatherings, which have been called especially for the purpose, problems of marketing and the advisability of reducing acreage and selling through co-operative associations.

However, while mills for the consumption of cotton are multiplying in this country, right in the southland, too, and with European spinners taking large quantities of the white staple, the planter has not so much to fear from the market side of the industry as on the plantation where the ravages of the boll weevil set at naught a large part of his labor, so far as the southwestern states are concerned. It is estimated that the boll weevil is traveling eastward at the rate of 75 miles a year. The conquering of this evil will bestow a great blessing on the southern planter.

Consumption of Sugar in America Equals Half Population's Weight

Probably the people of no other country have such a "sweet tooth" as those of the United States, each person consuming half his own weight in sugar in a year. To meet this demand for sugar, a large industry has grown up in the production of sugar from cane and beets. Sugar maple trees, from which maple sugar is obtained, also form an important part of the sugar industry, which, however, is confined to northern New England, Ohio, New York, and, most important of all, Vermont. Sugar cane growing is confined to Louisiana and a small section of southwest Texas.

Beet Sugar Outstrips Cane

The sugar beet industry is now the larger and bids fair to expand to greater proportions. Vast areas are planted in the Pacific coast states, Michigan, Wisconsin, Minnesota, and in a limited way in New York. The acreage devoted to sugar beets has been extended from year to year until it has become one of the important industries of the country. The sugar beet is readily grown on irrigated land and thrives in the great farming states of the middle west, notably Michigan.

Great possibilities are apparent in the further development of the sugar beet industry. Recent attention to the kind of seed selected has improved the yield of sugar per ton. As yet, compared with other industries, sugar beets are harvested in a primitive manner. Doubtless American inventive genius will come to the rescue of the farmer by designing machinery for chopping, pulling and topping the beets.

With production less than 800,000 tons, heavy imports are necessary. In the year ended June 30, 1908, imports were 1,502,-236 tons, long tons of 2,240 pounds; 1907, 1,957,256; and in 1906, 1,772,390 tons.

Sugar Production in United States

The following table, compiled from Willet & Gray's Statistical Sugar Trade Journal, shows the production of beet and cane sugar in the United States, in thousands of tons, three naughts (000) being omitted:

	Domestic production		
Fiscal years	Cane long tons	Beets long tons	Total long tons
1908–9	343	440	783
1907–8	347	420	767
1906–7	243	433	776
1905–6	342	284	626
1904–5	350	209	559
1899–0	149	95	244
1894–5	325	20	346
1889–0	136	2	139

Considerable numbers of cattle are brought into the United States each year for breeding purposes. In addition, rough beef cattle cross the line from Mexico into Texas. In recent years total imports of live cattle have averaged around 30,000 head annually; but in the fiscal year, 1908, imports were larger at 92,356, of which 89,168 were butcher cattle and paid duty.

Live Stock Industry is Changing as the West Becomes More Populous

Range Cattle Being Driven Out by the Arrival of Settlers—Present Day Demand is for High Grade, Healthy Stock—Native States and Cities Increase Restrictions

Each year marks progress in the evolution of the cattle raising industry, from the era of unbounded ranges and wild cattle in the early days, to the well-defined ownership of herds and definitely bounded pastures of the present. These were cheap lands, sometimes belonging to the government, sometimes to the state and often to the individual ranch man. Settlers are crowding into the west and crowding the cattle off the ranges, which are becoming highly cultivated farms. Grazing lands which in the recent past sold for a few dollars an acre are now "improved" farms.

Two Important Branches

There are two important branches of cattle raising, one looking to the production of beef and the other to the production of milk. Wisconsin, Michigan, New York and the New England states are prominent in the raising of dairy cattle. Illinois, Indiana, parts of Ohio, all of Iowa, Missouri, Kansas and Nebraska furnish the choice beef for home consumption and export. On the ranges of Montana, Wyoming, Colorado, Texas, parts of the Dakotas, Nebraska and Kansas are fattened the cheaper or grass-fed cattle. It is this branch of the cattle industry that is showing such a rapid change. In 1908 shipments of range cattle to the big stock yards exceeded those of 1907, and the shipments of 1907 were larger than those of 1906. Cattlemen say that the ranges are being cleaned up. Land is

DR HENRY PRENTISS ARMSBY

Dr Armsby is now Director of the Institute of Animal Nutrition of Pennsylvania State College. For a great many years he has been investigating animal feeding problems and his contributions to this field have been large, marked and important.

needed for agricultural purposes of a more highly cultivated type.

Modern civilization is demanding more of the cattle raiser of today than a decade ago. The beef of the country can no longer be made of animals ill or suspected of illness. Throughout the country a campaign has been waging for a higher grade of cattle. Everywhere the federal government keeps its watchful eye

Exports of Domestic Animals and Meat Products

[Stated in round millions]

Year ended June 30	Cattle, *No.	Hogs, *No.	Horses and Mules, *No.	Sheep, *No.	Canned beef, lbs.	Fresh beef, lbs.	Salted beef, lbs.	Tallow, lbs.	Bacon, lbs.	Hams, lbs.	Pickled pork, lbs.	Lard, lbs.	Oleomargarine, lbs.	Oleo oil, lbs.	Butter, lbs.	Cheese, lbs.	Eggs, *doz.
1908...	349	31	26	101	23	201	48	91	241	221	150	603	3	213	6	8	7590
1907...	423	24	40	135	16	282	64	128	250	209	166	628	5	195	13	17	6970
1906...	584	59	47	143	65	268	81	98	361	194	142	742	11	210	27	17	4952
1905...	567	44	40	268	67	237	56	63	262	203	119	610	8	153	10	10	2476
1904...	593	6	46	301	57	299	57	76	249	194	112	561	6	165	11	23	1776
1899...	389	33	52	143	38	282	47	107	563	226	137	711	6	142	20	38	3693
1894...	359	2	7	132	56	194	63	55	417	87	64	448	4	123	12	74	163
1889...	206	45	7	129	51	138	55	78	357	43	64	318	2	28	16	35	549

*Expressed in thousands.

on the live stock; state authorities co-operate and city authorities add restrictions, shutting out from their markets the undesirable animals which some years

Foreign Trade in Cattle

Imports and exports of live cattle for the years ended June 30 were as follows:

CATTLE IMPORTS

Year ended June 30	No.	Value
1908	92,356	$1,507,310
1907	32,404	565,122
1906	29,019	548,430
1905	27,855	464,572
1904	16,056	310,737
1903	66,166	1,161,548
1902	96,027	1,608,722
1901	146,022	1,931,433
1900	181,006	2,257,694
1899	199,752	2,320,362
1898	291,589	2,913,223
1897	328,977	2,589,857
1896	217,826	1,509,856
1895	149,781	765,853
1894	1,592	18,704

EXPORTS OF CATTLE

Year	No.	Value
1908	349,210	$29,339,134
1907	423,051	34,577,392
1906	584,239	42,081,170
1905	567,806	40,598,048
1904	593,409	42,256,291
1903	402,178	29,848,936
1902	392,884	29,902,202
1901	459,218	37,566,980
1900	397,286	30,535,153
1899	389,490	30,516,833
1898	439,255	37,827,500
1897	392,200	36,357,451
1896	372,461	34,560,672
1895	331,722	30,603,796
1894	359,278	33,461,922

ago were driven in indiscriminately. These restrictions are reflected in the high prices which have prevailed. Not since 1901 have native steers reached such a high price in Chicago as in 1908, and never before was the range of prices so narrow. The higher prices were again reflected in the export trade of the United States. After reaching the maximum of 593,409 head of cattle in 1904, exports had declined until in 1907 there were 423,-051 and in 1908, 349,210.

The decrease in the export trade was largely with the United Kingdom, which has always purchased the bulk of American export cattle. The United States' chief rival in the cattle export business is Argentina. Canada also supplies a good share of the European demand for beef cattle and dressed meats. One interesting development observed in the study of the movement of cattle in the United States in 1908 is that St Louis had a record year, while the other leading stock yards barely held their own compared with past years, Chicago being a possible exception.

Features of the year's cattle trade at Chicago were that there was a decrease in receipts of cattle, calves and horses, while sheep and hogs increased. The valuation of all the live stock received at the Union stock yards for the year was $306,566,000, not including $6,270,000 worth of hogs shipped direct to packers.

Top prices at Chicago and the dates of their making were as follows: Hogs, $7.60 on Sept 23; native steers, $6.60 Nov 10; feeders, $6.05 April 15; native sheep, $7.25 April 27; western sheep, $7.00 April 2; lambs $8.35 March 30.

Cattle Movement in the United States

As told by the Records of Stock Yards

In the following tables, compiled from reports and data furnished by courtesy of the various stock exchanges, receipts of cattle, stated in round thousands, are given for 1908 and for previous years:

AT FOUR LEADING WESTERN POINTS

	1908	1907	1906	1905	1904	1903	1902	1901	1900	1899	1898	1895
Chicago	*2617	3305	3329	3410	3259	3432	2942	3031	2729	2514	2481	2557
Kansas City	‡1885	2384	2296	1996	1996	2137	2083	2000	1970	1912	1758	1818
Omaha	†899	1158	1079	1026	944	1071	1011	818	828	838	812	811
St. Louis	a1068	1133	1121	1124	1074	1140	1113	892	698	684	684	803

*January 1 to November 16.
aJanuary 1 to November 30.   †January to November 15.   ‡January 1 to November 15.

IN CITIES OF THE MIDDLE WEST

	1908	1907	1906	1905	1904	1903	1902	1901	1900	1899	1898	1897
Cincinnati	*194	245	242	232	198	198	183	172	169	164	180	169
Indianapolis	*358	378	350	300	275	250	213	211	140	137	135	148
Cleveland	—	†45	68	—	—	80	39	38	32	27	23	21

*January 1 to November 15.   †January 1 to November 1.

EASTERN LIVE STOCK CENTERS

	1908	1907	1906	1905	1904	1903	1902	1901	1900	1899	1898	1897
New York	—	—	584	587	558	573	572	634	640	955	559	479
Boston	*162	249	240	211	202	161	160	180	178	189	193	229
Buffalo	173	‡212	265	220	611	599	520	640	620	500	440	707
Pittsburg	†521	*538	663	577	545	561	405	367	—	—	—	263

*January 1 to November 17.   †Estimated.   ‡January 1 to November 1.

IN OTHER IMPORTANT TRADE CENTERS

	1908	1907	1906	1905	1904	1903	1902	1901	1900	1899	1898	1897
St. Paul	*410	459	487	489	352	303	259	159	171	166	170	179
Sioux City	*334	410	374	402	326	379	399	308	300	348	201	306
St. Joseph	‡465	558	554	502	588	581	494	542	380	286	226	51
New Orleans	a157	b53	50	47	49	44	42	45	53	—	—	123
Denver	†332	558	329	165	265	249	324	227	240	283	288	251

*January 1 to November 15.   †January 1 to November 1.   ‡January 1 to November 18.
aJanuary 1 to November 25.   bYear ended September 1.

Hog Raising Story Told in Statistics

Record of 1908 and Previous Years of an Industry in which United States Excels

In the raising of hogs, the United States leads all other countries. Adapted both by climate and by large production of corn, this country produces 40% of the world's supply of pork, which plays an important part in the food supply of the human race. The United Kingdom, Germany, Belgium and the Netherlands depend upon the United States to a great extent for their pork and pork product. Exports of ham, bacon, lard and dressed pork to these countries amount to millions of dollars annually. These commodities form one of the most important groups of the exports of the United States.

Last year witnessed a high level of prices for live hogs. The range of heavy packing hogs in Chicago was $4.20 to $7.65. The higher price was exceeded only three times before in the history of the Chicago stock yards. In 1902 the

CHAMPION DODDIES AT INTERNATIONAL LIVE STOCK SHOW

These Angus cattle belonged to a herd which made an enviable record at the 1908 fairs. They were exhibited in Chicago by an Iowa stock raiser. The bull is Tenet Lad and the cow is Glenpyle Queen.

Range of Cattle Prices at Chicago

Year	Native steers 1200 to 1800 lbs.	Dry butcher cows and heifers	Stockers and feeders	Western range cattle
1908	$5.50@8.40	$3.25@6.10	$2.25@5.75	$3.15@6.25
1907	3.95@8.00	2.35@6.25	2.00@5.35	3.00@6.75
1906	3.85@7.40	1.85@6.25	2.00@5.00	2.50@5.85
1905	3.60@7.00	1.80@6.00	1.75@5.40	2.15@5.15
1904	3.75@7.00	2.00@5.75	2.00@5.10	2.25@5.00
1903	4.00@6.65	2.75@4.75	2.50@5.00	2.50@4.65
1902	3.60@9.00	3.25@8.25	1.90@6.00	2.00@7.40
1901	3.60@9.30	3.20@8.00	1.65@5.15	1.50@5.75
1900	3.90@8.50	3.20@6.00	2.10@5.25	3.20@5.35
1899	4.00@8.25	3.50@6.85	2.50@5.40	3.75@5.70
1898	3.80@6.25	3.20@5.40	2.40@4.75	3.25@5.00
1897	3.25@6.50	1.50@4.50	2.25@4.50	2.25@4.60
1896	3.40@6.50	1.25@4.50	1.90@4.10	2.10@5.50
1895	3.60@6.60	1.50@5.75	1.75@5.15	1.90@5.75
1894	3.00@6.60	1.00@4.40	1.75@4.15	1.50@5.50
1893	4.00@6.75	1.25@5.00	2.00@4.90	1.75@6.05
1892	3.75@7.00	1.00@4.00	1.50@4.10	1.50@5.20

high mark was $8.25; in 1903, $7.87½, and in 1893 the record price of the yard was made—$8.75 per 100 pounds.

Examination of the export statistics shows a shrinkage in the value and number of hogs shipped abroad. While the financial depression was no doubt in a measure the cause of restricted exports in hogs as well as other commodities, there is another probable cause which will

Range of Hog Prices at Chicago

Year	Heavy packing 260 to 460 lbs.	Mixed packing 200 to 250 lbs.	Light bacon 150 to 200 lbs.
1908...	$4.20@7.65	$5.10@7.40	$5.00@7.40
1907...	3.75@7.25	3.70@7.25	3.70@7.20
1906...	4.25@7.00	4.40@7.10	4.25@7.05
1905...	3.85@6.40	3.80@6.30	3.60@6.25
1904...	3.75@6.30	3.70@6.20	3.50@6.10
1903...	3.90@7.87½	3.85@5.70	3.90@7.55
1902...	5.70@8.25	5.65@8.20	5.40@7.95
1901...	4.80@7.37½	4.85@7.30	4.75@7.20
1900...	4.05@5.85	4.05@5.82½	4.00@5.75
1899...	3.35@4.80	3.40@5.00	3.30@5.00
1898...	3.10@4.80	3.10@4.70	3.10@4.65
1897...	2.50@4.50	2.90@4.60	3.00@4.65
1896...	2.40@4.45	2.75@4.45	2.80@4.45
1895...	3.25@5.45	3.25@5.55	3.25@5.70
1894...	3.90@6.75	3.90@6.65	3.50@6.45
1893...	3.80@8.75	4.25@8.65	4.40@8.50
1892...	3.70@7.00	3.65@6.70	3.60@6.85
1891...	3.25@5.70	3.25@5.75	3.15@5.95

outlive the panic effects. That is the increased production in Europe, particularly in Germany and Ireland, due to the eradication of the swine plague which devastated the herds there a decade ago.

The average price of hogs in Chicago in the year 1908 was $5.55 per 100 pounds, compared with $6.10 in 1907, and $6.30 in 1906, the banner year so far as prices are concerned.

Sheep Are Producers of a Double Revenue

Record of This Branch of Live Stock Industry Shows Gradual Development

Like cattle, the sheep industry is divided into two parts. Cattle are raised for beef and for milk production, swine for meat alone, and sheep for meat and wool. Large tracts of land, not very useful for other purposes, have been made to produce fortunes in sheep. Parts of Montana, Wyoming, Colorado and New Mexico have been especially adapted to sheep raising. Protected by an import duty, wool growers have found a good market at home. Until 1908, wool prices were on a high level. The summer of 1908 saw a decline. Drouth reduced the pasturages of New Mexico, Colorado, Arkansas and to a certain extent the more northern states, and the sheep-raising industry suffered a reverse.

At the leading stock yards of the country it has been noted that farmers in sections of the country who heretofore have been devoted entirely to other pursuits have taken an interest in sheep raising. This in a measure is counter-balancing the curtailment of the sheep-feeding industry in the western states. There was a remarkable range of prices in sheep in 1908. The Chicago market recorded sales of native sheep of poor quality selling as low as $2.50 per 100

Live Hog Traffic in the United States

(As told by the various Stock Yard Records)

Hog receipts from January 1, 1908, together with complete returns for previous years are shown in the following tables compiled from official statistics of the stock yards. The figures are in round thousands:

AT FOUR LEADING WESTERN POINTS

	1908	1907	1906	1905	1904	1903	1902	1901	1900	1899	1898	1897
Chicago	a6851	7201	7275	7726	7239	7325	7895	2890	8109	8178	8817	8364
Kansas City	*3161	2924	2676	2508	2227	1969	2279	3716	3094	3673	3351	
Omaha	*2119	2253	2394	2294	2299	2231	2247	2414	2201	2216	2101	1611
St. Louis	†2065	2301	1923	2026	1955	1700	1330	1924	1792	1801	1728	1627

aJanuary 1 to November 16.   *January 1 to November 15.   †January 1 to December 1.

AT THE HOG MARKETS OF THE MIDDLE WEST

	1908	1907	1906	1905	1904	1903	1902	1901	1900	1899	1898	1897
Cincinnati	*982	938	860	948	870	736	722	767	815	869	895	875
Indianapolis	a2061	1955	1869		1669	1530	1251	1487	1323	1681	1253	
Cleveland	—	‡883	1102	—		885	926	846	989	1098	918	652

aJanuary 1 to November 15.   *January 1 to November 17.   ‡January 1 to November 9.

IN THE HOG MARKETS OF THE EASTERN STATES

	1908	1907	1906	1905	1904	1903	1902	1901	1900	1899	1898	1897
New York	b1871	1802	1750	1822	1878	1518	1349	1681	1825	1825	1797	1578
Boston	*308	1300	1358	1312	1371	1266	1448	1401	1275	1681	1495	1420
Buffalo	1869	‡1523	1752	2279	2384	2440	2227	2040	2032	2160	2558	5621
Pittsburg	a2784	‡2197	2696	1194	1868	2008	1745	1125	—	—		1894

*January 1 to November 17.   aJanuary 1 to November 15.
‡January 1 to November 9.   bJanuary 1 to December 15.

STOCK CENTERS, SOUTH AND WEST

	1908	1907	1906	1905	1904	1903	1902	1901	1900	1899	1898	1897
St. Paul	*940	868	860	855	882	759	659	609	495	365	333	225
Sioux City	*1165	1289	1158	1209	1114	1008	1008	960	833	568	474	350
St. Joseph	‡2102	1923	1908	1900	1657	1700	1699	1105	1679	1402	1034	400
New Orleans	a30	†16	†19	†18	†17	‡11	11	17	19	—	—	18
Denver	*245	b218	193	162	162	117	87	109	116	120	82	75

*January 1 to November 15.   †Year ended September 1.   ‡January 1 to November 18.
aJanuary 1 to November 25.   bJanuary 1 to November 9.

Sheep and Lamb Prices at Chicago During 1908 and Previous Years

[Poor to best, per 100 pounds live weight]

Year	Native sheep	Native lambs	Western sheep	Tex. and Mex. sheep and lambs
1908................	$2.50@7.25	$3.50@12.50	$2.50@7.00	$3.50@10.00
1907................	2.00@7.60	4.00@ 8.60	2.00@7.25	4.00@ 9.25
1906................	2.75@6.50	4.50@ 8.75	3.00@6.40	2.75@ 7.85
1905................	2.50@6.35	4.00@ 8.60	2.75@6.30	2.50@ 7.50
1904................	2.25@6.00	3.50@ 7.10	2.50@5.50	2.40@ 7.25
1903................	2.25@7.00	2.50@ 8.00	2.75@3.75	2.50@ 7.50
1902................	1.25@6.50	2.00@ 7.25	1.25@6.30	2.50@ 7.60
1901................	1.40@5.25	2.00@ 6.25	1.50@5.25	2.75@ 5.90
1899................	2.25@5.65	3.50@ 7.45	2.50@5.55	4.00@ 7.00
1898................	2.00@5.25	3.50@ 7.10	3.00@5.25	3.75@ 6.75
1895................	1.75@5.50	1.75@ 6.35	1.50@5.25	1.00@ 5.15
1894................	1.50@5.40	1.00@ 6.00	1.10@5.40	1.00@ 4.50
1892................	2.25@6.90	3.00@ 8.25	3.00@6.75	2.25@ 6.35
1891................	2.00@7.00	3.25@ 8.50	3.25@6.85	2.05@ 5.75

pounds. In the same year the high price was $7.25. Native lambs also had a remarkable year, touching for a day the high price of $8.35 per 100 pounds.

It is conceded in the trade that the year 1908 was unfortunate for the feeders. Having purchased their sheep and lambs at high prices, the financial panic came on, smashing values of all commodities, feeders being obliged to unload their stock on dull markets. It is claimed that in Colorado alone the feeders lost $1,000,000.

That the reverses of 1908 coupled with the increased values of grazing land will have a marked effect on the future of the live mutton industry is assumed.

One of the problems with which feeders have to contend in years of prosperity is the question of labor. A few years ago herders received $30 to $35 a month. With the rapid industrial development came demand for labor and higher wages, until herders in 1908 commanded $50 a month. Another problem is the increase in railroad rates. Instead of $127 for a car of sheep from Idaho to Chicago, the charge was increased to $184.

The average price for the bulk of the native lambs sold in Chicago in 1908 was $5.50@6.40 per 100 pounds compared with $6.10@7.15 in 1907. The average price for the bulk of western lambs was $5.80@6.60 compared with $6.50@7.30 in 1907. Top price that year $9.25, the highest in the history of the yard.

Exports of live sheep which have shown a decline since 1904 reached the lowest mark since 1893, 101,000 head being exported in 1908 compared with 135,344 in 1907. Imports were not greatly encouraged by the prevailing prices, the total being 224,000 head compared with 221,000 in 1907.

Receipts of Sheep at the Leading Stock Yards

Receipts of sheep in the following tables are stated in round thousands of heads. Figures for 1908 were furnished by the stock yards and are for varying periods of the calendar year. Statistics for previous years were compiled from the official reports of the yards:

AT FOUR LEADING WESTERN MARKETS

	1908	1907	1906	1905	1904	1903	1902	1901	1900	1899	1898
Chicago............	a3699	4218	4805	4737	4505	4583	4516	4044	3549	3683	3589
Kansas City........	*1451	1582	1617	1319	1004	1152	1154	980	860	953	980
Omaha.............	*1880	2038	2165	1970	1754	1864	1743	1315	1277	1086	1085
St. Louis...........	†564	630	579	688	688	528	523	520	416	409	436

*January 1 to November 15.　†January 1 to December 1.　aJanuary 1 to November 16.

SHEEP TRADE OF THE MIDDLE WEST

	1908	1907	1906	1905	1904	1903	1902	1901	1900	1899	1898	1897
Cincinnati..........	318	305	324	323	370	394	356	332	283	387	427	469
Indianapolis........	*97	72	77	—	90	101	103	126	67	65	85	98
Cleveland..........	—	†243	294	—	—	194	187	143	130	96	70	73

*January 1 to November 15.　†January 1 to November 9.

IN THE EASTERN STATES STOCK MARKETS

	1908	1907	1906	1905	1904	1903	1902	1901	1900	1899	1898	1897
New York..........	†2002	1759	1532	1394	1761	1944	2038	2162	1953	1762	1883	1631
Boston.............	*1245	357	342	313	528	426	476	450	367	375	493	559
Pittsburg..........	‡1,113,424	a898	1301	1113	1295	1229	—	—	—	—	—	1011
Buffalo............	‡1264	a1232	1775	1689	2464	2440	1129	2061	1668	1712	1784	1878

*January 1 to November 17.　†January 1 to December 15.　
‡January 1 to November 15.　aJanuary 1 to November 9.

IN GROWING CENTERS SOUTH AND WEST

	1908	1907	1906	1905	1904	1903	1902	1901	1900	1899	1898	1897
St. Paul...........	a301	568	735	818	773	876	601	331	486	382	429	300
Sioux City.........	a42	65	64	57	28	42	61	67	61	36	21	10
St. Joseph.........	*544	765	826	981	794	599	561	526	390	258	121	14
New Orleans.......	b4	†4	†5	†7	†6	†6	13	12	12	—	—	13
Denver............	a498	660	826	519	519	318	317	226	306	221	284	306

*January 1 to November 18.　†Year ended September 1.　‡Year ended July 1.　
aJanuary 1 to November 15.　bJanuary 1 to November 25.

Story of the Horse
His Commercial Side

"Noblest Animal" Scarcely Held His Own as a Valuable Chattel in 1908

Receipts of horses at leading cities and large distributing centers throughout the United States in 1908 indicate that the "noblest animal" scarcely held his own. The export business, also, showed a decrease, and imports for the year ended June 30, 1908, were smaller than for two years previous. Exports for the past five years were as follows: 1908, 19,000 head; 1907, 33,882 head; 1906, 40,087 head; 1905, 34,822 head; 1904, 42,001 head. Imports were: 1908, 5,487 head; 1907, 6,080 head; 1906, 6,020 head; 1905, 5,180 head; 1904, 4,728 head.

In spite of the increased use of the automobile, both for pleasure riding and for commercial purposes, and the enactment

Number of Farm Animals in Each State

According to a report of the department of agriculture the number and average price of farm animals in the several states on January 1, 1908, were as follows:

[Number and total value expressed in thousands, three ciphers being omitted.]

State or Territory	HORSES Number	HORSES Aver. price per head	MILCH COWS Number	MILCH COWS Aver. price per head	OTHER CATTLE Number	OTHER CATTLE Aver. price per head	SHEEP Number	SHEEP Aver. price per head	SWINE Number	SWINE Aver. price per head
Me.	116	$106.00	183	$31.00	151	$16.00	267	$4.09	67	$8.75
N. H.	60	101.00	128	32.50	103	17.00	77	3.87	52	9.25
Vt.	93	101.00	291	30.00	221	14.00	223	4.16	99	8.15
Mass.	81	111.00	196	40.00	92	17.00	45	4.49	70	10.25
R. I.	14	121.00	26	42.50	10	19.00	8	4.40	13	10.00
Ct.	60	118.00	138	37.50	83	19.00	34	4.75	47	10.50
N. Y.	696	113.00	1,789	33.50	907	17.00	1,131	4.81	669	8.90
N. J.	102	113.00	190	43.00	82	21.00	44	4.99	155	10.00
Pa.	607	114.00	1,152	36.00	965	18.00	1,102	4.62	990	7.80
Del.	37	99.00	37	36.50	22	20.00	12	4.64	46	7.50
Md.	158	94.00	155	32.00	140	20.00	163	4.55	293	6.35
Va.	311	97.00	288	28.00	561	19.00	512	4.00	798	5.75
W. Va.	189	102.00	247	33.00	549	22.00	675	4.40	379	5.75
N. C.	190	107.00	294	24.00	450	12.00	220	2.62	1,357	5.60
S. C.	84	118.00	138	27.00	223	12.00	59	2.17	678	5.70
Ga.	139	111.00	308	25.00	680	11.00	269	2.01	1,599	5.50
Fla.	52	104.00	91	29.00	664	10.00	101	1.97	399	3.75
Ohio	949	111.00	928	36.00	1,050	21.00	3,110	4.48	2,559	6.50
Ind.	814	105.00	660	33.00	1,096	21.00	1,215	5.06	3,159	6.20
Ill.	1,591	107.00	1,184	35.00	2,164	22.00	793	5.01	4,672	6.60
Mich.	704	105.00	849	34.00	1,003	16.00	2,130	4.46	1,388	6.60
Wis.	643	105.00	1,392	30.50	1,137	13.00	1,044	4.15	1,910	7.00
Minn.	723	98.00	1,040	28.00	1,279	12.00	459	3.79	1,267	7.10
Iowa	1,419	99.00	1,555	30.50	3,881	21.00	718	4.97	8,413	6.50
Mo.	957	88.00	965	28.50	2,349	20.00	1,017	4.36	3,593	5.15
N. D.	616	97.00	224	27.50	642	16.00	627	3.56	233	7.50
S. D.	560	80.00	618	27.50	1,426	18.00	821	3.63	903	7.00
Neb.	1,015	87.00	879	29.00	3,265	19.00	431	3.76	4,243	6.25
Kan.	1,108	87.00	722	29.00	3,577	20.00	236	4.15	2,663	5.90
Ky.	391	95.00	398	27.50	714	18.00	1,071	4.22	1,274	4.60
Tenn.	315	97.00	331	23.00	595	12.00	348	3.39	1,502	4.65
Ala.	160	89.00	283	21.00	539	8.00	188	1.94	1,251	4.60
Miss.	260	77.00	330	20.00	589	8.00	181	1.80	1,316	4.50
La.	224	66.00	190	24.00	480	10.00	180	1.79	669	4.50
Texas	1,278	65.00	1,072	26.00	7,825	12.00	1,799	2.74	3,147	5.25
Okla.	744	73.00	338	26.00	1,814	16.00	98	2.88	1,588	5.33
Ark.	279	68.00	384	18.50	695	8.00	266	2.13	1,127	3.80
Mon.	292	73.00	69	36.00	879	20.00	5,524	3.90	66	10.00
Wyo.	117	60.00	23	38.00	838	24.00	5,885	4.15	18	9.25
Col.	262	71.00	144	37.00	1,454	20.00	1,695	3.33	150	8.00
N. M.	118	42.00	25	38.00	939	17.00	4,787	3.45	26	7.00
Ariz.	101	53.00	23	43.00	603	17.00	1,031	3.62	18	8.00
Utah	119	71.00	79	31.00	324	17.00	2,967	3.88	61	7.50
Nev.	102	77.00	17	45.00	367	20.00	1,586	3.79	15	10.00
Idaho	150	75.00	69	32.00	344	17.00	3,575	3.55	130	7.00
Wash.	311	98.00	184	37.00	389	18.00	824	3.73	182	7.75
Ore.	285	96.00	158	35.00	758	17.00	2,661	3.58	279	6.25
Cal.	396	94.00	410	36.00	1,155	19.00	2,422	3.47	551	7.20
Total	19,992	93.41	21,194	30.67	50,073	16.89	54,631	3.88	56,084	6.05

The total value of all farm animals in the U. S. on January 1, 1908, was $4,331,227, as follows: Horses, $1,867,530; mules, $416,039; milch cows, $650,939; other cattle, $845,938; sheep, $211,736; swine, $339,030.

of laws restricting the horse-racing business, the values of all classes of horses have been well maintained. The high water mark in values of horses in the United States was reached in 1907, when

In the Horse Markets

Receipts of horses at the leading distributing centers are shown in the following table, stated in thousands, three naughts (000) being omitted:

	1908	1907	1906	1905	1904	1894
Chicago	84	102	127	127	106	97
Kansas City	50	62	70	66	68	44
Omaha	*37	44	42	45	47	8
St. Louis	*100	117	166	178	181	13
Cincinnati	†16	24	26	26	20	4
Indianapolis	19	25	30	†30	31	8
Sioux City	*11	15	19	15	4	1
St. Paul	*14	15	9	6	6	‡
St. Joseph	*21	27	28	32	20	1
Denver	*10	—	—	13	13	6
Buffalo	—	†21	26	29	28	—

*Jan. 1 to Nov. 15.
†Jan. 1 to Nov. 18.
‡Less than 1,000 head.

the 22,518,000 head had an average worth of $106.30. The lowest average price in recent years was $31.51, which was recorded in 1897.

Something About Hide and Skin Industry in the United States

Involving commodities valued at millions of dollars, the hide and skin industry is carried on in the United States in such an unobtrusive way that its importance is not realized by the general public. Next in importance to the packing house hides come the country hides. Native steer hides weigh about 65 pounds, native cow hides about 55 pounds and western range steer hides 60 pounds.

In spite of the large quantity of hides produced in this country, the United States annually imports millions of dollars' worth and exports large quantities of leather goods, also some raw hides. Imports of hides in fiscal year 1908 were 98,353,249 pounds, equal to $12,044,435. Exports of raw hides and skins were 14,650,454 pounds and of leather goods $40,688,619. Exports of leather goods were valued at $40,000,000. For the preceding year $45,000,000. The year 1907-8 was the first in which an increase was not recorded in the leather goods export business.

United States Wool Clip in 1909 Increased—Growers' Co-operation

The total wool production in the United States for the year 1908 was 311,138,000 pounds. It was equivalent to 135,330,648 pounds of scoured wool and had an estimated value of $61,649,616. The clip was the largest since 1902, when

316,000,000 pounds were produced. But the banner year for wool production was 1893, when the high-water mark of 348,538,000 pounds was reached. Wyoming, the leading wool state, produced 36,000,000 pounds for 1908. Montana was second with 32,000,000 pounds, and Oregon and New Mexico were tied for the third place, with 16,500,000 pounds production. This increase in production is attributed by the national association of wool manufacturers, whose figures are used, to lighter shrinkage in the wools of 1908.

With the increased amount of wool produced came a recession in prices which was probably due to the financial panic as much as it was to increased production. The lower level of prices gave rise to the movement which resulted in the erection of a wool growers' storage house in Omaha. An effort to sell the wool at auction was not altogether successful, but the warehouses had the effect of temporarily reducing offerings on the open market and in the meantime prices advanced.

Encouraged by the Omaha experiment, the national wool growers' association has under way a plan for a warehouse in some central city which can handle 25,000,000 to 75,000,000 pounds of wool, this to be offered direct to manufacturers,

Prices of Fleece Wool

Wholesale prices of fleece wool at New York City, in cents per pound follow. The Ohio wools are washed clothing, the Kentucky and Indiana are unwashed.

Year	January Ohio X and XX	January Ohio, medium ‡ and ‖	January Ky. and Ind. ‡ and ‖	July Ohio X and XX	July Ohio, medium ‡ and ‖	July Ky. and Ind. ‡ and ‖
1908	35	34	35	33	26	25
1907	34	40	34	34	39	31
1906	35	39	34	34	38	34
1905	34	38	33	35	40	35
1904	33½	32	25	34½	33	30
1899	26½	29	21½	28½	31	23
1894	23	24	19	20	20	17

or, perhaps, sold as in London at auction. This was tried ten years ago in New York and failed. Imports of raw wool for the year ended June 30, 1908, were 125,980,524 pounds, compared with 203,847,545 pounds in 1907, and 201,688,668 pounds in 1906.

The total supply of all kinds of wool in the United States on January 1 was 64,571,100 pounds, compared with 99,745,060 pounds in 1908. The total supply of domestic wool was 50,556,100 pounds compared with 84,556,560 pounds in 1908 and 94,402,046 pounds in 1907.

Science and Practice of Feeding the Live Stock on the Farm

How to Mix Feeds for the Best Results—Elements Required for Proper Nourishment—Feeding Standards and Table Showing Elements in Different Feed Stuffs

The Balanced Ration

One of the best ways to prepare a ration is first to consider the feeds you have available at reasonable cost. Then combine these in such proportions as to give a good variety and a total amount of feed which the animal will readily consume. Then compare the ration fed with the standards on the next page and calculate its value as a ration. If your feed is deficient in some element, add a little of some of the feeds to bring it up to the standard, at the same time keeping in mind that it must have all of the features above mentioned.

To Calculate a Ration

From the table on the next page select the several feeds which you have available at reasonable cost; get the amounts of digestible nutrients as shown in the table, including dry matter, protein, carbohydrates and fats. These figures are the number of pounds per 100 pounds of the meal. By dividing by 100 you will get the number of pounds of nutrients in one pound of the feed. Then in each case multiply this by the number of pounds which you would reasonably expect to feed and you will have the

PRODUCER AND CONSUMER IN PERFECT ACCORD

amount of nutrients which your animal will get at each feed. It is best to calculate the ration on the basis of the amount of feed fed each day, paying no attention to the number of feeds.

You will need to select feeds that have enough bulk and roughness so that the animal can digest it readily without digestive troubles.

For instance, we decide to use 10 pounds of corn stover and 10 pounds of oat straw.

By comparing this ration with the standard we find that it is too low, so we will need one pound each of oil meal and corn meal to keep up the protein.

CORN STOVER.   OAT STRAW.
In 100 lbs. In 100 lbs. In 100 lbs. In 100 lbs.

Dry
  matter 59.5÷100×10=5.95   90.8÷100×10=9.08
Protein   1.7÷100×10=0.17   1.2÷100×10=.12
Carbohy-
  drates 32.4÷100×10=3.24   38.6÷100×10=3.86
Fats ...   .7÷100×10= .07   .8÷100×10=.08
Totals
Dry matter 15.03, Protein .29, Carbohydrates 7.10,
  Fats .15.
Standard
Dry matter 18. Protein .7, Carbohydrates 8. Fats .10

By combining the nutrients as before we now have a ration which is very near to the standard; in fact, near enough for all practical purposes.

Kind and Amount of Feed

The first essential of a balanced ration is palatability. The combination of feeds must be such that the animal will have a great desire to eat it and will be ready for each feed. Observation of the appearance of the animal will do much toward determining this factor. The feed must be adapted to the needs of the animal. The cost of feeds will also have a great deal to do with what feeds are used to make the balanced ration.

The quantity of food consumed by an animal is not always an indication that all is being used. Large amounts of feed are sometimes eaten by animals which have a stimulated appetite in an effort to satisfy it. A great deal of the feed taken passes through the body unchanged, and the animal actually starves while eating much feed. A variety and change of feed is especially necessary where dry feeding is followed. This is particularly true of dairy cows and swine. It is a good general principle to feed all of the roughage that an animal can conveniently devour and enough grain and other feed to keep the ration up to the standard.

In calculating a ration, it will seldom be possible to feed exactly the amount called for in a standard. Some animals will require more feed and others less.

Feeding Standards

Table showing the amount of dry matter and digestible nutrients required daily by farm animals per 1,000 pounds live weight.

	Dry matter, Lbs.	Protein Lbs	Carbohydrates, Lbs.	Fats, Lbs.
Ox at complete rest in stall	18.0	0.7	8.0	0.1
Fattening cattle (first period)	30.0	2.5	15.0	0.5
Milch cow (yielding 22 pounds daily)	29.0	2.5	13.0	0.5
Horse (medium work)	24.0	2.0	11.0	0.6

Composition of Feeding Stuffs

Corn and Wheat Products

Feed	Dry matter in 100 lbs. Lbs.	Protein Lbs.	Carbohydrates Lbs.	Fats Lbs.
Dent corn	89.4	7.8	66.7	4.3
Flint corn	88.7	8.0	66.2	4.3
Sweet corn	91.2	8.8	63.7	7.0
Corn cob	89.3	0.4	52.5	0.3
Corn and cob meal	84.9	4.4	60.0	2.9
Gluten meal	91.8	25.8	43.3	11.0
Germ meal	89.6	9.0	61.2	6.2
Starch refuse	91.8	11.4	58.4	6.5
Hominy chops	88.9	7.5	55.2	6.8
Starch feed wet	34.6	5.5	21.7	2.3
Wheat	89.5	10.2	69.2	1.7
High-grade flour	87.6	8.9	62.4	0.9
Low-grade flour	87.6	8.2	62.7	0.9
Wheat bran, spring wheat	88.5	12.9	40.1	3.4
Wheat bran, winter wheat	87.7	12.3	37.1	2.6
Wheat shorts	88.2	12.2	50.0	3.8
Wheat middlings	87.9	12.8	53.0	3.4
Wheat screenings	88.4	9.8	51.0	2.2

Other Grains and Seed Products

Feed				
Rye	88.4	9.9	67.6	1.1
Rye bran	88.4	11.5	50.3	2.0
Rye shorts	90.7	11.9	45.1	1.6
Barley	89.1	8.7	65.6	1.6
Malt sprouts	89.8	18.6	37.1	1.7
Brewers' grains wet	24.3	3.9	9.3	1.4
Brewers' grains dried	91.8	15.7	36.3	5.1
Oats	89.0	9.2	47.3	4.2
Oatmeal	92.1	11.5	52.1	5.9
Oat feed or shorts	92.3	12.5	46.9	2.8
Oat hulls	90.6	1.3	40.1	0.6
Rice	87.6	4.8	72.2	0.3
Rice hulls	91.8	1.6	44.5	0.6
Rice bran	90.3	5.3	45.1	7.3
Rice polish	90.0	9.0	56.4	6.5
Buckwheat	87.4	7.7	49.2	1.8
Buckwheat hulls	86.8	2.1	27.9	0.6
Buckwheat bran	89.5	7.4	30.4	1.9
Buckwheat shorts	88.9	21.1	33.5	5.5
Buckwheat middlings	87.3	22.0	33.4	5.4
Sorghum seed	87.2	7.0	52.1	3.1
Broom corn seed	85.9	7.4	48.3	2.9
Kafir corn	84.8	7.8	57.1	2.7
Millet	86.0	8.9	45.0	3.2
Flaxseed	90.8	20.6	17.1	29.0
Linseed meal, old process	90.8	29.3	32.7	7.0
Linseed meal, new process	89.9	28.2	40.1	2.8
Cottonseed	89.7	12.5	30.0	17.3
Cottonseed meal	91.8	37.2	16.9	12.2
Cottonseed hulls	88.9	0.3	33.1	1.7
Sunflower seed	92.5	12.1	20.8	29.0
Peanut meal	89.3	42.9	22.8	6.9
Peas	89.5	16.8	51.8	0.7
Soy beans	89.2	29.6	22.3	14.4
Cowpeas	85.2	18.3	54.2	1.1
Horse bean	85.7	22.4	49.3	1.2

Various Kinds of Silage

Feed				
Corn silage	20.9	0.9	11.3	0.7
Clover silage	28.0	2.0	13.5	1.0
Sorghum silage	23.9	0.6	14.9	0.2
Alfalfa silage	27.5	3.0	8.5	1.9
Cowpea vine silage	20.7	1.5	8.6	0.9
Soy bean silage	25.8	2.7	8.7	1.3
Corn soy bean silage	24.0	1.6	13.0	0.7

Forage Crops

Feed				
Fodder corn, green	20.7	1.0	11.6	0.4
Fodder corn, field cured	57.8	2.5	34.6	1.2
Pasture grasses (mixed)	20.0	2.5	10.2	0.5
Kentucky blue grass	34.9	3.0	19.8	0.8
Timothy	38.4	1.2	19.1	0.6
Orchard grass	27.0	1.5	11.4	0.5
Redtop	34.7	2.1	21.2	0.6
Sorghum	20.6	0.6	12.2	0.4
Meadow fescue	30.1	1.5	16.8	0.4
Hungarian grass	28.9	2.0	16.0	0.4
Green barley	21.0	1.9	10.2	0.4
Peas and oats	16.0	1.8	7.1	0.2
Peas and barley	16.0	1.7	7.2	0.2
Timothy hay	86.8	2.8	43.4	1.4

Orchard grass hay	90.1	4.9	42.3	1.4
Redtop hay	91.1	4.8	46.9	1.0
Mixed hay	87.1	5.9	40.9	1.2
Rowen (mixed)	83.4	7.9	40.1	1.5
Soy bean hay	88.7	10.8	38.7	1.5
Oat hay	91.1	4.3	46.4	1.5
Marsh or swamp hay	88.4	2.4	29.9	0.9
Wheat straw	90.4	0.4	36.3	0.4
Rye straw	92.9	0.6	40.6	0.4
Oat straw	90.8	1.2	38.6	0.8
Barley straw	85.8	0.7	41.2	0.6

Leguminous Crops

Red clover, green	29.2	2.9	14.8	0.7
Alsike, green	25.2	2.7	13.1	0.6
Crimson clover, green	19.1	2.4	9.1	0.5
Alfalfa, green	28.2	3.9	12.7	0.5
Cowpea, green	16.4	1.8	8.7	0.2
Soy bean, green	24.9	3.2	11.0	0.5
Red clover, medium hay	84.7	6.8	35.8	1.7
Mammoth clover hay	78.8	5.7	32.0	1.9
Alsike clover hay	90.3	8.4	42.5	1.5
White clover hay	90.3	11.5	42.2	1.5
Crimson clover hay	90.4	10.5	34.9	1.2
Alfalfa hay	91.6	11.0	39.6	1.2
Cowpea hay	89.3	10.8	38.6	1.1
Soy bean straw	89.9	2.3	40.0	1.0
Pea vine straw	86.4	4.3	32.3	0.8

Roots and By-Products

Potato	21.1	0.9	16.3	0.1
Beet, sugar	13.5	1.1	10.2	0.1
Beet, mangel	9.1	1.1	5.4	0.1
Rutabaga	11.4	1.0	8.1	0.2
Carrot	11.4	0.8	7.8	0.2
Artichoke	20.0	2.0	16.8	0.2
Cabbage	15.3	1.8	8.2	0.4
Sugar beet leaves	12.0	1.7	4.6	0.2
Pumpkin, field	9.1	1.0	5.8	0.3
Rape	14.0	1.5	8.1	0.2
Dried blood	91.5	52.3	.0	2.5
Meat scrap	89.3	66.2	.3	13.7
Beet pulp	10.2	0.6	7.3	—
Cow's milk	12.8	3.6	4.9	3.7
Skimmilk	9.6	3.1	4.7	0.8
Buttermilk	9.9	3.9	4.0	1.1
Whey	6.6	0.8	4.7	0.3

Greenland Mapped at Last

Not until the year 1908 was it known what were the exact boundaries of the world's largest island, Greenland. During the preceding two years Dr Mylius-Erichsen completed the work of mapping Greenland. He lost his life on the way home to Germany. The northeast coast is very different from what geographers had supposed. It had been marked on all previous maps as practically extending from about 78 degrees north latitude in a general northwest direction to the Independence bay of Peary. In fact, it extends for about 300 miles in a northeasterly direction till its most easterly point nearly touches 12 degrees west longitude from Greenwich.

Peary made surveys of the northwestern and northern parts of the Greenland coast, and at last Dr Mylius-Erichsen started out to complete the survey of the unknown coast from Cape Bismarck to the point where his survey could join that of Peary, this completing the map of the island. He sailed on the steamer Denmark from Copenhagen in June, 1906. He established

supply stations late that summer, and the next spring began a great sledge journey. Every mile of the coast was covered until the explorations joined those of Peary at Independence bay. The lives of the explorer and two of his companions were sacrificed to his mistaken notion of the shape of the northern coast of Greenland. Instead of being practically a smooth curve, as he thought, it presented great peninsulas and inlets, so that the food supply with them gave out before they could find their way back to the base of supply.

COMPLETED MAP OF GREENLAND

The body of a companion of the explorer, together with the notes of his survey, were found later by a search party.

The leader of the expedition died November 25, when only a few miles from the food supply. A companion had died 10 days before. Another companion, Bronlund, a Greenlander, reached the station, and when a search party arrived there in 1908 his body was found.

A bottle containing the survey sheets was found slung across the neck of Bronlund. It has been impossible to discover the bodies of Mylius-Erichsen, and Hagen, the other companion who lost his life.

Work of the Forest Service— Its Development, Scope, Methods

Management of the National Forests—Their Present Use and Value— Provision for the Future Generations—Important Experimental Work—Location and Extent of the Forests

The Forest Service is charged, under the direction of the secretary of agriculture, with the administration of the national forests. The forests, whose area on October 1, 1908, was 167,992,-208 acres, are of vital importance for their timber and grass and for the conservation of stream flow. They are so managed as to develop their permanent value as a resource by use. Opposition toward them, based on the belief that preservation would prevent use, has changed with the understanding of their real object to approval and support. The last valid objections to their establishment and maintenance have been removed by the agricultural settlement law of June 11, 1906, and by a clause in the agricultural appropriation act for the year 1906-7. By the first, agricultural land in national forests, if classified as chiefly valuable for agriculture, listed in the local land office, and opened by the secretary of the interior, may be taken up by home builders. Many small tracts of agricultural lands, scattered here and there along creeks and valleys, have unavoidably been included within forest boundaries, though the utmost care secured the elimination of all large bodies of such land when the boundaries were drawn. The need of such a law as that of June 11 was clearly seen, and its passage was secured.

The so-called "10% clause" of the agricultural appropriation bill provided that states having forests were to receive 10% of the gross receipts from the forests within their boundaries, to be distributed among the counties in which the forests lie and devoted to public schools and roads. Congress at the last session increased the percentage to 25%. Many counties have much of their area, in some cases more than half, in national forests, and this land is withdrawn from the possibility of private ownership and taxation. By the new law the loss to the counties from the withdrawal of taxable land is offset.

Business Management

The business management of the national forests is in itself a large undertaking. The business on the national forests is destined to grow rapidly and to assume far-reaching economic importance. In the fiscal year ending June 30, 1908, approximately $1,842,000 was received, chiefly from grazing and timber sales. The returns from timber sales

GIFFORD PINCHOT

As National Forester, Mr Pinchot has done more than any other one man to make successful the movement to save the nation's forests. He is chairman of the National Conservation Commission.

alone were nearly $850,000. Grazing, which formerly had been free, has brought in $2,340,000 under the permit system inaugurated in January, 1906.

The free use of timber and stone which, at the discretion of the secretary of agriculture, is granted to settlers and others who may not reasonably be required to purchase, as well as to school and road districts, churches, or co-operative organizations of settlers, very greatly aids the development of the regions in and near the forests.

It is the active policy of the forest service to manage the national forests upon a sound technical as well as business basis. Only improvement in the standard of the technical management can secure steady and constant increase in returns without depleting the forest. To this end careful investigation is essential. This includes special study of the habits and requirements of trees as a basis for the regulation of cutting of every kind. Special attention is given to finding new uses for species at present valueless or little used, as well as for the trees already classed as commercial and for timber killed by fire or insect attacks. Studies are made of damage by fire and the best means of preventing it, and, in co-operation with the bureau of entomology, of the prevention and control of insect ravages. In these and in many other ways the basis of knowledge necessary for the best forest work is being laid.

Practical Use

Aside from the care and perpetuation of the national forests, the forest service has to do with the practical uses of forests and forest trees in the United States, especially with the commercial management of forest tracts, woodlots, and forest plantations. It undertakes such forest studies as lie beyond the power or the means of individuals to carry on unaided. It stands ready to co-operate, to the limit of its resources, with all who seek assistance in the solution of practical forest problems, particularly where such co-operation will result in setting up object lessons to serve as encouraging examples for the general benefit.

Co-operative state studies are carried on with states which request the advice of the service. Examples of this work are the studies of forest conditions in New Hampshire, which appropriated $7,000 toward the total cost, and California, which appropriated $25,000. Maine, Massachusetts, Maryland, Rhode Island, Delaware, North Carolina, Kentucky, Tennessee, Missouri, and Mississippi have also called upon the service for expert assistance.

The fruits of its more important studies are published and distributed without charge upon request, or sold at a low price by the superintendent of documents.

Administration of the National Forests

For administrative purposes the national forests are grouped into 6 districts. The headquarters of the 6 districts, beginning January 1, 1909, are as follows: District 1, Missoula, Mont; district 2, Denver, Col; district 3, Albuquerque, N Mex; district 4, Ogden, Utah; district 5, San Francisco, Cal; district 6, Portland, Ore. Each district is in charge of a district forester, with an assistant district forester. The office of organization, at Washington, has direct oversight of personnel, equipment, and expenditures. It also examines all matters of an administrative character which are prepared in the other offices, and forms the central agency by which the administrative work of all branches is brought together and harmonized.

The permanent field force on the national forests now contains the grades of chief inspector, inspector, forest supervisor, deputy forest supervisor, forest assistant, planting assistant, lumberman, forest ranger, and forest guard.

Examinations for these positions are held as required in each state and territory in which national forests are situated. Applicants for the positions of ranger or supervisor must be legal residents, between the ages of 21 and 40. The restriction as to residence is not imposed upon applicants for the forest assistant examination, for which the age requirement is 20 to 40 years. Inspectors are appointed only from those who by their qualifications, training, and experience have gained great familiarity with national forest problems and unusual efficiency in the conduct of forest business. Their duties are to inspect the forests in their districts, see and report on existing conditions, recommend changes for the better in both the business and technical management and in personnel, and assist the local officers, by suggestions and advice, in all forest matters.

Forest Supervision

Appointments to the position of forest supervisor are made by the promotion of competent forest rangers or forest assistants, when they can be found in the state or territory in which the vacancies exist. Should there be none satisfactory, examinations of other applicants are held. The qualifications for the position of supervisor include all those required of rangers, as hereafter outlined, with superior technical, business, and administrative ability.

Supervisors have full charge of their forests, plan and direct all work, have entire disposition of rangers and other assistants, and are responsible for the efficiency of the local service. Under instructions from the forester, supervisors deal with the public in all business connected with the sale of timber, the control of grazing, the issuing of permits, and the enforcement of all regulations which govern the use, protection, and occupancy of national forests. Each supervisor is required to keep, at his own expense, one or more horses, and is allowed necessary traveling expenses.

Assistants

The position of forest assistant or planting assistant requires technical qualifications of high order. Forest assistants or planting assistants may be assigned to any part of the United States, and must be competent to handle technical lines of work, such as the preparation of working and planting plans, the investigation of the silvics and uses of commercial trees and the study of wood preservation. When assigned to a national forest, assistants are under the supervisor, from whom they receive their

orders and to whom they report. They are required to own and keep horses when necessary.

Lumbermen, technical men, and others temporarily assigned to forests are directly under the instructions of the supervisor and report to him on all forest matters.

To be eligible as ranger of any grade, the applicant must be, first of all, thoroughly sound and able-bodied, capable of enduring hardships and of performing severe labor under trying conditions. No one may expect to pass the examination who is not already able to take care of himself and his horses in regions remote from settlements and supplies. He must be able to build trails and cabins, shoot, ride, pack, and deal tactfully with all classes of people. He must know something of land surveying, estimating and scaling timber, logging, land laws, mining, and the live-stock business.

Duties

The examination of applicants is along the practical lines indicated above, and they are required to show that they can do these things by actually doing them. Where boats, saddle horses, or pack horses are necessary in the performance of their duty, rangers are required to own and maintain them. Rangers execute the work of the national forests un-

A LAWRENCE COUNTY (ILL) SPRAYING OUTFIT

This simple spraying outfit was made by G. W. Emerick, and is used in his orchard. It is placed in an ordinary farm wagon and can be easily driven among the trees. Its cheapness certainly commends it, and Mr Emerick says it is very effective.

der the direction of the supervisor. Their duties include patrol to prevent fire and trespassing; estimating, surveying, and marking timber; the supervision of cuttings, and other similar work. They issue minor permits, build cabins and trails, enforce grazing restrictions, investigate claims, report on applications, and arrest for violation of forest laws and regulations. In the absence of the supervisor, charge of the forest falls on one of the rangers or assistants.

Deputy rangers and assistant rangers have charge of definite districts, to which they are assigned by the supervisor. They supervise forest guards stationed within their districts, and may also be given temporary laborers when necessary.

In addition to the permanent classified force upon the forests, forest guards receiving from $60 to $75 per month are employed to fill vacancies for which the eligible list is inadequate or to supply additional men for patrol and protective work for not over 6 months at a time. Forest guards have the powers and duties of assistant forest rangers.

At Washington

The administrative head of the forest service is the forester at Washington. There is an associate forester, a law officer, editor, and dentrologist in the forester's office. The administration of the service at Washington covers branches under assistant foresters devoted to operation, grazing, silviculture, and products.

The following table shows the location and extent of the national forests. A table immediately following that also shows national monuments, which are objects of historic or scientific interest, within national forests.

The National Forests

State or Territory.	Forest.	Headquarters of supervisor.	Proclamation effective.	Area, Acres.	Total.
Arizona.....	Apache..............	Springerville...........	July 1, 1908	1,302,711	
	Chiricahua.............	Douglas................	July 2, 1908	287,520	
	Coconino.............	Flagstaff..............	July 2, 1908	3,689,982	
	Coronado.............	Benson................	July 2, 1908	966,368	
	Crook...............	Safford...............	July 1, 1908	788,624	
	Dixie	St. George, Utah......	May 22, 1908	626,800	
	Garces..............	Nogales...............	July 2, 1908	644,395	
	Kaibab..............	Kanab, Utah..........	July 2, 1908	1,080,000	
	Prescott.............	Prescott..............	July 2, 1908	1,465,268	
	Sitgreaves............	Snowflake.............	July 1, 1908	749,084	
	Tonto...............	Roosevelt.............	July 1, 1908	2,067,614	13,668,366
Arkansas....	Arkansas.............	Mena.................	Dec. 18, 1907	1,073,955	
	Ozark...............	Harrison..............	Mar. 6, 1908	917,944	1,991,899
California...	Angeles..............	Los Angeles...........	July 1, 1908	1,350,900	
	California.............	Willows...............	July 2, 1908	976,949	
	Cleveland.............	San Diego.............	July 2, 1908	1,904,826	
	Crater...............	Medford, Oreg........	July 1, 1908	58,614	
	Inyo................	Bishop................	July 2, 1908	1,458,444	
	Klamath.............	Yreka................	July 2, 1908	2,029,348	
	Lassen...............	Red Bluff.............	July 2, 1908	1,209,298	
	Modoc...............	Alturas...............	July 2, 1908	1,165,536	
	Mono................	Gardnerville, Nev......	July 2, 1908	658,106	
	Monterey.............	Salinas...............	July 2, 1908	514,477	
	Plumas..............	Quincy...............	July 2, 1908	1,354,158	
	San Luis.............	San Luis Obispo........	July 1, 1908	355,990	
	Santa Barbara.........	Santa Barbara.........	July 1, 1908	2,027,180	
	Sequoia..............	Hot Springs...........	July 2, 1908	3,051,782	
	Shasta...............	Sisson................	July 2, 1908	1,187,040	
	Sierra...............	Northfork.............	July 2, 1908	1,935,680	
	Siskiyou..............	Grants Pass, Oreg.....	July 1, 1908	37,814	
	Stanislaus.............	Sonora................	July 2, 1908	1,117,625	
	Tahoe...............	Nevada City..........	July 2, 1908	1,595,982	
	Trinity..............	Weaverville..........	July 2, 1908	1,753,033	25,742,782
Colorado....	Arapaho..............	Sulphur Springs........	July 1, 1908	796,815	
	Battlement............	Collbran..............	July 1, 1908	753,720	
	Cochetopa............	Saguache.............	July 1, 1908	932,890	
	Gunnison.............	Gunnison.............	July 1, 1908	945,350	
	Hayden:.............	Encampment, Wyo.....	July 1, 1908	84,000	
	Holy Cross...........	Glenwood Springs......	July 1, 1908	1,251,200	
	La Salle.............	Moab, Utah...........	July 2, 1908	29,502	
	Las Animas...........	La Veta...............	Mar. 1, 1907	196,140	
	Leadville.............	Leadville..............	July 1, 1908	1,184,730	
	Medicine Bow.........	Fort Collins..........	July 1, 1908	659,780	
	Montezuma...........	Mancos...............	July 1, 1908	1,175,811	
	Pike................	Denver................	July 1, 1908	1,457,524	
	Rio Grande...........	Monte Vista..........	July 1, 1908	1,262,158	
	Routt...............	Steamboat Springs......	July 1, 1908	1,049,686	
	San Isabel............	Westcliffe.............	July 2, 1908	560,848	
	San Juan.............	Durango..............	July 1, 1908	1,460,880	
	Uncompahgre..........	Delta................	July 1, 1908	921,243	
	White River...........	Meeker...............	May 21, 1904	970,880	15,693,157

Idaho.......	Beaverhead......	Dillon, Mont..........	July 1, 1908	304,140	
	Boise...............	Boise...............	July 1, 1908	1,147,360	
	Cache..............	Logan, Utah..........	July 1, 1908	276,640	
	Caribou............	Idaho Falls..........	Jan. 15, 1907	733,000	
	Challis.............	Challis.............	July 1, 1908	1,161,040	
	Clearwater.........	Kooskia.............	July 1, 1908	2,687,860	
	Coeur d'Alene.........	Wallace.............	July 1, 1908	1,543,844	
	Idaho..............	Meadows............	July 1, 1908	1,203,280	
	Kaniksu............	Newport, Wash......	July 1, 1908	544,220	
	Lemhi..............	Mackay.............	July 1, 1908	955,408	
	Minidoka...........	Oakley.............	July 2, 1908	619,204	
	Nezperce...........	Grangeville.........	July 1, 1908	1,946,340	
	Payette............	Emmett.............	July 1, 1908	844,240	
	Pend d'Oreille......	Sandpoint..........	July 1, 1908	913,364	
	Pocatello...........	Pocatello...........	July 1, 1908	288,148	
	Salmon.............	Salmon City.........	July 1, 1908	1,762,472	
	Sawtooth...........	Hailey.............	July 1, 1908	1,211,920	
	Targhee............	St. Anthony........	July 1, 1908	1,101,720	
	Weiser.............	Weiser.............	July 1, 1908	764,829	20,099,029
Kansas.....	Kansas.............	Garden City.........	May 15, 1908	302,387	302,387
Minnesota...	Minnesota..........	Cass Lake...........	May 23, 1908	294,752	294,752
Montana....	Absaroka...........	Livingston..........	July 1, 1908	980,440	
	Beartooth..........	Red Lodge..........	July 1, 1908	685,293	
	Beaverhead.........	Dillon.............	July 1, 1908	1,506,680	
	Bitterroot..........	Missoula...........	July 1, 1908	1,180,900	
	Blackfeet..........	Kalispell..........	July 1, 1908	1,956,340	
	Cabinet............	Thompson Falls.......	July 1, 1908	1,020,960	
	Custer.............	Ashland............	July 2, 1908	590,720	
	Deerlodge..........	Anaconda..........	July 1, 1908	1,080,220	
	Flathead...........	Kalispell..........	July 1, 1908	2,092,785	
	Gallatin...........	Bozeman...........	July 1, 1908	907,160	
	Helena.............	Helena.............	July 1, 1908	930,180	
	Jefferson...........	Great Falls.........	July 2, 1908	1,255,320	
	Kootenai...........	Libby..............	July 1, 1908	1,661,260	
	Lewis and Clark.......	Chouteau..........	July 1, 1908	844,136	
	Lolo...............	Missoula...........	Nov. 6, 1906	1,211,680	
	Madison............	Sheridan...........	July 1, 1908	1,102,860	
	Missoula...........	Missoula...........	July 1, 1908	1,237,509	
	Sioux..............	Camp Crook, S. Dak....	July 2, 1908	145,253	20,389,696
Nebraska...	Nebraska...........	Halsey.............	July 2, 1908	556,072	556,072
Nevada....	Humboldt..........	Elko...............	July 2, 1908	558,679	
	Inyo	Bishop, Cal........	July 2, 1908	62,573	
	Moapa.............	Las Vegas..........	July 2, 1908	345,005	
	Mono	Gardnerville........	July 2, 1908	1,440	
	Tahoe.............	Nevada City, Cal......	July 2, 1908	57,675	
	Toiyabe............	Austin.............	July 2, 1908	1,565,680	2,591,052
New Mexico	Alamo.............	Alamogordo..........	July 2, 1908	1,164,906	
	Carson.............	Santa Fe...........	July 1, 1908	946,480	
	Chiricahua	Douglas, Ariz.......	July 2, 1908	178,077	
	Datil..............	Magdalena..........	June 18, 1908	1,255,883	
	Gila...............	Silver City.........	June 18, 1908	1,790,698	
	Jemez.............	Santa Fe...........	July 1, 1908	944,085	
	Las Animas	La Veta, Colo.........	Mar. 1, 1907	480	
	Lincoln............	Capitan............	July 2, 1908	596,603	
	Magdalena..........	Magdalena..........	July 2, 1908	578,445	
	Manzano...........	Albuquerque.......	Apr. 16, 1908	587,110	
	Pecos..............	Santa Fe...........	July 2, 1908	430,880	8,474,547
Oklahoma...	Wichita............	Cache..............	May 29, 1906	60,800	60,800
Oregon.....	Cascade............	Eugene.............	July 1, 1908	1,767,370	
	Crater.............	Medford............	July 1, 1908	1,061,220	
	Deschutes..........	Prineville..........	July 14, 1908	1,504,207	
	Fremont............	Lakeview..........	July 14, 1908	1,260,320	
	Malheur............	John Day...........	July 1, 1908	1,167,400	
	Oregon.............	Portland...........	July 1, 1908	1,787,280	
	Siskiyou............	Grants Pass........	July 1, 1908	1,264,579	
	Siuslaw............	Eugene.............	July 1, 1908	821,794	
	Umatilla...........	Heppner...........	July 1, 1908	540,496	
	Umpqua............	Roseburg...........	July 1, 1908	1,567,500	
	Wallowa............	Wallowa............	July 2, 1908	1,750,240	
	Wenaha............	Walla Walla, Wash.....	Mar. 1, 1907	494,942	
	Whitman...........	Sumpter............	July 1, 1908	1,234,020	16,221,368
South Dakota....	Black Hills..........	Deadwood...........	July 1, 1908	1,163,160	
	Sioux..............	Camp Crook..........	July 2, 1908	100,560	1,263,720
Washington.	Chelan.............	Chelan.............	July 1, 1908	2,492,500	
	Columbia...........	Portland, Oreg........	July 1, 1908	941,440	
	Colville............	Republic...........	Mar. 1, 1907	869,520	
	Kaniksu	Newport...........	July 1, 1908	406,520	
	Olympic............	Hoodsport..........	Mar. 2, 1907	1,594,560	
	Rainier............	Orting.............	July 1, 1908	1,641,280	
	Snoqualmie.........	Seattle.............	July 1, 1908	961,120	
	Washington.........	Bellingham.........	July 1, 1908	1,419,040	
	Wenaha............	Walla Walla..........	Mar. 1, 1907	318,400	
	Wenatchee..........	Leavenworth..........	July 1, 1908	1,421,120	12,065,500

State or Territory.	Forest.	Headquarters of supervisor.	Proclamation effective.	Area, Acres.	Total.
Utah.......	Ashley..............	Vernal................	July 1, 1908	947,490	
	Cache..............	Logan..............	July 1, 1908	257,200	
	Dixie............	St. George............	May 22, 1908	464,320	
	Fillmore..........	Beaver............	July 1, 1908	578,459	
	Fishlake...........	Salina..........	July 2, 1908	537,233	
	La Salle...........	Moab...........	July 2, 1908	444,628	
	Manti..............	Ephraim........	Apr. 25, 1907	786,080	
	Minidoka..........	Oakley, Idaho........	July 2, 1908	117,203	
	Nebo..........	Payson.........	July 1, 1908	343,920	
	Pocatello	Pocatello, Idaho.......	July 1, 1908	10,720	
	Powell...........	Escalante............	July 2, 1908	726,159	
	Sevier..............	Panguitch.........	Jan. 17, 1906	710,920	
	Uinta.........	Provo.........	July 1, 1908	1,250,610	
	Wasatch...........	Salt Lake City........	July 2, 1908	249,840	
					7,424,782
Wyoming...	Ashley..............	Vernal, Utah..........	July 1, 1908	4,596	
	Bighorn..............	Big Horn............	July 2, 1908	1,151,680	
	Bonneville.............	Pinedale............	July 1, 1908	1,627,840	
	Caribou	Idaho Falls, Idaho......	Jan. 15, 1907	7,740	
	Cheyenne............	Saratoga............	July 1, 1908	617,932	
	Hayden.........	Encampment........	July 1, 1908	370,911	
	Shoshone...........	Cody..........	July 1, 1908	1,689,680	
	Sundance...........	Sundance........	July 1, 1908	183,224	
	Targhee	St. Anthony, Idaho.....	July 1, 1908	377,600	
	Teton...........	Jackson............	July 1, 1908	1,991,200	
	Wyoming.............	Afton............	July 1, 1908	976,320	
					8,998,723

Total of 139 National Forests in the United States..155,838,632

Alaska......	Chugach...............	Ketchikan...........	July 2, 1908	5,330,640	
	Tongass................	Ketchikan...........	July 2, 1908	6,756,986	
					12,087,626
Porto Rico..	Luquillo.............	....	Jan. 17, 1903	65,950	
					65,950

Grand total of 142 National Forests...167,992,208

National Monuments

Name.	National Forest.	State.	Date.	Area. Acres.
Cinder Cone..............	Lassen.............	California..........	May 6, 1907	5,120
Gila Cliff-Dwellings.........	Gila..............	New Mexico........	Nov. 16, 1907	160
Grand Canyon..............	Coconino and Kaibab	Arizona..........	Jan. 11, 1908	806,400
Jewel Cave................	Black Hills........	South Dakota	Feb. 7, 1908	1,280
Lassen Peak................	Lassen............	California..........	May 6, 1907	1,280
Pinnacles..................	Monterey...........	California..........	Jan. 16, 1908	2,080
Tonto..................	Tonto............	Arizona..........	Dec. 19, 1907	640

Total area of National Monuments within National Forests........................ 816,960

Forestry Here and in Other Countries

From "What Forestry Has Done," Circular 140, Forest Service, by Treadwell Cleveland, Jr.

Many people in this country think that forestry had never been tried until the government began to practice it upon the national forests. Yet forestry is practiced by every civilized country in the world, except China and Turkey. It gets results which can be got in no other way, and which are necessary to the general welfare. Forestry is not a new thing. It was discussed 2,000 years ago, and it has been studied and applied with increasing thoroughness ever since.

The more advanced and progressive countries arrive first and go farthest in forestry, as they do in other things. Indeed, we might almost take forestry as a yardstick with which to measure the height of a civilization. On the one hand, the nations which follow forestry most widely and systematically would be found to be the most enlightened nations. On the other hand, when we applied our yardstick to such countries as are without forestry, we could say with a good deal of assurance, by this test alone, "Here is a backward nation."

A singular and suggestive exception is England, which, though provided with mountain and heath lands capable of producing a large part of the wood for home consumption, has, with strange indifference, been leading all nations in volume of wood imports and depending mainly upon foreign sources for her supplies. England has hitherto been able to count with certainty upon outside aid from such near neighbors as Norway and Sweden. This policy has seemed satisfactory to the people, in spite of the examples of a more provident policy afforded by rival nations almost at her door. The geographical and economic positions of the country have permitted the government, for the time, at least, to ignore measures found necessary for the public welfare in other countries of the same rank.

The countries of Europe and Asia, taken together, have passed through all the stages of forest history and applied all the known principles of forestry. They are rich in forest experience. Their forest systems were built up gradually as the result of hardship. They did not first spin fine theories and then apply those theories by main force. On the contrary, they began by facing disagreeable facts. Every step of the way toward wise forest use, the world over, has been made at the sharp spur of want, suffering, or loss. As a result, the science of forestry is one of the most practical and most directly useful of all the sciences.

The United States, then, in attacking the problem of how best to use its great forest resources, is not in the position of a pioneer in the field. It has the experience of all other countries to go upon. The forest principles which hundreds of years of actual practice have proved right are at its command. The only question is, how should these be modified or extended to best meet American conditions. In the management of the national forests the government is not working in the dark. Nor is it slavishly copying European countries. It is putting into practice, in America, and for Americans, principles tried and found correct, which will insure to all the people alike the fullest and best use of all forest resources.

Foreign Forestry Lessons

In the history of what forestry has done in other countries, two things stand out. One is that those countries which have gone furthest in the practice of forestry are the ones which to-day are most prosperous, which have the least proportion of waste land, and which have the most promising futures. The other is that those countries which spend most upon their forests receive from them the greatest net returns.

France and Germany together have a population of 100,000,000, in round numbers, against our probable 85,000,-000, and state forests of 14,500,000 acres, against our 160,000,000 acres of national forests; but France and Germany spend on their forests $11,000,000 a year and get from them in net returns $30,000,000 a year, while the United States spent $1,400,000 in 1907 and got a net return of less than $130,000.

Austria has 24,000,000 acres of forest, of which only 7% belongs to the state and 58% is private land. Communal and entailed forests make up the remainder. Of the private forests 34% is in estates ranging from 20,000 to 350,000 acres in area, and for the last 50 years at least 75% of the total forest area has been held in large, compact bodies. These large blocks are naturally favorable to forest management. Private forestry is further encouraged by the system of forest taxation, which relieves forests in which forestry is practiced. In the United States there are many enormous private forest holdings on which forestry would unquestionably be practiced were it not that excessive or ill-devised forest taxation effectually discourages it.

New Law of Heredity

The scientific world is on the point of giving full recognition to a new, strange law. Briefly, the law, which seems to touch the ultimate mysteries of heredity, is this: When pure stock or strains are crossed, it is found that a certain list of qualities remain, so to speak, indestructible, and appear uncontaminated in a definite proportion of the offspring of all generations after the first.

Some concrete examples show the practical effect of the law. When the tall variety of sweet pea and the short variety of sweet pea are crossed, the first generation are all tall. Tallness is the dominant quality over shortness, which is called recessive. But in the second generation it is found that just one-quarter are dwarf, and not only are they dwarf, but they will remain pure dwarf, without any reversion, and when crossed with dwarf will never again show signs of tallness. The other three-quarters will be tall, and of these tall again just one-quarter will be pure tall, and never again show signs of dwarfness. The remaining two-quarters will be impure, but again when crossed with their like will give both pure talls, pure dwarf, and mongrels in due proportion. So that we find in all grandchildren, so to speak, of pure strains the proportion 1:2:1 has a mystic application; that is, one-quarter of these grandchildren will be exact or pure reproductions in one quality of their grandmother, one-quarter will be pure reproductions of their grandfather, and two-quarters, though resembling one grandparent, will have latent in them the qualities of both.

Natural Resources Conservation Begun in Earnest by the Nation

The Great Movement Started by President Roosevelt's Conference of Governors—How It Has Been Organized—Getting at the Facts—Waterways First

An important movement was inaugurated by Pres Roosevelt during the year 1908 in the interest of the conservation of the natural resources of the United States. A conference of the governors of the states, 44 of whom were present, and 500 other persons, including high officials and experts in all lines of industry, met at the White House in May. The conference adopted a declaration of principles in part as follows:

The Governor's Declaration

We, the governors of the states and territories of the United States of America, in conference assembled, do hereby declare the conviction that the great prosperity of our country rests upon the abundant resources of the land chosen by our forefathers for their homes, and where they laid the foundation of this great nation.

We look upon these resources as a heritage to be made use of in establishing and promoting the comfort, prosperity and happiness of the American people, but not to be wasted, deteriorated nor needlessly destroyed.

We agree that our country's future is involved in this; that the great natural resources supply the material basis upon which our civilization must continue to depend, and upon which the perpetuity of the nation itself rests.

We declare our firm conviction that conservation of our natural resources is a subject of transcendent importance, which should engage unremittingly the attention of the nation, the states and the people in earnest co-operation. These natural resources include the land on which we live and which yields our food; the living waters which fertilize the soil, supply power and form great avenues of commerce; the forests which yield the materials for our homes, prevent erosion of the soil and conserve the navigation and other uses of our streams, and the minerals which form the basis of our industrial life and supply us with heat, light and power.

We commend the wise forethought of the president in sounding the note of warning as to the waste and exhaustion of the natural resources of the country, and signify our high appreciation of his action in calling this conference to consider the same and to seek remedies therefor through co-operation of the nation and the states.

Co-operation Between States and Congress

We agree that this co-operation should find expression in suitable action by the congress within the limits of, and co-extensive with, the national jurisdiction of the subject, and complementary thereto, by

NEW LIVE STOCK BUILDING, MINNESOTA STATE FAIR

the legislatures of the several states within the limits of, and co-extensive with, their jurisdiction. We declare the conviction that in the use of the natural resources our independent states are interdependent and bound together by ties of mutual benefits, responsibilities and duties.

We agree in the wisdom of future conferences between the president, members of congress and the governors of the states regarding the conservation of our natural resources, with the view of continued co-operation and action on the lines suggested. And to this end we advise that from time to time, as in his judgment may seem wise, the president call the governors of the states, members of congress and others into conference.

We agree that further action is advisable to ascertain the present condition of our natural resources and to promote the conservation of the same. And to that end we recommend the appointment by each state of a commission on the conservation of natural resources to co-operate with each other and with any similar commission on behalf of the federal government.

We urge the continuation and extension of forest policies adapted to secure the husbanding and renewal of our diminishing timber supply, the prevention of soil erosion, the protection of headwaters and the maintenance of the purity and navigability of our streams. We recognize that the private ownership of forest lands entails responsibilities in the interests of all the people, and we favor the enactment of laws looking to the protection and replacement of privately owned forests.

We recognize in our waters a most valuable asset of the people of the United States, and we recommend the enactment of laws looking to the conservation of water resources for irrigation, water supply, power and navigation, to the end that navigable and source streams may be brought under complete control and fully utilized for every purpose. We especially urge on the federal congress the immediate adoption of a wise, active and thorough waterway policy, providing for the prompt improvement of our streams and conservation of their watersheds required for the uses of commerce and the protection of the interests of our people.

We recommend the enactment of laws looking to the prevention of waste in the mining and extraction of coal, oil, gas and other minerals with a view to their wise conservation for the use of the people and to the protection of human life in the mines.

Let us conserve the foundation of our prosperity.

Organization

The first step was organization. The president appointed a national conservation commission, organized in four sections, to consider the four great classes of natural resources: Water, forest, lands, and minerals. A chairman and a secretary were designated for each section, and the chairmen and secretaries of the sections act as the executive committee. The chairman of the executive committee acts as chairman of the entire committee. He is Gifford Pinchot, the national forester. The secretary of the commission is Thomas R. Shipp of Washington, D C. The inland waterways commission forms the section of waters of the national conservation com-

mission. The four sections of the conservation commission are organized with the following make-up:

Waters—Theodore E. Burton, Ohio, chairman; Dr. W. J. McGee, bureau of soils, secretary; Francis G. Newlands, Nevada; William Warner, Missouri; John H. Bankhead, Alabama; F. H. Newell, reclamation service; Gifford Pinchot, forest service; Herbert Knox Smith, bureau of corporations; Joseph E. Ransdell, Louisiana; Dr. George F. Swain, institute of technology, Massachusetts, and Brig.-Gen. William L. Marshall, chief of engineers, U. S. A.

Forests—Reed Smoot, Utah, chairman; Overton W. Price, forest service, secretary; Albert J. Beveridge, Indiana; Charles A. Culberson, Texas; Charles F. Scott, Kansas; Champ Clark, Missouri; Prof. Henry S. Graves, Yale forest school; William Irvine, Wisconsin; Newton C. Blanchard, Louisiana; Charles L. Pack, New Jersey; Gustav H. Schwab, national council of commerce, New York.

Lands—Knute Nelson, Minnesota, chairman; George W. Woodruff, assistant attorney-general, secretary; Francis E. Warren, Wyoming; Swager Sherley, Kentucky; Herbert Parsons, New York; N. B. Broward, Florida; James J. Hill, Minnesota; George C. Pardee, California; Charles Macdonald, American society of civil engineers, New York; Murdo Mackenzie, Colorado; T. C. Chamberlin, university of Chicago; Frank C. Goudy, Colorado.

Minerals—John Dalzell, Pennsylvania, chairman; Dr. Joseph A. Holmes, geological survey, secretary; Joseph M. Dixon, Montana; Frank P. Flint, California; Philo Hall, South Dakota; James L. Slayden, Texas; Andrew Carnegie, New York; Dr. Charles R. Van Hise, Wisconsin; John Mitchell, Illinois; John Hays Hammond, Massachusetts; Dr. Irving Fisher, Yale university, Connecticut; Dr. I. C. White, West Virginia.

The inland waterways commission was first appointed by the president in March, 1907. Its creation was due to the demands of commercial organizations throughout the Mississippi valley. The project desired to be accomplished was a deep waterway from the Great Lakes to the Gulf of Mexico, including incidentally deepening the channel of the Mississippi, Missouri, and Ohio rivers. From this beginning it has been proposed to extend the development of navigation to many other rivers and lakes, including the great river systems of the south, of the Atlantic and Pacific coasts.

Congress neglected to extend the life of the waterways commission. The president, however, believing that the work originally committed to the commission ought to be continued without delay, reappointed the commission after congress had adjourned in June, 1908. Arrangements were made for co-operation between the national conservation commission and the executive departments of the federal government. Co-operation between the national committee and state commissions appointed by the governors was secured. Not all states thus co-operated, but many did.

Getting Facts

Organization being accomplished, the next step was to secure information. An elaborate schedule of inquiries was prepared, covering the four classes of resources under consideration. The questions about lands included requests for information regarding present conditions and what is needed under these topics: Public land laws, tenure, agricultural production, the public range, and swamp and overflow lands. The questions about waters covered irrigation, navigation, power and floods. The questions about forests included timber, pulpwood supply, public and private forest administration, waste through fire and careless logging, forest planting. Under minerals, information was sought with reference to existing resources, present production and use, waste, and the protection of life and property in mining. Information was also sought with reference to conserving our resources of live stock, fish and game.

Waterways Have Call

. The results of the investigation were reported to the president and were to be embodied in a special message to congress which had not been issued at the time this article was prepared. At the conference in December, special emphasis was made upon the development of our inland waterways for navigation purposes. Approval was expressed for the lakes-to-the-gulf deep waterway project, but especially was it urged by President Roosevelt and President-elect Taft, who apparently voiced the sentiments of a majority of those interested, that a comprehensive plan should be prepared for inland waterway development of the nation. As forests and waterways cannot be separated in a successful treatment of either, forest protection was given serious consideration.

In order to provide funds for waterway development, it was proposed that a bond issue be authorized by the government, so that these permanent improvements should not be paid for out of current income. The scheme proposed is similar to that employed in financing the Panama canal.

The rivers and harbors congress held at Washington at the same time that the December conference was held went on record as desiring to have congress authorize the issue of $500,000,000 of bonds. the proceeds to be used for river and harbor work. A permanent commission to study waterway projects in the United States and abroad was urged.

The Country Life Commission

A national commission was appointed by Pres Roosevelt in the latter part of the summer of 1908 with a view to securing better social, sanitary, and economic conditions on the farms of the United States. The commission included the following: Chairman, Prof L. H. Bailey of the New York college of agriculture, Henry Wallace of Wallaces' Farmer, Des Moines, Ia, Pres Kenyon

PROF L. H. BAILEY

Prof Bailey has long been in the public eye as the champion of the farmer and country avocations. He is director of the New York State College of Agriculture, and at the present time heads Pres Roosevelt's Country Life Commission. Prof Bailey has done much to popularize agriculture as a profession for young men.

L. Butterfield of the Massachusetts agricultural college, Gifford Pinchot, United States forester, and Walter H. Page, editor of the World's Work, New York.

The commission made a tour of the country, held public meetings, and through mail inquiries secured data and suggestions from thousands of people interested in farm life. The report of the commission will be made the basis of a message to the 60th congress.

New Era in Turkey Brings Many Reforms

Proclamation of Constitution Followed by Secession of Turkish Provinces—Results of the New Turk Movement

Turkey entered a new era in the autumn of 1908. A reform political organization known as the Young Turks accomplished a bloodless revolution. Frightened at the prospect of trouble for himself, the sultan, Abdul Hamid II, restored the constitution granted and revoked 30 years ago. Under the constitution, the Turkish parliament convened in December, 1908. Important reforms in the administration of the government were inaugurated, and many social and other reforms have begun throughout the empire. Even the doors of the imperial harem have been thrown open, and

TURKEY IN EUROPE

This map shows the Balkan peninsula, where war was threatened in 1908, and Crete, which calmly pulled down the Turkish flag and raised the Grecian. Within a single week Bulgaria, which includes eastern Roumelia, declared itself an independent empire, Bosnia and Herzegovina were annexed by Austria-Hungary, and Crete seceded. The dismemberment of Turkey has long been feared, because of the expected quarrel over the division of the spoils among the European nations.

Turkish women at last appear in public unveiled, as in other lands.

Impressed with what appeared to be a confession of weakness on the part of the sultan, Bosnia and Herzegovina, which have been Turkish provinces, were annexed by Austria-Hungary. The latter has exercised practical control over the provinces for years. Bulgaria, including eastern Roumelia, declared itself an independent empire, and another Turkish province, the island of Crete, seceded and proclaimed its union with Greece. A conference of the powers will be held during 1909 to decide whether or not Turkey shall lose these provinces, and if so, upon what terms. The separation is apparently final.

The Ottoman Empire

Turkey and its tributary states which make up the Ottoman empire include a population of 40,000,000 people and has an area of 1,600,000 square miles. The fundamental laws of the empire are based upon the Koran. The will of the sultan has been absolute, so far as not opposed to the Mohammedan religion. The immediate possessions of the empire lie in Europe, Asia, and Africa. These include what is known as European Turkey, Asia Minor, Armenia, Kurdistan, Mesopotamia, Syria, Arabia, and Tripoli. The tributary states are Bulgaria, Bosnia, Crete, Cyprus, Samos, and Egypt. Probably Germany has more influence with the Turkish government at present than any other of the great powers, with the exception of England, which practically rules Egypt.

Disastrous Italian Earthquake

A terrible earthquake devastated southern Italy and the portion of the island of Sicily near by. About 200,000 lives were lost and 20 cities and villages were destroyed. Reggio, in Calabria, Italy, and Messina, in Sicily, were the two largest cities destroyed. The earthquake caused a tidal wave in the strait of Messina. This inundated a number of towns and sunk many ships. Throughout most of the coast region of Calabria and northeastern Sicily there was great distress among the survivors. Thousands of persons were left without shelter and food or means of livelihood, and thousands were injured in the ruined towns.

A great work needed to be done to rescue the injured from the ruins and to care for them. The Italian government, led by the king and queen, who went personally to the scene of the disaster, at once undertook the relief work. Aid was given by foreign warships near enough to be sent to the scene of the trouble, and throughout the world relief funds and supplies have been gathered to be forwarded to the sufferers.

Relief From United States

In this relief work the United States has taken a prominent part. Through the Red Cross society liberal contributions were made. President Roosevelt ordered ships of the battleship fleet, which is on its way around the world and was about to enter Suez canal, to take on supplies and rush to the Italian sufferers; $300,000 worth of supplies were thus taken. Congress, upon the president's request, passed an appropriation to cover this, and appropriated $500,000 more.

After the first shock intermittent earthquake shocks followed for a number of days, adding to the damage and increasing the fright of the stricken people. The relief parties found hundreds of little children whose parents had been killed, and very many families were broken up and separated. The distress in the cities was greatly increased by the fact that water systems were demolished, and there was serious lack of drinking water.

Robbers pillaged the ruins until driven away or shot down by soldiers, sailors and others who finally came to the rescue. The population of Messina was 100,000, and that of Reggio 45,000. Not more that 10,000 or 15,000 altogether escaped from these cities.

American Consul Killed

Among those who lost their lives in the earthquake at Messina was an American consul, A. S. Cheney, and his wife; also Joseph H. Pierce, former vice-consul, and several members of his family.

The earthquake was supposed to have been caused by volcanic action between Mt Etna in Sicily and Mt Vesuvius, so that a slip occurred in the earth's crust. The sea bottom in the strait of Messina has changed considerably, rising in some parts and sinking in others, the action being similar to that which is supposed to have occurred when Sicily was separated from the mainland.

Little Helen, at the close of her evening prayer, said: "And, O God, make me a good girl. I asked you to yesterday, but you didn't."

A Year of Progress in Aerial Navigation

The Achievements of the Wright Brothers in France and America—Count Zeppelin's Giant Dirigible Balloon

Great progress was made in aerial navigation during the year 1908. The motor-driven aeroplane of the brothers Orville and Wilbur Wright of Dayton, O, was the most successful flying machine produced. Orville Wright was engaged in flights at Fort Meyer, Va, preliminary to undertaking official government tests when his machine broke and fell 150 feet. Mr Wright was seriously injured and Lieut Thomas E. Selfridge of the United States

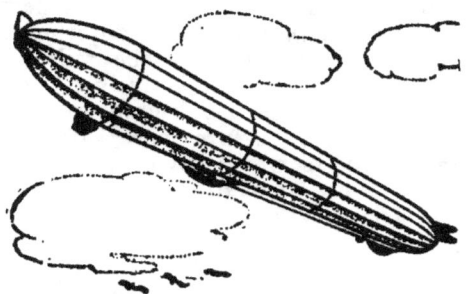

COUNT ZEPPELIN'S DIRIGIBLE BALLOON

army signal corps, who accompanied him, was killed. The United States army is in the market for aeroplanes and war balloons. It is expected that a Wright aeroplane will be submitted to the government during 1909. The accident occurred in September, 1908.

At that time Wilbur Wright was breaking aeroplane records, with a similar machine, in France. He made a flight of 61 miles in 1 hour, 31 minutes and 51

WRIGHT BROTHERS' AEROPLANE

seconds. The patent rights on the Wright aeroplane in France were sold for $100,-000, a company was formed for their manufacture and 50 were ordered at once. The Wright brothers plan to build a fac-

tory for the manufacture of improved types of their aeroplanes at Detroit, Mich.

The Wright aeroplane is 40 feet wide, its greatest height 8 feet. It weighs 800 pounds. It is propelled by a 4-cylinder, 30-horse-power motor. There are 2 propellers, a vertical rudder behind the ma-

FARMAN AEROPLANE

chine, and a horizontal rudder in front. The surface of the planes consists of 500 square feet of taut muslin.

Dirigible balloons have reached their highest type in that of Count von Zeppelin of Germany. He constructed one in 1908 that was 420 feet long and 40 feet in diameter. He went up with it July 1 and sailed around 12 hours, at an average speed of 34 miles an hour, keeping it always under control. On August 6, while at anchor, it caught fire and was destroyed. The people of Germany raised nearly

AEROPLANE OF SANTOS-DUMONT

$1,000,000, with which Count Zeppelin built another like the one he lost.

Others have made important progress in the construction of airships, but the achievements of the Wrights and Count von Zeppelin seem to have been most noteworthy.

Mardi Gras is the name frequently applied to Shrove Tuesday, the day before Ash Wednesday, the first day of Lent. It is the last day of the carnival and is noted for its merrymakings throughout Italy and the south of France. The carnival of Mardi Gras in New Orleans is celebrated with great splendor.

The 60th Congress—First Session's Work

Currency Measure Provides for Emergency—Other Laws Passed—The Agricultural Appropriations

The 60th congress, during its first session in 1908, made appropriations exceeding $1,000,000,000, thereby breaking all records in time of peace. Aside from the appropriations, the most noteworthy achievement of the session was the enactment of the Aldrich-Vreeland currency measure, which is a temporary provision for financial emergencies and expires according to its term 6 years from July 1, 1908. It provides that national banks to the number of not less than 10 in any city, or in contiguous territory outside the large cities, having together a capital and surplus of at least $5,000,000, may organize as national currency associations; any bank thus allied with a circulation of not less than 40% of its sound stock capital, with a surplus of not less than 20%, may deposit with its association securities representing actual commercial transactions with at least 2 responsible names and not exceeding 4 months to run, and upon application by the association to the treasury department, circulation may be issued not over 75% of the cash value of such securities.

Any such banks may directly offer municipal bonds and may take out circulation 90% of the market value not to exceed par of such bonds. The total amount of circulation issued upon commercial paper and municipal bond shall not exceed $500,000,000 for the whole United States, which shall be fairly distributed by the secretary of the treasury. Such circulation must pay a tax of 5% per annum for the first month, and for each succeeding at the rate of 1% additional until the rate of 10% is reached and 10% shall thereafter apply until the notes are retired. This compromise currency measure carries provision for an investigation and report on the currency situation with a view to permanent legislation for currency reform. The members of the commission are Senators Aldrich, Burrows, Hale, Knox, Daniel, Teller, Money and Bailey, and Representatives Vreeland, Overstreet, Burrows, Weeks, Bonynge, Smith, republicans; Padgett, Burgess, Pujo, democrats.

Other Measures

Other legislation included the following: The employers' and government liability bills, which make the railroads and the government respectively responsible for injuries to their employees; several measures to secure more safety to railway and steamboat travelers and employees; a law regulating child labor in the District of Columbia; a bill providing for investigation of mining conditions and bringing the Alaska coal fields under stringent government control, so that a private coal monopoly is made illegal; a measure enlarging the medical and nursing branches of the army and navy and perfecting the military system; increasing the pay of soldiers and sailors of the navy, and providing pensions for the widows of soldiers and sailors; prohibiting race-track betting in the District of Columbia; limiting appeals on habeas corpus proceedings; repealing the statute applying the American coast-

THE NATIONAL CAPITOL AT WASHINGTON

wise shipping laws to the Philippines; voting to return the surplus boxer indemnity to China, and, on the part of the senate, the confirmation of important international agreements.

The latter included arbitration treaties with Great Britain, Japan, France, Italy, Spain, Switzerland, Mexico, Portugal, The Netherlands, Sweden, and Denmark. These treaties resulted from The Hague peace conference of 1907. A century and a quarter of disputes over Canadian questions was brought to satisfactory adjustment by an agreement concerning the boundary and fisheries. The senate at this session ratified more treaties than during the last preceding 20 years.

The Agricultural Appropriations

Salaries, office of the secretary..	$132,900
Weather bureau....................	1,662,260
Bureau of animal industry......	1,080,860
Bureau of plant industry......	1,341,676
Forest service....................	3,896,200
Bureau of chemistry............	826,720
Bureau of soils................	234,700
Bureau of entomology..........	184,960
Bureau of biological survey.....	62,000
Division of accounts and disbursements	46,690
Division of publications........	179,710
Bureau of statistics............	221,640
Department library.............	33,580
Contingent expenses............	86,200
Office of experiment stations....	1,034,620
Office of public roads..........	87,390
Paper making tests.............	10,000
Naval stores industry..........	10,000
National bison range...........	40,000
Prevention of spread of moths..	250,000
Eradicating cattle ticks........	250,000
Total in agricultural appropriation bill	$11,672,106
Meat inspection	3,000,000
Printing and binding..........	460,000
Grand total	$15,132,106

The Pension Roll

The number of pensioners on the government roll at the close of the fiscal year 1908 was 951,686; one year before there were 967,371. During the fiscal year 1908, 38,682 were added to the roll and 54,366 were dropped, a net loss of 15,684. The most pensioners at any one time was 1,004,196, on January 31, 1905. The net decrease since then is 52,509.

The survivors of the Civil War on the pension roll June 30, 1908, numbered 620,985, a decrease during the year of 23,383. The number dropped because of death during the year was 34,333.

The amount paid out for pensions during the fiscal year 1908 was the largest since the organization of the pension bureau, $153,093,086, with the exception of

the year 1893 when $156,906,637 was paid. The cost of administration in 1908 was the least since 1883. $2,800,000

The following table shows the number of pensioners on the roll in each class at the close of the fiscal years 1908 and 1907:

	1908	1907
Revolutionary war:		
Daughters...........................	2	3
War of 1812:		
Widows........................	471	558
Indian wars:		
Survivors.......................	1,820	2,007
Widows.........................	3,018	3,201
War with Mexico:		
Survivors.......................	2,932	3,485
Widows.........................	6,914	7,214
Civil war:		
General law—		
Invalids.....................	142,044	178,816
Widows......................	75,515	75,629
Minor children...............	541	599
Mothers......................	3,688	4,578
Fathers......................	656	873
Brothers, sisters, sons and daughters	240	224
Helpless children................	528	489
Act of June 27, 1890—		
Invalids.....................	140,600	349,283
Widows......................		180,539
Minor children...............	3,954	4,032
Helpless children................	295	292
Act of February 6, 1907........	338,341	116,239
Act of April 19, 1908, widows....	188,445	......
Army nurses....................	510	542
War with Spain:		
Invalids.....................	20,548	19,031
Widows......................	1,145	1,100
Minor children...............	331	316
Mothers......................	3,096	3,090
Fathers......................	536	527
Brothers and sisters..............	7	11
Helpless children...............	2	2
Regular establishment:		
Invalids.....................	11,786	11,076
Widows......................	2,580	2,526
Minor children...............	120	122
Mothers......................	871	821
Fathers......................	139	133
Brothers and sisters..............	5	5
Helpless children...............	7	8
Total........................	951,687	967,371
Net loss......................	15,684	

The last surviving pensioned soldier of the War of 1812 was Hiram Cronk, 105, of Ava, N Y, who died May 13, 1905.

McCumber Service Pension Law

In 1904 Pres Roosevelt directed that when a claimant has passed 62 years pension examiners should consider him disabled one-half in ability to perform manual labor and he should be rated as entitled to $6 a month. It was further ordered that at 65 the rate should be $8, after 68 $10 and after 70 $12, the full rate for disability. The McCumber service pension law goes a little further. It abolishes all classes and awards a pension of $12 after 62 years, $15 after 70 and $20 after 75.

The man who always stops to think before speaking may not say very much, but he seldom has occasion to take any of it back.

The Presidential Succession

The presidential succession is fixed by chapter 4 of the acts of the 49th congress, first session. In case of the removal, death, resignation or inability of both the president and vice-president, then the secretary of state shall act as president until the disability of the president or vice-president is removed or a president is elected. If there be no secretary of state, then the secretary of the treasury will act; and the remainder of the order of succession is as follows: Secretary of war, attorney-general, postmaster-general, secretary of the navy and secretary of the interior. The secretary of agriculture and the secretary of commerce and labor were added by subsequent enactment. The acting president must, upon taking office, convene congress, if not at the time in session, in extraordinary session, giving 20 days' notice. This act applies to such cabinet officers as shall have been confirmed by the senate and are eligible.

David R. Atchison, United States senator from Missouri, was legal president of the United States, Sunday, March 4, 1849, as Gen Taylor, the president-elect, was not sworn into office to succeed Pres Polk until the following day.

Values of Foreign Coins

Proclaimed by the Secretary of the Treasury, Washington, October 1, 1906.

Country	Standard	Monetary unit		Value in terms of U. S. gold dollar
Argentine Republic	Gold	Peso		$0.965
Austria-Hungary	Gold	Crown		.203
Belgium	Gold	Franc		.193
Bolivia	Silver	Boliviano		.382
Brazil	Gold	Milreis		.546
British Possessions, N. A. (except Newf'nd).	Gold	Dollar		1.000
Central American States—				
Costa Rica	Gold	Colon		.465
British Honduras	Gold	Dollar		1.000
Guatemala				
Honduras				
Nicaragua	Silver	Peso		.382
Salvador				
Chile	Gold	Peso		.365
China	Silver	Tael	Canton	.624
			Haikwan (Customs)	.637
			Peking	.610
			Shanghai	.572
			Hongkong	.412
		Dollar	British	.412
			Mexican	.415
Colombia	Gold	Dollar		1.000
Denmark	Gold	Crown		.268
Ecuador	Gold	Sucre		.487
Egypt	Gold	Pound (100 piastres)		4.943
Finland	Gold	Mark		.193
France	Gold	Franc		.193
German Empire	Gold	Mark		.238
Great Britain	Gold	Pound sterling		4.866½
Greece	Gold	Drachma		.193
Haiti	Gold	Gourde		.965
India [British]	Gold	Pound sterling*		4.866½
Italy	Gold	Lira		.193
Japan	Gold	Yen		.498
Liberia	Gold	Dollar		1.000
Mexico	Gold	Peso†		.498
Netherlands	Gold	Florin		.402
Newfoundland	Gold	Dollar		1.014
Norway	Gold	Crown		.268
Panama	Gold	Balboa		1.000
Persia	Silver	Kran		.070
Peru	Gold	Libra		4.866½
Philippine Islands	Gold	Peso		.500
Portugal	Gold	Milreis		1.080
Russia	Gold	Ruble		.515
Spain	Gold	Peseta		.193
Straits Settlements	Gold	Pound sterling‡		4.866½
Sweden	Gold	Crown		.268
Switzerland	Gold	Franc		.193
Turkey	Gold	Piastre		.044
Uruguay	Gold	Peso		1.034
Venezuela	Gold	Bolivar		.193

NOTE.—The coins of silver-standard countries are valued by their pure silver contents, at the average market price of silver for the three months preceding the date of this circular.

*The sovereign is the standard coin of India, but the rupee ($0.3244) is the current coin, valued at 15 to the sovereign.

†Seventy-five centigrams fine gold.

‡The current coin of the Straits Settlements is the silver dollar issued on government account and which has been given a tentative value of $0.5677584.

Home Economics and Household Helps

**Practical Suggestions for Home Makers—What to Do in Emergencies
When a Doctor is Not at Hand—Ideas for Beautifying the
Home—Ready Reference Aids for the Housewife**

Before the Doctor Comes

PROMPT action in emergencies often means the saving of human life. One cannot always wait for the doctor, but must take preliminary steps that his arrival be not too late. Here are some emergency suggestions:

Accidents

In case of arterial hemorrhage, apply pressure at a point where there is a bone under it on the artery supplying the wound. For hemorrhage in arm, apply pressure just back of collar bone, and against the first rib, using thumb or any hard pad. If below the elbow, place a stick in the elbow joint, bending the arm forcibly against it and tying arm in that position. For hemorrhage in leg, apply pressure on the hip bone or at back of the knee, or at outer side of heel, the upper part of thigh or in front of ankle, according to location of wound. A tourniquet is also useful in case of hemorrhage.

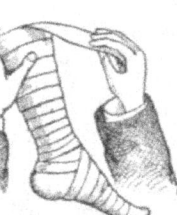

FIG 1—BANDAGING LEG AND SHOWING REVERSE

Elevating the wounded part lessens the blood pressure. Hold a wounded hand or forearm above the head. Lie on the back and hold a wounded foot or leg above the body.

Small hemorrhages are often controlled by a stream of cold water, an ice bag, or piece of ice.

For nose bleed, apply ice or cold water on surface of nose. Hold thumb on notch in the jaw bone on the side of the bleeding nostril, through which the artery passes.

Persistent bleeding from the cavity of an extracted tooth may be stopped by introducing a bit of antiseptic gauze, putting a bit of cork on top and placing the teeth upon it. If necessary, support the under jaw by a bandage over the head.

Bandaging is an art which should be acquired by everyone. A little practice at home, when there is no immediate necessity for it, will give the skill which in an emergency may mean so much in the prompt dressing of wounds. A well-placed bandage never slips.

Dressing Wounds

Be sure that the hands are thoroughly clean before attempting to dress a wound. They should be dipped for a short time in an antiseptic solution. Thorough sterilization of everything coming in direct contact with the surface of the wound should be attended to. Remove the foreign bodies, dirt, pebbles, bits of clothing, etc, and cleanse thoroughly with an antiseptic solution. Bring edges of wound together in as natural a position as possible. A

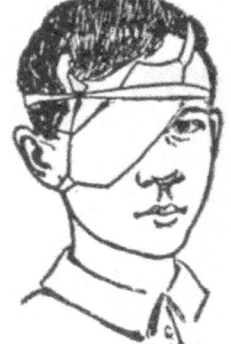

FIG 2—RIGHT WAY TO BANDAGE EYE

wound is often successfully closed by putting a wide, long piece of adhesive plaster on each side and a little away from the edges of the wound and then sewing the edges of the plaster and drawing up the threads until the edges of the wound are brought smoothly in contact. Cover with several layers of antiseptic gauze. Secure by bandages of strips of adhesive plaster. There should be a compress over the wound. Use absorbent cotton, weave the gauze if there is much hemorrhage. Use splints with bad cuts of fingers or hands to hold them steady.

For burns or scalds, cover with cooking soda and lay wet cloths over. White of egg and olive oil, or olive or linseed oil, plain or mixed with whiting or chalk, are good.

Ways to Bandage

A bandage should consist of but one piece and be free from seams and selvage. Every house should have a supply of antiseptic bandages. In applying a bandage, place the outside of the free end upon the part; hold in position with the left hand until a few turns can be made with the bandage held by the thumb and fingers of the right hand. Held thus, the bandage unrolls into the hand and is always easy to control. A limb should be bandaged in position it is afterward to occupy. Each turn of a bandage should be made with even pressure. A circular bandage will answer where the part is of uniform size, but where it tapers, the "reverse" is used, otherwise the lower edge of each turn will be loose. The reverse is made by holding the thumb of one hand on the bandage and simply turning the bandage over, so as to bring the opposite side next to the surface of the limb. At each turn the bandage is reversed.

The foot and ankle are bandaged by applying the outside of the bandage to the ankle, taking two turns around it and carrying the bandage diagonally over the top of the instep, around the foot two or three times toward the toes, then two or three times toward the ankle and back around the ankle and up by means of the reverse.

FIG 3—HEAD BANDAGE

To bandage the head, place the free end of the bandage on top of the head and carry under the chin to top of head and so around two or three times. Then bring it down even with the top of the ear and make a reverse, thus carrying the bandage around the forehead and back of head, crossing the other and pinning where it crosses. This will hold tight to any part of the head covered. To successfully bandage an eye, place the flannel part of bandage over the eye, carrying one tape under the ear and the other over the head, crossing them on the back and bringing them around above the ears and tying in front, as shown in illustration. Never bandage too tightly.

In Time of Shock

Patients suffering from sudden shock, due to injury or mental excitement, should be given hot stimulants if they can swallow. The quantity is to be measured by effects and not by drops or grains. In severe cases, the head may be lowered and the limbs bandaged and elevated to ease the load on the heart. Hot water bags should surround the body, which should be well covered with warmed blankets.

To Revive from Drowning

The first essentials are the restoration of breathing and body heat, and the maintenance of the latter. Loosen or cut apart all neck and wrist bands, so as to remove obstructions to breathing. Remember that every moment is precious.

Turn the patient on his face with head lower than body, grasping body about the middle and raising as far as you can without lifting the head off the ground; give the body a sharp jerk to remove mucus from throat and water from windpipe. Holding the body suspended, slowly count 1, 2, 3, and then repeat the jerk gently two or three times.

Place body on ground again, face downward; stand astride the body, holding it by the joints of the shoulders. Raise the chest as far as possible without lifting the head from the ground; hold it long enough to slowly count three. Replace the body on the ground with forehead resting on the arm that has been bent at the elbow. Be sure that nose and mouth are free, so that they can take in air.

With elbows against knees for a leverage, press downward and inward with increasing force against the sides of the victim's chest and over the lower ribs long enough to count two, then let go suddenly. Grasp the shoulders as before and raise the chest, remembering to leave the forehead resting on the ground. Replace body on the ground, pressing downward and inward against the sides of the chest, let go suddenly and, grasping the shoulders, raise the chest and press upon the ribs. If necessary, repeat these alternate movements for an hour at least at the rate of 10 to 15 times a minute unless breathing is restored sooner.

As soon as breathing is restored, attend to the restoration of body heat. Warmed blankets should be wrapped around the body, and hot water bottles or hot bricks applied. The head should be warmed nearly as fast as the body to avoid convulsions. As soon as the patient can swallow, give hot tea, coffee or milk. Spirits given too freely at this time might cause depression.

Don't get discouraged. Work persistently. People have been restored to consciousness by persistent effort so long after being taken from the water that it seemed impossible that the spark of life could still exist. Follow the above directions explicitly, and if there is still life you will surely succeed in restoring the victim; just keep at it.

Hygiene and Health—
Home Treatment
of Tuberculosis

THE battle of medical science against the "great white plague," tuberculosis, has been persistent, unyielding, vigorous. The success met with has been more and more encouraging, so that now the disease is fought with an understanding of its nature, which makes a successful and complete cure almost a certainty if the disease is taken in hand in its early stages.

A consensus of opinion among the leading investigators is that consumption or tuberculosis is not hereditary, but is contagious, and is curable. In the majority of cases where children develop consumption, it may easily be proved that the disease developed from contagion and was not hereditary. A predisposition to weak lungs may favor the progress of the disease, once the germs find lodgment.

On the first sign of the disease a physician should be consulted. If he recommends treatment at a sanitarium, and this cannot be afforded, home treatment may possibly be very successfully employed. The patient's bedroom should be the largest, sunniest and best ventilated in the house. Carpets and curtains should be as scarce as possible, compatible with the esthetic sense. The bed should be at least a foot from the wall. A patient should not breathe hot air. There should be as much rest out of doors as possible. It is vital that the patient shall every moment breathe fresh air, and be in the sunshine all the time possible. Pure air and sunshine are fatal to the tubercle bacillus.

Consumptives should be in the fresh air 24 hours out of the 24. Sleeping in tents is often found beneficial. Rain and dampness are to be avoided, especially during the hours of sleep. Sleeping on porches and piazzas, properly protected by awnings, has been very successful in the treatment of many.

Good, wholesome foods, well cooked, are essential. In general, such foods as will give the largest amount of nutrition with the smallest amount of labor for the alimentary tract, are best, such as roast or boiled beef, mutton, lamb, fresh vegetables and fruits, cereals, mixed liberally with cream, plenty of sugar and good butter, and at or between meals, six or more eggs, and from 2 to 3 quarts of milk, distributed within the 24 hours.

Indigestible things, such as sweets, pasties, etc., must always be carefully avoided.

The use of medicine should not be countenanced, except under a physician's immediate direction. Much actual harm is done by the indiscriminate use of medicines recommended by friends. Always consult a physician before using a medicine. The five fundamental principles in the treatment of consumption are as follows: The proper disposition of infective material; rest; fresh air and sunshine; pure food and drink in abundance; use of medicines as directed by a physician.

The first practical consideration is the disposition of the infective material. The most common form of infection is through the dried sputum, which is taken up in the dust of the atmosphere. Infants become infected from the dust of the floor, or by putting in their mouths objects on which sputum may have been deposited. Consumptive persons should never kiss children. Frequently the germs are transferred in this way, and the expression of love becomes the curse of fatal disease. The consumptive invariably, while indoors, should have some cloth or tissue paper before the face during coughing or sneezing. He should always spit into a receptacle made expressly for the purpose, or a moist cloth.

You Betcher
Pudding and Milk,
Mothers Know,
Is just what makes
Little girls grow!

Disinfectants and Their Use

BEAR in mind the difference between a disinfectant and deodorant; the former kills germs, while the latter destroys odors but does not kill germs. Many so-called disinfectants have very little germ-killing power, and are simply deodorants.

Cleanliness is the foremost agency in destroying disease germs. Sunlight is a valuable destroyer of bacteria; its importance cannot be overestimated. Moisture is favorable to the growth of bacteria. Germs thrive in damp corners.

Boiling water is one of the most efficient disinfectants; one-half hour boiling destroys all disease germs.

Bichloride of mercury or corrosive sublimate in solution, one part to 1,000 parts of water, may be used for wiping infected woodwork, etc. It should not be used on metal. It is extremely poisonous, and has no color or odor; great care must be exercised in its use.

Carbolic acid in 3% or 5% solution in water is most effective. This is poisonous, but its strong odor reveals its presence. It may be mixed with typhoid discharges; soiled bedding may be soaked in this solution before boiling.

Chloride of lime, 6 ounces to a gallon of water, is excellent in scrubbing floors.

For disinfecting a room after the patient has left, formaldehyde is most efficient. This does not injure fabrics, furniture or hangings. This is usually used under the direction of an officer of the board of health. It is exceedingly poisonous.

Do not buy disinfectants about which you know nothing.

Sulpho-naphthol is excellent to use about the kitchen sink. A little in the water in which wounds, scratches, etc, are washed, will prevent infection.

Diagnosed

The Ostrich—Oh, oh, doctor, I've such a pain!
Dr Pelican—What have you been eating?
The Ostrich—Only some needles and a few spools of thread.
Dr Pelican—Er—um. You've undoubtedly got a stitch in the side.

The Home Beautiful

Ideas for Making it Attractive—Decorations and Suggestions for Color Harmony—Stenciling and Its Application to Home Arts and Crafts

IN no single phase of the household does the average person so often fail of obtaining their ideals as in the matter of decoration. It is a problem which comparatively few are really equipped to work out for themselves. They may instinctively recognize good taste and charm in home decoration, yet be wholly at a loss as to how these results are to be attained. Simplicity should be the keynote of all home decorations, color harmony in wall papers, the carpet and the furnishing of each room should be studied. Clashing colors invariably jar. They introduce a false note in the harmony of the household life. You cannot get away from them. They persist through everything. The old days of much bric-a-brac and fancywork of all kinds have given way to saner, more wholesome and altogether more harmonious ideas. Today some of the most beautiful results are attained at a minimum of expense.

If the ceiling of a room is low, a striped paper from baseboard to ceiling will give the effect of height. A ceiling which appears too high may be lowered to the eye by allowing the ceiling paper or kalsomine to extend a foot or more on to the side walls, putting the picture molding at the junction with the side wall paper. A little study of draperies will enable one to change the apparent height of windows.

Harmony in Colors

Color harmony is a matter for careful study. In this matter it is a good plan, if you know that you are deficient in color sense, to consult someone whom you know to have a thorough knowledge of color values and harmony. There are a few cardinal points to bear in mind. A room naturally bright and full of sunshine needs to have its brightness modified in cold blues or greens, while a sunless room, as a north room, for instance, becomes more beautiful by the use of yellow and gold in the colors of its walls and draperies; golden browns and rich reds can be used to good advantage.

Always the strongest tones of color belong to the base. By this rule the floor covering should carry the strongest tones, the color showing a gradation of the walls to the ceiling, which should be lightest of all.

Always remember that the room as a whole, floor, walls and ceiling, is but a setting for the furnishings, and the color scheme should be chosen accordingly.

In the framing and hanging of pictures, to say nothing of the choice of pictures, are all the elements of harmony and of discord. Pictures in a brown tone are usually best framed in a simple, plain wood of a brown finish. As a rule, only expensive oil paintings call for a gilt frame. A frame is but a setting for a picture, and should never, by its elaborateness, distract the eye and draw it away from the picture.

In the hanging of pictures avoid straight lines. There is nothing which conveys a greater sense of stiffness and formality than a room in which all the pictures are hung at exactly the same height. Study the effect of light on each picture as you hang it. Each one is entitled to the best light it can have to bring out its real beauty.

The choice of draperies is another subject for individual study, and perhaps advice, from one who has had experience in such matters. In general, portieres should follow the color schemes of the walls. If the walls are plain, it is well to have figured portieres, and vice versa.

Carefully chosen window hangings soften the lines and take away the bareness and stiffness from the room. Nevertheless, the use to which a room is to be put must govern the selection. Draperies which shut out the light and air and sunshine from the living-room are not only a menace to health, but they even affect the mental atmosphere.

Of late, stenciling as a means of home decoration has come into vogue rapidly. Unlike many fads which have prevailed, this art has in it that which should give it a permanent standing in home arts and crafts. It makes possible at the least possible expenditure the most attractive decorations imaginable. It is applicable to many materials from a frieze or dado on a wall to portieres, window curtains, and sofa pillows, to say nothing of smaller pieces, such as table and bureau scarfs, etc. A decorative scheme in perfect harmony may be carried out at a minimum of expense.

Stenciling—A Home Art

The use of the stencil for beautifying the home is constantly growing. Stencil-ing is so practical, so inexpensive, so easily and quickly done and so satisfying to the artistic temperament that it has unquestionably come to stay. It is applicable to all parts of the home—the walls of the nursery, the curtains of the bedroom, the portieres of the parlor, the sofa pillows of the piazza, the table runner of the library, and to many materials. Its possibilities are almost without end.

What a Stencil Is

A stencil is a piece of thin metal, cardboard, celluloid or manila paper in which have been cut conventional designs for the purpose of transferring through the openings thus made, in color or plain black, the design to the object to be decorated. For practical purposes stencil cardboard or heavy manila card are plenty good enough.

How to Use the Stencil

The necessary equipment consists of several round stencil brushes, such as shown in the accompanying illustrations, tubes of oil paints in the four primary colors, and such other colors as you may desire, a small can of turpentine for thinning the paint, and some old plates or pieces of glass on which to spread the colors.

Having decided upon the material which you are to decorate and the pattern and colors which you will use, spread the material on a smooth surface, and place the

BRUSH   stencil in position, being sure that it is perfectly flat and absolutely true. It is a good plan to hold it down at the edges with a flatiron, or other weight. Squeeze out a little of the

paint of the color you are to use first upon the piece of glass or plate, and holding one of the stencil brushes, as shown in the illustration, rub the paint thoroughly into the brush by a circular motion. Do not be afraid to grind it thoroughly. Remember that stenciling is not painting, and you do not want your color to lie on the surface of the material. Now add to the paint a few drops of turpentine, and continue to rub the

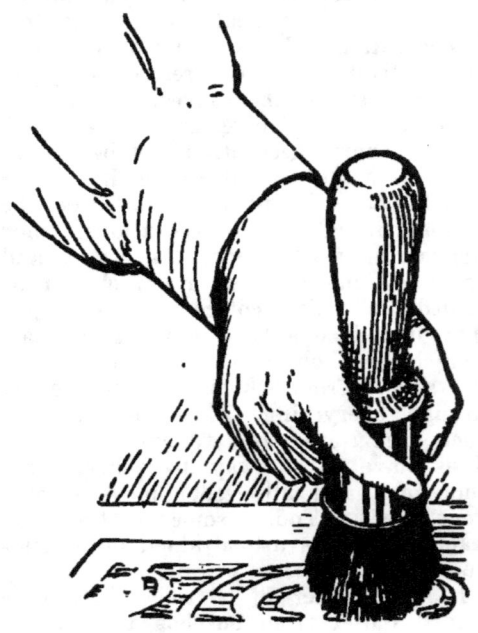

CORRECT WAY OF HOLDING BRUSH

brush in this way until your paint is absolutely smooth, with no possibility of streakiness. You can judge for yourself how much to thin it in this way. Try it out with a stencil on an old piece of cloth before attempting to stencil your fine goods.

Having gotten your color to a satisfactory condition, apply to those parts of the stencil which are to be of this color, covering the other parts of the stencil with a piece of paper, if necessary. In applying the color do not be afraid to use some strength. Do not try to paint it in the commonly accepted sense of the term. If the opening is large, bear down hard and use a circular motion. You want to grind your color into the fabric. If the opening of the stencil is narrow, and there are some delicate parts which might be broken, make your strokes lengthwise of the opening.

At first an equal application of color may be easiest, but later if you desire to shade, simply bear on a given portion which you want darker a little longer and

New stencil designs. Prices of cut stencils, beginning at upper left-hand corner: No 223 GH 50 cents; No 224 GH, upper right-hand corner, 60 cents; No 225 GH, middle, 30 cents; No 226 GH, bottom, 75 cents. See next page.

a little harder than on the part which you wish to shade off. If there are a number of small parts to the design in close conjunction it is often of advantage to use a small brush, a No 2 paintbrush, for these parts, thus avoiding running over the edges into other parts of the design. Use a separate brush for each color.

You will find out after one or two attempts that this stenciling is the easiest thing in the world. The most beautiful effects can be obtained. It is adapted to the decoration of all sorts of materials. Scrim, burlap, crash, denim, monk's cloth, all of these and many other materials stencil beautifully. The simplest window curtains can be made things of real art by a little thought in the selection of delicate and appropriate colors and attractive designs. Beautiful sofa pillows can be made in almost no time at all. The actual work of stenciling a sofa pillow need not be over five minutes. Shirtwaist boxes, boxes for the baby's toys, picture frames, table scarfs, portieres, all these and many more artistic things for the home are adapted to the stencil.

How to Get the Stencils *

One who has some talent in drawing or copying may make up their own designs, but the better way is to buy either the patterns or the cut stencils. The illustration at the opening of this article is an excellent example of what is meant by a conventionalized flower design. Such a design as this in green for leaves and stem and soft yellow for the blossoms makes a most charming and effective border if repeated.

* The patterns shown on the preceding page are for sofa pillows and curtain ends. The cut stencils may be obtained at the prices quoted of the Stencil Department, Orange Judd Co., Springfield, Mass., New York, N. Y., Chicago, Ill. A stenciling outfit of five brushes, four tubes of colors, turpentine can and two repeat cut stencil designs will be sent postpaid for $1.25.

NO. 1.   NO. 2.   N° 3. THE RAT COMPLETE.

GO AHEAD JUST AS YOU DID FOR THE RAT.

HOW TO DRAW AN OWL.

FOR The Young Artists.

Helpful Household Secrets

Knowledge of Which will be of Daily Use—Tested Ways of Doing Things and Suggestions for Cleaning Various Material~

A CARPET may be brightened by wiping with a clean cloth wrung out of ammonia and water, 2 tablespoons ammonia to 4 quarts of tepid water. Rinse the cloth frequently in this water.

In cleaning rugs lay them face down on the grass and beat; then turn face upward and brush. Hang on the line to dry. Don't shake a rug from one end, for this loosens the threads at the upper ends, causing ravelings of the border.

The less water used on straw matting the better. Alkalies and soap are apt to discolor it and should not be used. Soiled spots should be scrubbed with hot water without soap. A thin paste of fuller's earth and cold water spread thickly over a place where grease or oil has been spilled will remove the stain. Lay a paper over it and let it remain 2 or 3 days before brushing up.

Care of Marble

Marble is easily dissolved by acids; even the mildest acid will remove the high polish. If an acid is spilled on the marble mantel or table, or even if a lemon is allowed to lie there a few minutes, an alkali such as ammonia, soda or borax should be applied at once. This will neutralize the acid. If the acid has remained long enough to roughen the surface, rub with pumice-stone and water. Apply a generous amount of pumice-stone wet with water and rub with a large flat stone. It will take time and muscle, but it will restore the polish.

Should the marble become stained, remember that acid must be avoided as a removing agent. One of the alkalies mentioned above may be used with safety. Recent oil or grease spots can sometimes be removed by the application of fuller's earth or French chalk. Make a strong, hot solution of sal-soda and enough fuller's earth to make a thin paste. Let this remain on the spot for 24 hours.

Marble should be kept clean by washing with a soft cloth and soap and water.

Windows, Copper, etc

Windows may be washed with clear water, or water in which is a little ammonia, soda, or soap.. Plenty of clean, soft cloths and some soft paper to polish with after the dirt has been removed, together with plenty of elbow grease, will insure clean windows.

Copper and brass are easily cleaned by means of acids. Remember, however, that tarnish will quickly follow unless every trace of the acid is removed. If a cleaning preparation contains an acid, as most preparations on the market do, follow with an application of whiting to neutralize the acid and preserve the surface from tarnishing. Brasses cleaned with oil and rotten stone, or with tripoli, will have a rich yellow tone. Acids and naphtha produce a tone less rich.

Common salt and oxalic acid, or vinegar, are very satisfactory agents. Use a soft cloth, and rub the surface until all tarnish is removed, then wash in plenty of water and wipe perfectly dry. If the article is badly tarnished, try a solution of sal-soda to remove all grease, then rub with oxalic acid, lemon juice, or strong vinegar. Wash thoroughly and wipe dry. Polish with rotten stone, or tripoli, and sweet oil, using a woolen cloth and rubbing hard. Wipe off this scouring mixture and go over the surface with a dry flannel and dry tripoli, or rotten stone. Finish with a soft, clean cloth. Brass and copper will keep untarnished a long time if in a dry place.

Woodwork

When cleaning woodwork, bear in mind that paint is softened by wet alkali, and if the solution is strong enough, it will dissolve the paint. Potash and sal-soda are particularly caustic, while borax is the least caustic. Of course, the stronger the solution the more quickly it acts on the paint. It is obvious, therefore, that no strong alkali should be applied to painted or varnished surfaces, and no strong caustic soaps should be used on them.

Should an alkali be spilled on such a surface, oil applied instantly will neutralize the alkali and save the surface.

Whiting is an excellent agent for cleaning painted woodwork. Mix it with cold water to the consistency of thick cream; have two pails half filled with hot water, and a woolen cloth in each pail; keep a third woolen cloth dry. Wring nearly all the water from one of the cloths, dip in the whiting mixture, and rub hard the surface to be cleaned, following the slight grain left by the painter's brush. Wash off the whiting with the cloth and water from the second pail. Rinse the cloth dry and wipe the surface with this, then rub perfectly dry with a dry woolen cloth.

An enamel finished surface requires different treatment. First wipe off the dust, then follow with a clean woolen cloth dipped in hot water and wrung as dry as possible. Rub dry with a second woolen or cotton cloth. This dry rubbing gives brilliancy to the surface; on soiled places which the damp cloth will not clean, rub a little with powdered tripoli until the stain disappears. Don't press too hard. Powdered pumice-stone will answer the same purpose.

Avoid as much as possible using water on natural wood finish. Use oil and turpentine, or alcohol. If a place becomes very much soiled, clean by rubbing with a woolen cloth wet with turpentine or kerosene. The turpentine removes the gloss and should be followed with oil. A little tripoli moistened with oil is excellent for a bad spot. Once a year go over the woodwork with a mixture of paraffine and turpentine, using equal parts. If the finish looks especially dry and cracked, rub in pure oil. On dark wood, where the finish is dry, cracked or faded looking, boiled linseed oil is better than paraffine oil. The frequent application of oil, well rubbed in, will keep woodwork exposed to heat, sun or moisture, in fairly good condition. Outside doors should be oiled with pure boiled linseed oil several times each year.

Silver, Cut Glass and Steel

Silver will not often require cleaning if every time it is used it is washed in plenty of soap and hot water and rubbed dry with clean, soft towels. If silver tarnishes quickly, it indicates that there is some gas in the house. Therefore, gas pipes and drain pipes should be inspected. Sifted fine French whiting, wet with diluted alcohol or ammonia and applied with a soft cloth, afterward polishing with chamois, will keep silver in handsome condition.

Sawdust from a resinous wood, such as box or bass-wood, and free from any hard substance, is excellent for cut glass. It absorbs the moisture which cannot be reached with a towel. After wiping the glass, bury it in a bed of sawdust for half an hour or more, then brush with a soft brush and polish with a soft cloth.

Remember that to scratch glass is to weaken that part so that a little heat or cold will sometimes cause a break at that point. The grain of sand at the bottom of the dish-pan or on the dish-cloth may be the cause of splitting a beautiful dish.

Use little soap on gilt china. To polish pewter, britannia and blocked tinware use powdered rotten stone and oil, or oil and whiting.

Keep steel from rusting by covering with sweet oil or mutton tallow and wrapping with soft paper.

To remove rust, use oil and quicklime on the article. After several days rub with oil and rotten stone, or bristol brick.

The Care of Gloves

Gloves require care to obtain their full wearing value. A cheap glove is dear at any price. When the glove is put on for the first time, talcum powder should be sifted over the hands. The glove fingers should be worked on easily from the tip to the hand, keeping the seams straight.

Always see that gloves are dry before putting them away. They keep best in a box especially for them. Lay them flat in this after they have been aired and drawn into proper shape.

Naphtha or gasoline may be used to clean white and light colored gloves. Both are *dangerous explosives* and must not be exposed to fire. They are best used out of doors, or in a room with the window wide open. To cleanse the glove slip it on one hand and dip a clean piece of white flannel in the cleansing fluid, wet the glove all over and then rub it nearly dry with a second piece of clean flannel. Keep on the hand until dry, in order to retain the shape. Sprinkle talcum powder on them and hang in the air until the odor has left. Chamois gloves for summer may be washed on the hands, using a lather of white soap and water. Badly soiled spots may be cleansed by rubbing them with magnesia. Before washing, rinse them in warm water and then in cold. Keep the gloves on the hand until nearly dry, then pull them off carefully in their proper shape, and hang to dry.

To blacken kid gloves which have become white at the seams or at the finger tips, dip a feather in a little olive oil containing a few drops of black ink and brush lightly over the white places. Light colored suede or undressed kid gloves may be cleaned with corn meal or dry bread crumbs, dusting off with a piece of clean white flannel.

Cleaning Fabrics

A pan of gasoline and a velvet brush are all that are needed for cleaning light velvets that are much soiled. Dip the velvet in the pan and brush the soiled places with the velvet brush. Hang up while dripping to air until all the vapor is gone.

Steaming will take the creases out of velvet. Stand a large hot iron on end on a cold stove or asbestos matting. Wring a piece of cheesecloth or thin muslin out of water and spread smoothly over the iron. Holding the velvet with both hands, pass the back over the iron, holding the velvet so that you pull it on the straight, either selvage way or across. Just as fast as the muslin dries pull a fresh piece over the iron. Finally pass the back of the velvet over the bare iron to dry it off and effectually raise the pile.

A badly marred piece of velvet can be "mirrored" by placing it flat on the ironing table face up and passing the iron over it down the nap. Don't stop in the middle of a pass or you will leave the shape of the iron, which can be removed only by steaming. Velvet ribbons may be renovated in the same way.

Restoring Silks and Laces

To clean silks other than wash silks, brush well in a gasoline bath. Some of the worse spots will yield to ether. Often creases can be pressed out of silk without dampening it. This, of course, after the gasoline has entirely evaporated.

Black silk ribbon may be steamed with ammoniated water; if very dusty they may be wiped over with a cloth wrung out of cold tea, or water in which a few raw potatoes have been standing a few hours. Alcohol will brighten black silk.

To clean white lace, if all silk, use gasoline, afterward pressing on the wrong side with muslin over it. Black silk lace may be treated in the same way. Press on a thickly covered board.

White laces may be soaked in several waters prepared with good suds of pure soap. Shake the lace frequently in this and pat or brush gently any very dirty spots. Do not rub. Rinse in several waters, and pin out on the neatly covered ironing table, brushing out and pinning the little loops of the edge. Pin right side up. When dry, turn and press out the pattern with the round end of an orange stick or ivory penholder.

Real laces should never be cut. They are usually made in patterns, which may be ripped apart and joined again if used in a different way. The joining should be made with the finest silk or cotton.

For lace that is very soft and needs stiffening, dissolve a little gum arabic and put in with the last rinsing water. Some laces, like Valenciennes, may be washed and ironed wet.

White laces may be tinted by coloring the last rinsing water with coffee to obtain an ecrue shade; tea will give a cream, and a yellow tint can be obtained by using saffron steeped in boiling water and strained.

Miscellaneous

For wax, cover with absorbent paper and press with warm iron.

For mildew, try lemon juice, followed by bleaching in direct sunlight.

There is nothing better for beaver hats and cloth than a corn-meal bath.

For iron rust, sprinkle with salt, moisten with lemon juice, and lay in the sun.

For blood stains, wash in cold water, then rub with naphtha soap and soak in warm water.

Coffee stains will yield to boiling water. Spread stained part over a cup or bowl, and pour water through.

Paint is best removed with benzine or turpentine. If delicate goods are stained, use chloroform or naphtha.

Wagon grease yields to either oil or lard rubbed on stain, followed by careful washing with warm water and soap.

Chocolate stains may be removed by soaking in cold water, then sprinkling with borax and washing in boiling water.

For fresh ink, try milk. Allow stained portion to stand in milk until the latter is discolored, then use more. Salt and lemon juice will remove dried ink stains. However, this is apt to take out color of goods also.

Aigrettes, ostrich feathers and Paradise plumes should be cleaned in gasoline, shaken out and dried in the wind. Wings and breasts can be cleaned by shaking and rubbing gently with corn-meal in a cardboard box; then shake thoroughly in the air. Light furs can be cleaned in the same way.

It is economical to have two pair of shoes which can be worn alternately. This plan is of advantage from a sanitary standpoint also. A shoe worn every other day does not settle to the foot and lose its good appearance as does a shoe worn every day. Moreover, the shoes worn every other day are better aired, which is necessary for comfort. Shoes should be kept in a ventilated box.

The Kitchen; Its Various Problems

The Composition of Food—Meats and Their Various Cuts—Tables for the Guidance of the Cook—Household Pests

Composition of Food

Food is required to increase or repair the materials of the body, to keep it warm and to endow it with a renewal of working power. Sixteen elements, arranged in many compound substances, form the materials of the human body. These same elements, usually arranged in similar compounds, are found in food.

Nutrients or food substances fall into 2 groups, the incombustible and the combustible. In the former are water and salts; fat, sugar, starch, etc, are in the latter.

By means of the oxygen taken into the lungs combustible nutrients are burnt more or less completely within the body. In this way is produced the actual energy of heat and mechanical power of heat, maintaining warmth of the body and furnishing the motive force for work. The daily waste and work of the body must be met by a daily supply of nutrients.

Diet of Adult and What it Should Contain

According to one of the best authorities on diets and dietetics, the average daily diet for an adult should contain the elements of the following table, in the proportions given:

Nutrients	In 100 parts	Each 24 hours		
		lb	oz	gr
Water	81.5	5	8	329
Albuminoids	3.9	0	4	110
Starch, sugar, etc	10.6	0	11	178
Fat	3.0	0	3	337
Common salt	0.7	0	0	325
Phosphates, potash, salts, etc	0.3	0	0	170

The person doing active physical work requires a larger percentage of flesh-forming and heat-giving nutrients in his food. The harder he works, the more necessity

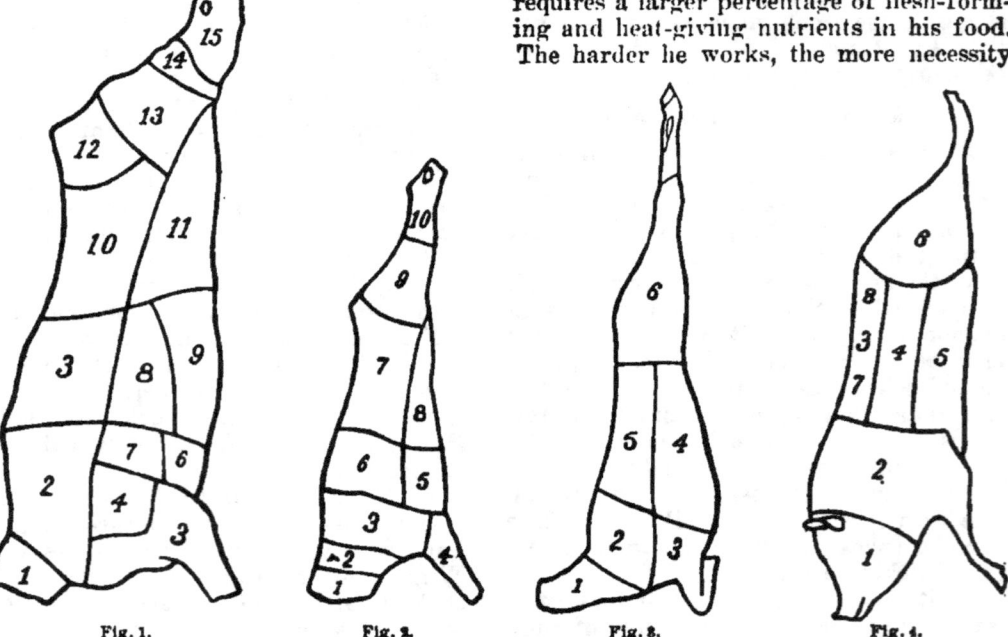

Fig. 1. Fig. 2. Fig. 8. Fig. 4.

A CHART SHOWING CUTS OF MEAT

Fig 1, Cuts of Beef—1, neck; 2, chuck; 3, ribs; 4, shoulder clod; 5, fore shank; 6, brisket; 7, crossribs; 8, plate; 9, navel; 10, loin; 11, flank; 12, rump; 13, round; 14, second cut round; 15, hind shank.

Fig 2, Cuts of Veal—1, neck; 2, chuck; 3, shoulder; 4, fore shank; 5, breast; 6, ribs; 7, loins; 8, flank; 9, leg; 10, hind shank.

Fig 3, Cuts of Lamb—1, neck; 2, chuck; 3, shoulder; 4, flank; 5, loin; 6, leg.

Fig 4, Cuts of Pork—1, head; 2, shoulder; 3, back; 4, middle cut; 5, belly; 6, ham; 7, ribs; 8, loin.

for increased quantities of nutritious compounds. In view of the fact that no one article alone can furnish the exact quantities of the elements demanded, a mixed diet is usually best.

Weights and Measures for the Cook

For the benefit of those housewives who have not at hand the means to readily weigh their ingredients, the following generally accepted table is given:

2	cups	butter (packed solidly)......1 pound
4	"	flour (pastry)...............1	"
2	"	granulated sugar............1	"
2 2-3	"	powdered sugar..............1	"
3 1-2	"	confectioners' sugar.........1	"
2 2-3	"	brown sugar.................1	"
2 2-3	"	oatmeal.....................1	"
4 3-4	"	rolled oats.................1	"
2 2-3	"	granulated corn meal 1	"
4 1-3	"	rye meal....................1	"
1 7-8	"	rice........................1	"
4 1-2	"	graham flour 1	"
3 7-8	"	entire wheat flour..........1	"
4 1-3	"	coffee......................1	"
2	"	finely chopped meat.........1	"
9 large eggs...........................1	"
1-3 cup almonds blanched and chopped..1 ounce
A few grains is less than one-eighth teaspoon.
3 teaspoons dry materials.............1 tablespoon
16 tablespoons.......................1 cup
2 tablespoons butter..................1 ounce
4 tablespoons flour...................1 ounce

Cooking Lore

Young housewives are often perplexed as to how long various forms of food should be cooked. The following table for cooking various common foods has been prepared by a noted cook:

BOILING

Articles	Hours	Minutes
Leg of Lamb, according to size	2 to 3	
Ham of 12 to 14 pounds ....4 to 5		
Tongue ....3 to 4		
Chicken of 3 pounds...........1 to 1¼		
Turkey of 9 pounds............2 to 3		
Salmon of 2 to 3 pounds		30 to 35
Asparagus		20 to 30
String-beans...................1 to 2½		
Cabbage.......................		35 to 60
Turnips.......................		30 to 45
Onions.......................		45 to 60
Green corn		12 to 20
Peas.........................		20 to 25
Parsnips.....................		30 to 45
Cauliflower...................		20 to 25

BROILING

Articles	Hours	Minutes
1-inch steak.................		4 to 6
1½-inch steak		8 to 10
Lamb chops...................		6 to 8
Chicken.....................		20
Liver and tripe..............		4 to 5
Shad, bluefish and whitefish.....		5 to 20
Sliced halibut, swordfish........		2 to 15

ROASTING

Articles	Hours	Minutes
Saddle of mutton..............1¼ to 1½		
Leg of lamb..................1¼ to 1¾		
Leg of veal..................3½ to 4		
Loin of veal.................2 to 3		
9-pound turkey...............2½ to 3		
9-pound goose...............2		
Duck........................1 to 1¼		
Chicken.....................1 to 1½		
Pork (chine or spare rib).......3 to 3½		
10 pounds of rump beef, rare....1		35

These tables give an approximate idea of the length of time required for cooking meats and vegetables. In cooking meats it must be remembered that it is not so much the weight as the amount of surface exposed to heat which counts. Thus, a flat roast would cook through quicker than the same amount of meat in a more compact form.

Care of a Refrigerator

Selection of and care of a refrigerator are matters of vital moment to the family health. It is not simply a matter of keeping food cool, but of keeping it hygienic. A properly constructed refrigerator produces a continuous circulation of cold, dry air. Usually it is attained by putting ice in the upper chamber, so constructed that as the air chills it will sink into the food chambers below, forcing the warmer air there to rise, and in turn to chill and sink.

The old-time charcoal filling of the ice box or chest has largely given way to mineral wool, in addition to a liberal air space, air being the best known non-conductor of heat. Galvanized iron has largely replaced zinc as a lining.

In buying a refrigerator, first make sure of the circulation. A simple test is to stand a lighted candle in the provision chamber, or hold a match there; note if there is a sufficient current to make the flame waver. Of course the refrigerator must have ice in the box.

Wood racks are subject to mold and mustiness. They will harbor germs. Metal and glass shelves are better. A refrigerator should be washed out thoroughly at least once a week. It should be kept spick and span inside.

Put nothing edible directly on ice.

Edibles should never touch the shelf surface.

Never allow dishes to slop over.

Wash ice clean before putting it in.

Remember that milk and other foods will take the odor of strong smelling fruits, vegetables or cooked foods. Keep such things closely covered, and put in bottom of food chamber. Milk and butter should be kept on top shelf. There should be no dampness inside the food box.

Dishes should be perfectly clean before putting in the food chamber.

Worth Knowing

To make window glass opaque, dissolve 1 tablespoon epsom salts in 1 glass beer or ale, which should be warm. Smear over the window, where it will form into a lacelike crystal, through which it is impossible to see. By using a little more salts, a denser mixture can be made.

To soften hard leather shoes, first wash them over with warm water and then rub in castor oil. Any oil will answer, but the castor oil is best. The shoes, after this application, will become soft and pliable.

In pressing ribbons with a hot iron, lay them between two sheets of manila paper, and they will come out in better shape, and minus the usual objectionable gloss.

For staircase pads, pieces of old blankets may be folded neatly and placed on each step separately, taking care to cover the edges. Pads make the stair carpet wear double the time.

One of the best things for cleaning tinware is washing soda. Use a strong solution of the soda, rinse in clean, hot water, wipe and dry in sun.

To clean copper, take a handful of salt and enough vinegar and flour to make a paste. After using the paste, wash the copper in hot water, rinse in cold, and dry.

To clean marble, take 2 parts washing soda, 1 part pumice stone and 1 part salt. Sift all through a fine sieve, mix with water and rub over the stains. Rinse and wipe dry.

Getting Rid of Household Pests

Water Bugs

A weak solution of turpentine poured into the water pipes once a week for a few weeks will usually drive away water bugs. Use ½ pint turpentine to 3 pints of water. This will drive the bugs from their hiding places and they should then be killed whenever seen. Preparations effective in exterminating cockroaches will usually destroy water bugs.

Cockroaches

Garbage left standing is one of the surest means of attracting cockroaches. Dissolve 1 pound of alum in 3 pints hot water and force a hot solution into all cracks and openings where the roaches are likely to be. Afterward spread borax about the places where the roaches have been in the habit of coming.

Another way is to blow insect powder into all cracks. Brush up and burn the dead insects and powder, and blow in a second dose. Brush up as before and spread powdered borax about the cracks and holes.

Ants

These little pests dislike oil of pennyroyal. Pour this oil on bits of cotton batting and spread about the places where the ants appear. If fresh pennyroyal can be procured, spread the leaves about the infested places.

Bedbugs

Naphtha is a quick, clean and sure exterminator of this pest. Open all the windows in the room and shake and examine minutely all the bedding. Hang out sheets, blankets, etc; saturate mattresses, pillows, etc, with naphtha and put them out of doors if possible. Brush the walls of the room, paying special attention to every crack, groove and corner. Examine the backs of all framed pictures. These pests will often hide in the cracks in a picture frame. If there are any ribbons used for decorative purposes, such as hanging photographs, etc, examine the knots closely. Take up the carpet, take the bedstead apart and lay on the floor with the grooved sides up; saturate with naphtha. Fill with naphtha any cracks or breaks in the walls or floors, then leave the room and lock the door, allowing no one to enter it. If much naphtha is used it is safer to keep the door shut and the windows open for the greater part of the day. Remember that *no fire* must be allowed anywhere near naphtha.

The first application will kill all the living insects, but not the eggs. In three or four days repeat the operation and this will destroy all the bugs which have hatched since the first killing.

Moths and Buffalo Bugs

The above treatment is equally good for these pests. Of course, the eggs will not be destroyed, and these are gotten rid of by brushing and shaking before the naphtha is applied. Washing out closets spring and fall with a weak solution of carbolic acid is an excellent means of keeping them free from insects.

Rats and Mice

Trapping is the best means of getting rid of these pests. Leave no holes for them. When you find a hole, sprinkle chloride of lime in the hole, fill with broken glass and seal with mortar or plaster of paris. There are many kinds of traps and when one becomes ineffective, try another. Pests of this sort quickly become educated to a source of danger.

Making and Using a Fireless Cooker

Why It is Economy to Have One—Some Foods Better Cooked—Recipes Tried and Found Good

THE rapid growth in popularity of the fireless cooker has resulted in a great number of cookers being put on the market. Perhaps there is no kitchen convenience which means more in the saving of time and labor to the farmer's wife than one of these. While the patented articles are most convenient in arrangement, and very handy, they actually do the work for which they are designed little, if any, better than the so-called " hay box," which any enterprising woman can make for herself.

Making the Box

The hay box consists of a thoroughly sound box, large enough to hold one or more kettles. This box is filled with hay, nests being made for each kettle. This is done by putting the kettles in place and packing the hay tightly around them. Of course there must be a liberal amount underneath them. This hay is for insulation purposes, and must be packed as tight as possible. It is an excellent plan to line these nests with cloth, by packing the cloth on the edges of the box, the kettle having been taken out. The kettles should have tight-fitting lids, and over all a pillow or thick cushion should be placed, and the lid closed down tightly. An old trunk makes an excellent hay box.

The principle involved is very simple. The food is first started cooking in the kettle which is to go into the hay box, the cover being kept tightly in place to retain the steam. After boiling has progressed for a little while varying from a few minutes up to twenty minutes, according to the article being cooked, the kettle is placed in the hay box, a cushion put on, and the cover fastened. The insulation, by means of the hay, prevents the escape of the heat, and the cooking process goes on for hours. This, of course, means a tremendous saving in fire, and incidentally very many things are much better for being cooked in this way. Excelsior or asbestos wool may be used in place of the hay for filling.

To test the efficiency of the cooker, place in it a kettle of boiling water; if at the end of fifteen hours the temperature is below 104 Fahrenheit, the cooker has not been packed correctly.

What Can and Cannot Be Done

Broiling or frying is, of course, impossible in such a cooker. However, some of the fireless cookers roast splendidly. Roasting is accomplished by first searing the meat on top of the stove or in the oven; it is then placed in a kettle with a tight cover, and this is in turn placed in a larger kettle having in the bottom a standard of some sort on which the inner kettle can rest. The second kettle has boiling water coming up around the inner kettle or pail; the whole, with a tight cover, is placed on the stove and boiled vigorously for perhaps half an hour, and then is transferred to the hay box and left for several hours, according to the size of the roast. Boiled vegetables may be placed in the receptacle holding the water.

The fireless cooker is most useful, however, for stewing, pot roasts, soups, and the cooking of various vegetables, steamed puddings, etc. The long cooking in low temperature results in a better tasting dish, because none of the volatile flavors have been boiled away. Meat is more tender; cheap cuts become tender without being stringy, and are well flavored because cooked below the boiling point. This means, of course, that economical cuts of meat can be utilized and made into deliciously flavored dishes, where their very toughness made them undesirable cooked in any other way.

Hay Box Recipes

Boiled Turkey

An old turkey may be cooked in from six to eight hours. If especially old it may be taken out and reheated after about four hours. In this case the seasoning may be added at the second heating. The turkey should be stuffed as usual, then put in the kettle and boiling water poured upon it until it is completely covered. Then take it out and replace the boiling water with tepid water and put the turkey in—heat slowly, and when boiling put it in the cooker. When reheating, add one tablespoon of salt and a few peppercorns. When tender, put on a platter, spread all over with butter, and put in oven until all the butter is melted. Serve with a drawn butter sauce filled with oysters, or with the liver and giblets chopped fine.

Vanilla Sauce

Cream ½ a cup of butter, add one cup of powdered sugar, and cream again. Put bowl into boiling water and keep the water boiling while you add ½ cup of rich milk or cream. Beat until smooth and creamy, take from fire, add one tablespoon vanilla.

Leg of Mutton

That very homely dish, a boiled leg of mutton, is excellent cooked in the cooker. Trim the leg well and put it in the big kettle of boiling water. Let it boil about ten minutes, then add one tablespoon of salt, and two or three peppercorns. If you like the flavor, a tiny piece of mace, of bay leaf, and two or three cloves may be added. Now put kettle in cooker. In five or six hours it should be ready to be served with caper sauce.

Caper Sauce

Two tablespoons of butter, three tablespoons of flour, ¼ teaspoon each of salt and paprika, and ½ teaspoon of onion juice. When this is cooked add one and one-half cups of boiling water, one tablespoon of butter cut in bits, one tablespoon of parsley, and four tablespoons of capers.

Fresh Fruit Pudding

This may be cooked in a cooker. Cream ¼ cup butter, add ½ cup of sugar, and the yolk of one beaten egg. Clean and dredge with flour two cups of fresh fruit—currants, cherries, gooseberries or raspberries—and have them dry. Now add to the sugar and butter mixture a little at a time, alternately, one cup of milk and two cups of flour. Two scant teaspoons of baking powder and ½ teaspoon of salt should have been sifted with the flour. Now fold in the egg white, turn into a buttered mold and put cover on very tight, and plunge into boiling water in bottom of one of the large kettles. Put in the things to be steamed in the several receptacles while the kettle is still on the stove. Put on cover and let boil about ten minutes, then put in cooker for four or six hours. When serving the pudding, pass a cold boiled custard or a hot sauce if the day be chilly.

Tripe Stew

Wash the tripe well, about one pound of it, cut into strips, put it in the small kettle, barely cover with cold water and heat slowly. When boiling put it in the big kettle and then in the cooker and let it cook five or six hours or more, if convenient. When the tripe is done put it upon a hot dish and add to one tablespoon of butter and two of flour, cooked together, one cup of the liquid, ½ cup of tomatoes, one slice of onion, ½ teaspoon salt, and ¼ teaspoon paprika; strain over the hot tripe. One tablespoon of parsley and the same amount of cooked red or green peppers, minced fine, may be put over the tripe as an acceptable garnish.

Veal Savory

Cut two pounds of veal in small pieces and saute slightly in bacon fat in which a slice of onion, two of carrot and one tablespoon of parsley have been cooked and skimmed out. Put this meat into the small kettle, pour one cup of boiling water into the saute pan, and strain it over the meat. When this is boiling put it into the large kettle of boiling water and put into the cooker. After six hours or less, take it out and put meat on a dish while you brown one tablespoon of butter with two of flour; add one teaspoon of curry, if liked, or one tablespoon parsley instead. When this has boiled five minutes pour it over the meat and serve with a border of rice garnished with dice of pimentoes. The rice may be steamed in the cooker, as well as some prunes, to eat with the meat as a relish and a pudding.

A beef stew may be cooked in the same way. The method is the same, but a different gravy is used; boiled or mashed potatoes may take the place of the rice. The gravy should be made as follows: To the browned flour and butter add ½ teaspoon salt, one teaspoon Worcestershire sauce, ½ teaspoon onion juice, and a dash each of nutmeg and cayenne. To steam the prunes let them soak in very little water for several hours, then drain them out, saving the water for the sauce. Put the prunes in the cooker, sprinkle with lemon juice and pieces of lemon and steam about four hours or more. When ready to serve, add a very little sugar to the prune water, cook a few moments, and if to serve as a meat relish add a few drops of vinegar and pour over the prunes.

Games and Ways of Entertaining

Fun for Old and Young—Suggestions Which Can be Varied to Suit the Occasion—Ideas for Parties of All Kinds

" Bright Idea "

A quick-witted, good-natured person is sent from the room to be the first victim in this game, which is called "Bright Idea." A familiar object, such as a picture on the wall, or an ornament in the room, is then selected, and the one who has been sent out is recalled. " I have a bright idea," says one of the members, and the one who has returned and is the guesser, responds, " What's it like?" "Like you," is the answer. "In what way?" asks the guesser. " It is ornamental," is the reply. Then another says, " I have a bright idea," and the same formula is repeated, resemblances, real and absurd, being given. The person whose answer leads to a correct guess is the next one to be sent from the room.

Shadow Pictures

Try shadow pictures for an impromptu evening's entertainment. They are simply shadows made by hanging up a large sheet, setting a lamp behind it, and permitting the actors to walk between the lamp and the sheet, making gestures in illustration of a story or poem, which is at the same time read aloud by someone behind the scenes. The audience sees, of course, nothing but the shadows. Choose such tales as have in them plenty of action. The results are sometimes very funny, indeed. Some of the Mother Goose and simple nursery ballads are excellent for this purpose. Given people with a lively sense of humor, and some ability as actors, and there is no end of fun. A studied effort should be made to have all shadows thrown sharply on the sheet. Everything should be in profile, of course.

The Sober Whistler

Choose sides and form two lines opposite each other. A member of one side now starts to whistle a given tune, and all his opponents join forces in trying to make the whistler laugh. If he laughs before his tune is completed, one person from his side can be chosen by the other side. If, on the other hand, he maintains a sober face and whistles to the end, his side wins from the opposing team. When one side has obtained all the play-

ers from the other side, the game is finished. This game is particularly interesting when both sexes take part. A variation of this is as follows: Let a given number of ladies stand at one end of a hall, facing an equal number of gentlemen; at a given signal, the men race across the hall, and each whistles a bar of music into his partner's ear. She in turn jots down upon a piece of paper the name of the tune he has whistled, and this paper he must carry back to the starting place. The one to arrive there first wins.

A Party of Deaf and Dumb

In this game the problem is to *show* your partner what it is you wish to do, for the lips must be kept tightly closed. Gestures, signs and finger play must be depended upon for information. Slips of paper, bearing all kinds of commands, such as " Go out and count the stars with your partner," or " Get your partner to promenade on the veranda with you," are given the guests. When a couple have done the bidding of both their slips, they exchange them for two more, and continue to do this until they have been the rounds of all the slips.

" The Little Dwarf "

This game is one which mystifies and delights the children. During the evening they are called into a room where the lights are dim, to meet a dear friend of the hostess. There they are presented to a little old lady dwarf, about two feet high. She shakes hands with them, speaks to them, sings a song in a cracked voice, and may even dance a jig. This is the way it is done: The old lady is made up of two players. A curtain is draped across one corner of the room, and in front of that is placed a box or table, also draped to the floor, and close to the wall. One player stands behind the table in front of the curtain, and is made up to look like a little old lady, with spectacles, cap, false curls and nose. Her arms are encased in stockings, and her hands thrust into a pair of child's shoes, to represent feet. These move about realistically under a skirt as short as a small baby's, and tied around just under the arms of the player. A shawl draped over the shoulders hides any deficiencies. Behind the curtain and close to the first player, stands the second, with arms thrust through slits in the curtain, and encircling the waist of the little lady. These form the little old lady's hands, which should be covered with mittens if possible. Pin the shawl to the curtain wherever necessary, and be sure that the lights in the room are dim.

Definitions

This is an old game with an ever new zest. Provide each guest with a pencil and paper, and then give out some familiar word, each one of the party being required to write a definition of it. After ten words have been given out, the definitions should be collected and read aloud, or submitted to a committee of judges, who decide the prize winners. It is surprising how difficult of exact definition becomes the most commonplace and simple word. Try it.

Distances

Place two saucers, bottom up, on the floor about a foot apart, and ask some one to place a third saucer for the apex of the triangle, having the distance between the inner edge of that saucer and the inner edge of each of the others equal to the distance between the outer edges of the two saucers first placed. This is a very good test of the judgment of distances, for nine people out of ten will place the third saucer much too near the others.

The Old Game of Proverbs

Boys and girls who like noise, and most of them do, always enjoy Proverbs. One player retires until all the rest decide upon one particular motto, then each in turn is given a word from this proverb to shout upon a certain signal. The one who is to guess gives the signal, and each shouts his word. It is astonishing what a noise will result, and how very difficult it often is to guess the proverb.

The Inquisition

Get two packs of cards; deal one all around, and place the other face downward before the dealer. The latter takes up the first card and places it face up on the table, first asking, "Who is the prettiest girl in the room?" or "Who overate himself at dinner?" or any other such question. The one who has the cor-responding card places it face up in front of him and says, "I." The dealer continues to ask questions, and turn up cards until he has been through the whole pack. A dealer who is ingenious and original in the matter of questions will produce no end of fun. It is surprising how often the questions seem to fit the holder of the corresponding card.

White Elephant

There is no end of fun in this game for either young or old. Each one is sent an invitation to the party asking him to take advantage of the occasion to get rid of some "white elephant." Any article which is undesirable to the owner, or, as is more often the case, a ridiculous article bought for the occasion, is brought, neatly wrapped up and made as deceptive as possible in outward appearance. At a given signal, each exchanges his white elephant for one belonging to someone else, and takes to a secluded spot and opens it. If he does not care to keep it, he carefully does it up again, and starts out to make a fresh exchange. This continues until the host or hostess gives another signal, when each one retains the article then in his or her possession. One who is nimble-fingered and quick at doing up bundles may make a great number of exchanges in the time allotted.

Brushing the Coin

A quarter of a dollar and a whisk broom are provided for this "stunt." The coin is placed in the little depression in the middle of the palm of the hand, and the motion of the broom must be purely a brushing one. It is not fair to dig up the quarter with the end of the broom. It is astonishing how difficult it is to brush that quarter out. This never fails to interest and amuse all who try it.

Cat and Mouse

This game is especially good for the conclusion of a children's party. Two rows of chairs are placed opposite each other and facing, to form an aisle. The chairs must be far enough apart to allow a person to pass between. All take their seats with the exception of two who are blindfolded, and placed one at each end of the aisle. One is the cat and the other is the mouse. They may go anywhere up and down the aisle, between and around the chairs. They must not be helped by the other players. To insure perfect quietness of movement, they may remove their shoes. Absolute silence must reign. Players speak to each other only by signs and not by words. The "mouse," when caught, becomes the cat, and a new player is chosen for the mouse. There is no end of fun in the game, for the antics of the cat and mouse, the one endeavoring to steal upon the other, and the latter trying to avoid the former, are very funny.

Statuary

The following account of this game, as played by a party of young married people, tells exactly how it can be played, and suggests its possibilities:

Two boys retired behind closed doors, taking me with them as their first victim. "Now," said the leader to his companion and me, "assume some ridiculous position." We finally compromised on a tragic pose, in which he was kneeling before me with outstretched hands, while I disdained him haughtily. Then the door was opened a crack, and a couple was called in (a man and a woman), a couple as poorly matched as possible. They were greeted by the leader, who, with a feather duster or baton, delivered an eloquent little speech, to the effect that a group of statuary had just been delivered to him, with which he was slightly dissatisfied. He pointed to us, and by this time we were having difficult work to keep our composure. Could these two friends suggest some slight change? Something, for instance, which would give a more tender effect. Of course, the newcomers, in accordance with human nature, were only too willing to change the position of arm or head in some grotesque fashion, until they declared themselves satisfied. "Then," said the leader, "you may take their places." We were released, and with shouts of laughter at their expense, we forced our critics to assume the exact position they had recommended. Merriment grew greater and greater as the entire company were called into the room, two by two, and the shouts of laughter that arose at each final sentence, "Then you take their places," kept up the curiosity of the uninitiated.

Telephone Mind Reading

This is most mysterious to the uninitiated. Tell your guests that there is a mind reader in town who will read any card that they may pick out from a pack. When the card has been selected, give the telephone number of this friend of yours who knows the trick. This party is then to be called up by the one who selected the card. The name that you give this one to ask for depends on what card has been selected. For instance, if the three of hearts has been chosen, the last name of your mind reader must begin with the letter H, the first letter of the suit selected. The number of the card determines the letter of the alphabet which shall begin the first name. If the three of hearts be chosen, any name beginning with the third letter of the alphabet will serve, as Charles Harris, or Charlotte Harmon. By making your arrangements previously, your mind reader can very quickly tell which card has been chosen, and your guests are bound to be mystified. In a country town, or in the farming district, where arrangements can be made between parties a long distance apart, there is no end of fun in this comparatively simple trick.

The Cheap Jag

This is funny alike for players and onlookers. In some cases, it is hilariously funny. It really is not as bad as it sounds, being so harmless that the most sober and staid may indulge without shock to conscience. All that is needed is a broom or stick of the height of an ordinary broom. Holding this upright, one end on the floor, rest the forehead on the top of the other end, closing the eyes. Now walk three times around the broom or stick, keeping the forehead on the upper end and the eyes closed. On completion of the third round, open the eyes, drop the broom and endeavor to walk across the room to a given point. This is usually enough to give the victim as thorough a harmless "jag" as can be desired. If one of the company fails to respond to three times around, make it five. It may seem easy to walk straight, but just try it.

Mind Reading Made Simple

Two confederates are required for this. One asserts that he has the powers of a mind reader, and that if he leaves the room, and all of the people there will concentrate their thoughts upon one member of the group, he will on his return be able to tell which one it is. The mind reader leaves the room, and it is agreed among the others which one they are to think of. Of course, the mind reader's confederate knows which one it is, and he places his hands in the exact position of the one selected. The mind reader, returning to the room, notes the position of his confederate's hands, and then, rapidly scanning the entire company, notes who it is who has his hands in the same position as his confederates, thus, of course, knowing instantly who it is that the circle has thought of.

Progressive Stories

This may be played by any number of people, and if there chance to be some of vivid imaginations, the outcome is sometimes astonishing. The story is started in the usual " Once upon a time " manner, and then is taken up by the second player, who carries it along a little way and passes it on to the next and so on. The last of all must give it a suitable ending. Of course, it is utterly impossible to see where such a story is going to lead. If one or two children are among the players they often add much to the humor and interest of the story, for the childish imagination is often quicker than that of older people.

A Left-Handed Party

The host and hostess have their right hands fastened behind them when receiving their guests. Every guest giving his right hand in greeting pays a forfeit later in the evening. Potato races form a part of the entertainment. Potatoes are placed at one end of the room in a row. Contestants start from the other end of the room and take up the potatoes in spoons held in the left hand, and return to the other end of the room in a given time. A large bowl of peanuts is placed on the table and each guest removes as many as he can on the back of his left hand. Everything, even to partaking of refreshments, is done with the left hand.

The Eye Test

All the girls are sent to one room and the boys to an adjoining room, with a door between. In this doorway is hung a curtain having in it two holes just large enough to show the eyes. The light is put out in the room where the boys are, and they come to the curtain in turn, and place their eyes to the holes, while each girl, without audibly expressing her opinion, writes the name of the boy to whom she thinks the eyes belong. A correct list in the order of appearance is kept in the room where the boys are. After each boy has had his turn the lights in the other room are extinguished, and the boys are allowed to guess the girls' eyes in the same way. Prizes may be given for the most successful lists.

What Worried the Gump

" See here ! " Thus to the gump exclaimed the gentle jangaree,
" Why you sit staring at that glass is more than I can see ! "
" Because," the gump made answer, as another squint he took,
" I'm wondering if I really am as handsome as I look ! "

United States Weather Bureau and the Great Work It is Doing

Purposes and Methods of a Useful, Scientific Organization—Upper Air Research—Long-Range Forecasting—How Information is Gathered and How Warnings are Issued

A service for forecasting weather conditions throughout the United States has been maintained by the federal government since 1873. It was originally designed for the benefit of navigation alone. Recognition of the practical utility of such service has led to its continued extension in the interest of both agriculture and commerce. During the first 20 years the work was conducted by the signal corps of the army under the war department, but a demand for a strictly scientific bureau unhampered by military regulations resulted in a reorganization in 1891, and the establishment of the present weather bureau as a branch of the department of agriculture. The act providing for the reorganization, approved Oct 1, 1890, summarizes the functions of the bureau as follows:

Functions of Bureau

"The Chief of the Weather Bureau shall have charge of forecasting the weather; the issue of storm warnings; the display of weather and flood signals for the benefit of agriculture, commerce, and navigation; the gaging and reporting of rivers; the maintenance and operation of sea coast telegraph lines and the collection and transmission of marine intelligence for the benefit of commerce and navigation; the reporting of temperature and rainfall conditions for the cotton interests; the display of frost, cold wave, and other signals; the distribution of meteorological information in the interest of agriculture and commerce, and the taking of such meteorological observations as may be necessary to establish and record the climatic conditions of the United States, or are essential for the proper execution of the foregoing duties."

The public knows the weather bureau best through its daily forecast and weather maps. These forecasts are based upon simultaneous observations of local weather conditions taken daily at 8 a m and 8 p m, 75° meridian time, at about 200 stations scattered throughout the United States and West Indies. Each station has one or more trained observers and is equipped with mercurial barometers, thermometers, wind-vanes, rain and snow gages and anemometers, and many of them with sunshine recorders, barographs, thermographs, and other devices, which register automatically a continuous record of the local weather condition and changes as they occur. The result of the twice-a-day observations are immediately telegraphed to the central office, at Washington, D C, where they are charted for study and interpretation by experts trained to foretell the weather conditions which may be expected to prevail during the following 36 and 48 hours. From the data furnished, the forecaster, by comparison with preceding reports, is able to trace the paths of storm areas from the time of their appearance to the moment of observations and approximately forecast their subsequent courses and weather conditions.

Upper Air Research

Heretofore the weather forecasts made by the Weather Bureau have been based entirely on the conditions of the air as measured at the surface of the earth. Now, the Weather Bureau has established a research station to investigate all problems of the physics of the air, and especially to send meteorological instruments high into the upper levels at Mt Weather, Va. The data secured every day from altitudes of from 1 to over 4 miles mark an epoch in meteorological science. At times the temperature at over a few thousand feet high has been found to be as much as 15° warmer than at the surface of the earth. Again, the kites carrying the instruments would pass through swiftly moving air at the surface, only to encounter a stagnant condition at higher levels. At other times, what appeared to be a deep covering of cloudiness would be found to be only a thin layer of cloud, while in other cases the kites have traveled through a depth of cloudy weathe. of over 1 mile in thickness.

These data have an important bearing on the forecasts for the middle Atlantic and New England states, and are telegraphed to Washington each evening and considered by the forecaster. At this research institution the electric and magnetic conditions of the earth and air are also continuously recorded, and studies are being made of the amount of heat in the earth's atmosphere, as well of the temperature next the earth, and of the various forms of energy that reach us from the sun, and an effort is being made to find the relation between these conditions and our weather, although the Weather Bureau has not yet learned how best to interpret and apply all the data now being secured. This institution is officered by the highest trained experts that can be secured anywhere in the world.

Long-Time Predictions

Another important departure that has been made possible by the study of the science that is back of the art of weather forecasting is the making of long-range weather predictions, which were begun for practice purposes, but not for publication, in 1907. The accuracy of these became so marked that during 1908 these predictions have been given to the public. An instance of their value, which has been demonstrated many times before, is found in the case of the drouth which covered the greater part of the country from the Rocky mountains eastward during August and September, 1908. September 22 the Bureau announced that early in the following week general rains would set in over the Rocky mountain plateau and extend eastward. This prediction was fully verified.

The Bureau states that at times the conditions of the atmosphere may be so uncertain that long-range predictions can not be made; but it has demonstrated that in the majority of cases it can forecast the general character of the weather a week in advance with a high degree of accuracy. The value of these forecasts to the agricultural and other industries of the nation can hardly be measured. This gratifying result has been accomplished not only by getting observations from the higher levels, but more especially by the securing of daily telegraphic reports covering the entire northern hemisphere and the making of a meteorological chart such as is accomplished nowhere else in the world.

Reports From Distant Stations

Weather service similar to that of the United States is maintained by Canada and Mexico, and by a system of interchange, daily reports are received (To Page 78)

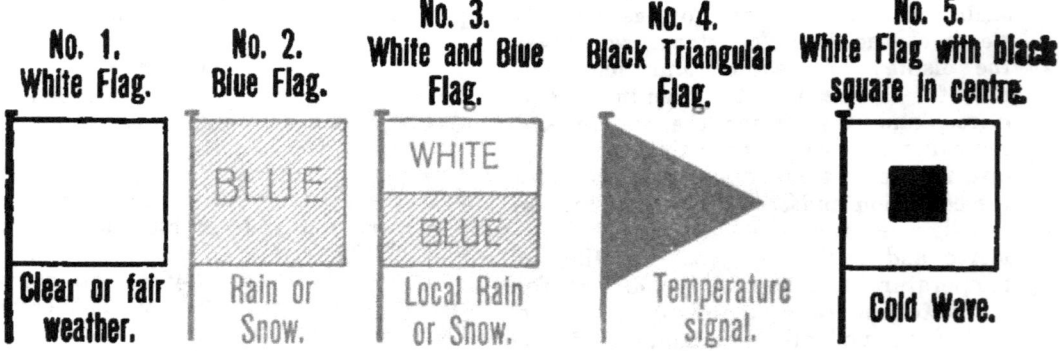

Flag Signals of the Weather Bureau

No 1—Clear or fair weather.

No 2—Rain or snow.

No 3—Local rain or snow.

No 4—Temperature; placed above 1, 2, 3, indicates warm weather; placed below, cold weather; no display, stationary temperature.

No 5—A cold wave or sudden fall in temperature.

A special storm flag, red with black square in center, is prescribed for use in North Dakota, South Dakota, Minnesota, Iowa, Nebraska, Wyoming, Montana, Colorado, Kansas, Indian Territory, Oklahoma and Texas to indicate high winds accompanied by snow.

When the signs are displayed on poles the signals should be arranged to read downward.

En Voyage

Whichever way the wind doth blow,
Some heart is glad to have it so;
Then blow it east, or blow it west,
The wind that blows, that wind is best.

My little craft sails not alone;
A thousand fleets from every zone
Are out upon a thousand seas;
And what for me were favoring breeze
Might dash another with the shock
Of doom upon some hidden rock.

And so I do not dare to pray
For winds to waft me on my way,
But leave it to a Higher Will
To stay or speed me, trusting still
That all is well, and sure that He
Who launched my bark will sail with me.

Then, whatsoever wind doth blow,
My heart is glad to have it so;
And blow it east, or blow it west,
The wind that blows, that wind is best.
—Caroline A. Mason.

Who Bides His Time

Who bides his time, and day by day
  Faces defeat full patiently,
And lifts a mirthful roundelay,
  However poor his fortunes be—
He will not fail in any qualm
  Of poverty—the paltry dime;
It will grow golden in his palm,
  Who bides his time.

Who bides his time—he tastes the sweet
  Of honey in the saltest tear;
And though he fares with slowest feet,
  Joy runs to meet him, drawing near;

The birds are heralds of his cause;
  And like a never-ending rhyme,
The roadsides bloom in his applause,
  Who bides his time.

Who bides his time and fevers not
  In the hot race that none achieves,
Shall wear cool-wreathen laurels, wrought
  With crimson berries in the leaves;
And he shall reign a goodly king,
  And sway his hand o'er every clime,
With peace writ on his signet ring,
  Who bides his time.
—James Whitcomb Riley.

United States Weather Bureau

(*From Page 76*)

from a number of stations in these countries; daily observations are also reported from Alaska, Honolulu, the British Isles, Germany, France, Portugal, Russia, Siberia, Iceland, and the Azores. The field represented by the daily report is extended over the greater portion of the North American continent having telegraphic communication, the North Atlantic ocean, the west coast of Europe and northern Europe and Asia.

Forecast centers are established at Chicago, Boston, New Orleans, Louisville, Denver, San Francisco, and Portland, Ore. The Chicago district comprises the upper Mississippi valley and the northwest; Boston—The New England states; New Orleans—Louisiana, Texas, Arkansas, Oklahoma; Louisville—Kentucky and Tennessee; Denver—Wyoming, Utah, Colorado, New Mexico and Arizona; San Francisco—California, and Nevada; Portland—Washington, Oregon, and Idaho. In addition to the general weather forecasts issued at Washington, forecasts for all not included in the named districts are issued.

Distribution of Forecasts

Within two hours after the morning observations have been taken the forecasts are telegraphed from the forecast centers to more than 2,100 distributing points, whence they are further distributed by telegraph, telephone and mail. The forecasts reach nearly 160,000 addresses by mail, and more than 1,000,000 telephone subscribers, mainly in the rural districts, receive the forecasts within an hour of the time the prediction is made. This system is at government expense and in addition to the service given through the press. The weather map is mailed immediately after the morning forecast is telegraphed. Conditions throughout the country are clearly represented on this map.

A feature of the Weather Bureau is its climatological service divided into 44 local sections, each section as a rule representing a state; each collects temperature and rainfall observations from more than 3,600 co-operative stations. Monthly reports of the data thus covered are given widespread distribution.

Crop Season Bulletins

During the crop-growing season from April to September inclusive, each section also receives weekly mailed reports of weather conditions from numerous correspondents, and publishes them in weekly weather bulletins. During the growing season in cotton, corn, wheat, sugar and rice producing sections, designated centers receive telegraphic reports of rainfall and daily extremes of temperature for publication in bulletin form, each local center receiving the reports from all others. During the crop season the central office from Washington issues weekly national weather bulletins containing a series of charts illustrating currents and conditions of temperature and rainfall for the entire country, and a general summary of the weather from each state. During the winter the central office issues every Tuesday a snow and ice bulletin; each shows the area covered by snow in depth and the thickness of ice in rivers, etc. This publication is of special value to those interested in the winter wheat crop, to ice dealers, and manufacturers of rubber goods, the sale of which is largely affected by the presence or absence of snow and ice. The annual report of the Chief of the Bureau presents a full summary of climatic data for the United States.

Storm, Frost and Flood Warnings

Warnings of especially severe and injurious weather conditions, such as storms and hurricanes, cold waves, frosts, floods, heavy rains and snows, are given in a special manner in addition to the distribution of weather data above referred to. Storm signals for the benefit of maritime interests are displayed from nearly 300 points along the Atlantic, Pacific and Gulf coast, and the shores of the Great Lakes, including every port and harbor of importance. So perfect has this service become that scarcely a storm of marked danger has occurred for years of which ample warnings have not been issued from 12 to 24 hours in advance. The sailings of great numbers of vessels are largely determined by these warnings.

Cold wave warnings are next in importance, and are issued from 24 to 36 hours in advance throughout threatened regions by means of flags displayed at regular weather bureau and sub-display stations. Warnings are also given by telegraph, telephone, and mailed to all places receiving the daily forecast and to a large number of special addresses in addition. The warnings of frost and freez-

(*To Page 80*)

Money Musk

In shirt of check and tallowed hair
The fiddler sits in the bulrush chair,
Like Moses' basket stranded there
    On the brink of Father Nile.
He feels the fiddle's slender neck,
Picks out the notes with thrum and check,
And times the tune with nod and beck,
    And thinks it a weary while.
All ready! Now he gives the call,
Cries, "Honor to the ladies!"   All
The jolly tides of laughter fall
    And ebb in a happy smile.

D-o-w-n comes the bow on every string,
"First couple join right hands and swing!"
And light as any bluebird's wing
    "Swing once and a half times round!"
Whirls Mary Martin all in blue—
Calico gown and stockings new,
And tinted eyes that tell you true,
    Dance all to the dancing sound.

She flits about big Moses Brown,
Who holds her hands, to keep her down,
And thinks her hair a golden crown,
    And his heart turns over once!
His cheek with Mary's breath is wet,
It gives a second somerset!

He means to win the maiden yet—
    Alas, for the awkward dunce!
"Your stogo boot has crushed my toe!
I'd rather dance with one-legged Joe,
You clumsy fellow!" "Pass below!"
    And the first pair dance apart.
Then "Forward six!" advance, retreat,
Like midges gay in sunbeam street,
'Tis Money Musk by merry feet
    And the Money Musk by heart!

"Three quarters round your partner swing!"
"Across the set!" The rafters ring,
The girls and boys have taken wing
    And have brought their roses out!
'Tis "Forward six!" with rustic grace,
Ah, rarer far then—"Swing to place!"
Than golden clouds of old point lace,
    They bring the dance about.

Then clasping hands all—"Right and left!"
All swiftly weaves the measure deft
Across the woof in living weft,
    And the Money Musk is done!
Oh, dancers of the rustling husk,
Good-night, sweethearts, 'tis growing dusk,
Good-night for aye to Money Musk,
    For the heavy march begun!
                    —Benjamin F. Taylor.

United States Weather Bureau

(From Page 78)

ing weather are of great value to the fruit, sugar, tobacco, cranberry and market gardening interests. Growers of oranges and other fruit in Florida and California, for instance, have invested large sums in tents and screens, in heating, smudging and irrigating apparatus for the protection of groves and orchards, which they use when notified by the Weather Bureau of expected low temperatures.

The commerce of our rivers is greatly aided, and lives and property in regions subject to overflow protected, by the publication of the river stages and the issue of river and flood forecasts based on reports received from about 500 special river and rainfall stations.

The annual cost of the Weather Bureau amounts to about $1,300,000.

Instruments

Thermometers and instruments employed by the Weather Bureau are purchased under annual contracts with responsible manufacturers; the addresses of such manufacturers may be obtained through the Weather Bureau by private parties who desire similar instruments. Persons who wish the errors of the weather thermometers accurately determined are referred to the Director of the National Bureau of Standards, Washington, D C, who does such work for a small fee.

The increasing demands on the Bureau for obtaining and supplying various meteorological data requires almost yearly the invention of new apparatus. During the past year these demands have been met in part by the preparation of the following:

(1) A chart and instrument kiosk, or booth, which is an instrument shelter to be located on the street level at places accessible to the public in the business and news centers of every large city.

(2) A recording hygrometer, for stations, that automatically registers the moisture contents of the air, which is one of the most important meteorological elements requiring observation.

(3) A tele-thermoscope, a device which installed in a Weather Bureau office indicates the temperature of the outside air.

Water Resources

A demand for a better knowledge of the water resources of the United States has become so urgent as to make it advisable to put forth special efforts to supply the necessary data to the public. In the arid and semi-arid regions of the west these consist primarily in securing the amount of precipitation in the high levels of the mountains, from which come the waters to the storage basins and the irrigation projects now undergoing rapid development. It is a difficult problem to secure regular and accurate observations in the remote regions of the mountains, but special effort will be made by the Weather Bureau.

Evaporation Study

It is necessary to determine the amount of evaporation in the lower levels, where the storage basins are located. Evaporation in the driest portions of the country, as in the Colorado desert, may amount to as much as 8 feet of water annually, although it differs greatly according to circumstances. More reliable information on this subject is needed. The formation of the Salton Sea in the Colorado desert, by the overflow of the Colorado river during 1906, affords a favorable laboratory on a large scale for research on evaporation.

A preliminary study on this subject was conducted by Prof Bigelow, of the Weather Bureau, in the summer of 1907, at Reno, Nev, for the purpose of securing adequate knowledge to permit a proper planning of the campaign at the Salton Sea. The necessary plant was installed at the Salton Sea during the summer of 1908, and it is hoped that by continuing the observations for 2 or 3 years a satisfactory law covering evaporation generally may be secured.

The plan of co-operation with the other bureaus of the government has been enlarged to include the Reclamation Service, and the Water Resources Branch of the Geological Survey, which are specially interested in evaporation at the reservoirs, not only of the arid west, but in the eastern districts of the country. During the summer of 1908, several plants for the measurement of evaporation were installed at the reservoirs of the Reclamation Service, and if practicable some other reservoirs in the central and eastern districts will be equipped. It is important to measure the evaporation in different climates on a uniform plan in order that a comprehensive law may be deduced.

The care of the body and the care of the soul are not two duties, but two parts of one duty.—[Phillips Brooks.

If We Knew

If we knew the cares and crosses
  Crowded 'round our neighbor's way,
If we knew the little losses
  Sorely grievous day by day,
Would we then so often chide him
  For the lack of thrift and gain,
Leaving on his heart a shadow,
  Leaving on his heart a stain?

If we knew that clouds above us,
  Held by gentle blessings there,
Would we turn away all trembling
  In our blind and weak despair?
Would we shrink from little shadows
  Lying on the dewy grass,
While 'tis only birds of Eden
  Just in mercy fleeting past?

If we knew the silent story
  Quivering through the heart of pain,
Would our womanhood dare doom them
  Back to haunts of vice and shame?
Life has many a tangled crossing,
  Joy has many a break of woe,
And the cheeks tear-washed are whitest—
  This the blessed angels know.

Let us reach within our bosoms
  For the key to others' lives,
And, with love to erring nature,
  Cherish good that still abides;
So that when our shrouded spirits
  Soar to the realms of light again,
We may say, "Dear Father, judge us
  As we judge our fellow-men."

The Nail of Destiny

"For the want of a nail the shoe was lost,
For the want of the shoe the horse was
    lost,
For the want of the horse the rider was
    lost,
For the want of the rider the battle was
    lost,
For the want of the battle the kingdom
    was lost,
  And all for the want of a horseshoe nail!"

But this, my child, as you doubtless know,
Was a number of hundred years ago.
Brought down to date, the facts are these:
The general chooses to ride at ease
In a new six-cylinder automobile,
And he punctures the tire of his off-front
    wheel;
So the battle goes bump—not for the lack,
But because of a smallish carpet tack.
                                —Puck.

Legal Holidays

January 1, New Year's Day—In all the states (including the District of Columbia), except Colorado, Kentucky, Massachusetts, Mississippi, New Hampshire and Rhode Island.

January 8, Anniversary of the Battle of New Orleans—In Louisiana.

January 19, Lee's Birthday—In Florida, Georgia, North Carolina, South Carolina and Virginia.

February 12, Lincoln's Birthday—In Connecticut, Illinois, Minnesota, New Jersey, New York, North Dakota, Pennsylvania and Washington (state).

February 22, Washington's Birthday—In all the states (including the District of Columbia), except Colorado, Iowa, Mississippi and New Mexico.

March 2, Anniversary of Texan Independence—In Texas.

March 4, Fireman's Anniversary—In New Orleans, La; also, Inauguration Day (every four years), in Washington, D C.

April 6, Confederate Memorial Day—In Louisiana.

April 19, Patriots' Day—In Massachusetts.

April 21, Anniversary of the Battle of San Jacinto—In Texas.

April 26, Confederate Memorial Day—In Alabama, Florida and Georgia.

May 10, Confederate Memorial Day—In North Carolina and South Carolina.

May 20, Anniversary of the Signing of the Mecklenburg Declaration of Independence—In North Carolina.

May 30, Memorial Day—In Arizona, California, Colorado, Connecticut, Delaware, District of Columbia, Illinois, Indiana, Iowa, Kansas, Kentucky, Maine, Maryland, Massachusetts, Michigan, Minnesota, Missouri, Montana, Nebraska, Nevada, New Hampshire, New Jersey, New York, North Dakota, Ohio, Oklahoma, Oregon, Pennsylvania, Rhode Island, South Dakota, Tennessee, Utah, Vermont, Washington, Wisconsin and Wyoming.

June 3, Jefferson Davis' Birthday—In Alabama, Florida and Georgia.

June 17, Bunker Hill Day—In Boston, Mass (a local, not legal, holiday).

July 4, Independence Day—In all the states and the District of Columbia.

July 24, Pioneers' Day—In Utah.

August 16, Bennington Battle Day—In Vermont.

September 6, 1909, Labor Day—In Alabama, California, Colorado, Connecticut, Delaware, District of Columbia, Florida, Georgia, Idaho, Illinois, Indiana, Iowa, Kansas, Maine, Maryland, Massachusetts, Michigan, Minnesota, Missouri, Montana, Nebraska, New Hampshire, New Jersey, New Mexico, New York, Ohio, Oklahoma, Oregon, Pennsylvania, Rhode Island, South Carolina, South Dakota, Tennessee, Texas, Utah, Virginia, Washington, Wisconsin and Wyoming.

September 9, Admission Day—In California.

September 12, Defenders' Day—In Baltimore, Md.

October 31, Admission Day—In Nevada.

November 1, All Saints' Day—In Louisiana.

November 2, 1909, Election Day—The first Tuesday after the first Monday in November, in general election years, is observed as a holiday in Arizona, California, Colorado, Delaware, Florida, Idaho, Indiana, Iowa, Kansas, Louisiana, Maryland, Michigan, Minnesota, Missouri, Montana, Nevada, New Hampshire, New Jersey, New Mexico, New York, North Dakota, Ohio, Oklahoma, Oregon, Pennsylvania, Rhode Island, South Carolina, South Dakota, Tennessee, Texas, West Virginia, Washington, Wisconsin and Wyoming.

November 25, 1909, Thanksgiving Day—The fourth Thursday in November is observed in all the states and the District of Columbia, although in some states it is not a statutory holiday.

December 25, Christmas Day—In all the states and in the District of Columbia.

Shrove Tuesday is a legal holiday in Alabama and New Orleans. Good Friday is a holiday in Alabama, Louisiana, Maryland, Minnesota, Pennsylvania and Tennessee.

Sundays and fast days are legal holidays in all the states which designate them as such.

There are no statutory holidays in Mississippi, Kansas and Nevada, but by common consent the Fourth of July, Thanksgiving and Christmas are observed as holidays in Mississippi, and Decoration Day, Labor Day and Arbor Day in addition in Kansas.

Every Saturday after 12 o'clock noon is a legal holiday in New York, New Jersey, Pennsylvania, Maryland, Tennessee, Virginia and the District of Columbia, the city of New Orleans and in New-castle county, Delaware, except in St George's Hundred; in Louisiana and Missouri, in cities of 100,000 or more inhab-

(To Page 84)

The Mother-Look

You take the finest woman, with th' roses
in her cheeks,
An' all the birds a-singin' in her voice each
time she speaks;
Her hair all black an' gleamin' or a glowin'
mass o' gold—
An' still th' tale o' beauty isn't more th'n
half-way told,
There ain't a word that tells it; all description it defies—
Th' mother-look that lingers in a happy
woman's eyes.

A woman's eyes will sparkle in her innocence an' fun,
Or snap a warnin' message to th' ones she
wants to shun.
In pleasure or in anger there is always
han'someness,
But still there is a beauty that was surely
made to bless—
A beauty that grows sweeter an' that all
but glorifies—
Th' mother-look that sometimes comes into
a woman's eyes.

It ain't a smile, exactly—yet it's brimmin'
full o' joy,
An' meltin' into sunshine when she bends
above her boy,
Or girl, when it's a-sleepin', with its dreams
told in its face;
She smooths its hair, an' pets it as she lif's
it to its place,
It leads all th' expressions, whether grave
or gay, or wise—
Th' mother-look that glimmers in a lovin'
woman's eyes.

There ain't a picture of it! If there was,
they'd have to paint
A picture of a woman mostly angel an'
some saint,
An' make it still be human—an' they'd have
to blend the whole.
There ain't a picture of it, for no one can
paint a soul!
No one can paint th' glory comin' straight
from paradise—
Th' mother-look that lingers in a happy
woman's eyes.

—Wilbur D. Nesbit.

Legal Holidays

(From Page 82)

itants; in Ohio, in cities of 50,000 or more inhabitants; and June 1 to September 30 in Denver, Col. In Connecticut, Maine and West Virginia, banks close at 12 noon on Saturdays.

May 18—International Peace Day is gradually coming to be observed, to celebrate the anniversary of the opening of the first Peace Conference of The Hague in 1899.

June 14, Flag Day—Not a legal holiday, but its observance, especially in the public schools. is urged by several governors and various patriotic societies, to stimulate devotion to the flag and patriotism. All anniversaries of important historical events attended by patriotic observance in any state should be observed as Flag Days in the public schools of that state by the display of the national flag. June 14 is the anniversary of the day in 1777 when the Continental Congress adopted the United States flag with 13 stripes and 13 stars, symbolic of the original states. Since 1818 a star has been added for each state admitted to the Union.

Arbor Days

Alabama—February 22.

Arizona—Friday following first day of April, also Friday following first day of February.

Arkansas—First Saturday in March.

California—Observed by separate counties, but not generally.

Colorado—Third Friday in April.

Connecticut—Appointed by governor. last Friday in April or first in May.

Delaware—Appointed by governor. usually in April.

District of Columbia—Not observed.

Florida—First Friday in February.

Georgia—First Friday in December.

Idaho—Last Monday in April.

Illinois—Date fixed by governor and superintendent of public instruction.

Indiana—Last Friday in October.

Iowa—Date fixed by governor.

Kansas—Date fixed by governor.

Kentucky—Not regularly observed.

Maine—Date fixed by governor, usually early in May.

Maryland—In April; date fixed by governor.

Massachusetts—Last Saturday in April.

Michigan—Last Friday in April.

Minnesota—Date fixed by governor; usually last of April or first of May.

Missouri—Friday after first Tuesday in April.

Mississippi—December 10.

Montana—Second Tuesday in May.

Nebraska—April 22.

Nevada—Date fixed by governor; usually in April.

New Hampshire—No date fixed; usually in May.

New Jersey—Usually third Friday in April; fixed by governor.

New Mexico—Second Friday in March.

New York—Friday after first of May.

North Carolina—October 12 usually observed.

North Dakota—First Friday in May.

Ohio—Second or third Friday in April.

Oklahoma—Second Friday in April.

Oregon—Second Friday in April.

Pennsylvania—In October; fixed by superintendent of instruction.

Rhode Island—Second Friday in May.

South Carolina—Third Friday in November.

South Dakota—Date fixed by governor.

Tennessee—Date fixed annually in November.

Texas—February 22.

Utah—April 15.

Vermont—Latter part of April or first of May. Date fixed by governor.

Virginia—Not regularly observed.

Washington — Irregularly observed; date set by governor; different dates east and west of the Cascades.

West Virginia—Third Friday in April and third Friday in November.

Wisconsin—Date fixed by governor.

Wyoming—Date fixed by governor.

Centenaries in 1909

The year 1909 will be noteworthy for the large number of famous people who were born just a century before. The birth roll for 1809 includes a long list of statesmen, soldiers, musicians, authors. inventors and scientists. The centenaries of many of these great men will be celebrated with important exercises. Among the anniversaries are the following:

January 19, Edgar Allan Poe.

February 3, Mendelssohn, the musician, and Joseph Johnston, the Confederate soldier.

February 12, Abraham Lincoln and Charles Darwin.

February 15, Cyrus McCormick, inventor of the reaping machine.

March 3, Chopin, the musician.

August 6, Alfred Tennyson.

August 29, Oliver Wendell Holmes.

December 29, William E. Gladstone.

There is no Death

There is no death! The stars go down,
  To rise upon some fairer shore;
And bright in heaven's jeweled crown
  They shine forever more.

There is no death! The dust we tread
  Shall change beneath the summer showers
To golden grain or mellowed fruit,
  Or rainbow-tinted flowers.

The granite rocks disorganize,
  And feed the hungry moss they bear;
The forest leaves drink daily life,
  From out the viewless air.

There is no death! The leaves may fall,
  And flowers may fade and pass away;
They only wait through wintry hours,
  The coming of the May.

There is no death! An angel form
  Walks o'er the earth with silent tread;
He bears our best loved things away,
  And then we call them "dead."

He leaves our hearts all desolate,
  He plucks our fairest, sweetest flowers;
Transplanted into bliss, they now
  Adorn immortal bowers.

The birdlike voice, whose joyous tones,
  Made glad these scenes of sin and strife,
Sings now an everlasting song,
  Around the tree of life.

Where'er he sees a smile too bright,
  Or heart too pure for taint and vice,
He bears it to that world of light,
  To dwell in paradise.

Born unto that undying life,
  They leave us but to come again;
With joy we welcome them the same,
  Except their sin and pain.

And ever near us, though unseen,
  The dear immortal spirits tread;
For all the boundless universe
  In life—there are no dead.

—J. L. McCreery.

The Bankruptcy Law

Extracts from the United States Bankruptcy Act of July 1, 1898

Who May Become Bankrupts

SEC 4. (a) Any person who owes debts, except a corporation, shall be entitled to the benefits of this act as a voluntary bankrupt.

(b) Any natural person (except a wage-earner or a person engaged chiefly in farming or the tillage of the soil), any unincorporated company, and any corporation engaged principally in manufacturing, trading, printing, publishing, or mercantile pursuits, owing debts to the amount of one thousand dollars or over, may be adjudged an involuntary bankrupt upon default or an impartial trial, and shall be subject to the provisions and entitled to the benefits of this act. Private bankers, but not national banks or banks incorporated under state or territorial laws, may be adjudged involuntary bankrupts.

Duties of Bankrupts

SEC 7 (a) The bankrupt shall (1) attend the first meeting of his creditors, if directed by the court or a judge thereof to do so, and the hearing upon his application for a discharge, if filed; (2) comply with all lawful orders of the court; (3) examine the correctness of all proofs of claims filed against his estate; (4) execute and deliver such papers as shall be ordered by the court; (5) execute to his trustee transfers of all his property in foreign countries; (6) immediately inform his trustee of any attempt, by his creditors or other persons, to evade the provisions of this act, coming to his knowledge; (7) in case of any person having to his knowledge proved a false claim against his estate, disclose the fact immediately to his trustee; (8) prepare, make oath to and file in court within ten days, unless further time is granted, after the adjudication of an involuntary bankrupt, and with the petition of a voluntary bankrupt, a schedule of his property, showing the amount and kind of property, the location thereof, its money value in detail, and a list of his creditors, showing their residences, if known (if unknown, that fact to be stated), the amount due each of them, the consideration thereof, the security held by them, if any, and a claim for such exemptions as he may be entitled to, all in triplicate, one copy of each for the clerk, one for the referee

and one for the trustee; and (9) when present at the first meeting of the creditors, and at such other times as the court shall order, submit to an examination concerning the conducting of his business, the cause of his bankruptcy, his dealings with his creditors and other persons, the amount, kind and whereabouts of his property, and in addition, all matters which may affect the administration and settlement of his estate; but no testimony given by him shall be offered in evidence against him in any criminal proceedings.

Providing, however, that he shall not be required to attend a meeting of his creditors, or at or for any examinations at a place more than one hundred and fifty miles distant from his home or principal place of business, or to examine claims except when presented to him, unless ordered by the court, or a judge thereof, for cause shown, and the bankrupt shall be paid his actual expenses from the estate when examined or required to attend at any place other than the city, town or village of his residence.

Mile Records on Land

Electric locomotive, 27 seconds, 1903.
Automobile, 28½ seconds, 1906.
Steam locomotive, 32 seconds, 1893.
Motor-paced bicycle, 1 minute 6 1-5 seconds, 1904.
Bicycle, unpaced, 1 minute 49 2-5 seconds, 1904.
Running horse, 1 minute 35½ seconds, 1890.
Pacing horse, 1 minute 55 seconds, 1906.
Trotting horse, 1 minute 58½ seconds, 1905.
Man skating, 2 minutes 36 seconds, 1896.
Man running, 4 minutes 12¾ seconds, 1887.
Man walking, 6 minutes 23 seconds, 1890.

Remembering His Catechism

Willie was discovered beating the rug in the yard and eagerly watching the dust ascend.

"What are you doing, Willie?" asked his mother.

"I'm sending up some dust so God can make a lot of new people," was the reply.

Before insisting upon a square deal, be certain you know how to play the game.

Be good, but don't overdo it.

The Isle of Long Ago

Oh, a wonderful stream is the River Time,
  As it runs through the realm of tears,
With a faultless rhythm and a musical
    rhyme,
And a boundless sweep and a surge sub-
    lime,
  As it blends with the Ocean of Years.

How the winters are drifting, like flakes of
    snow,
  And the summers like buds between,
And the year in the sheaf; so they come
    and they go,
On the river's breast, with its ebb and flow,
  As it glides in the shadow and sheen.

There's a magical isle up the River Time,
  Where the softest of airs are playing;
There's a cloudless sky and a tropical clime,
And a song as sweet as a vesper chime,
  And the Junes with the roses are stray-
    ing.

And the name of that isle is the Long Ago,
  And we bury our treasures there;
There are brows of beauty and bosoms of
    snow;

There are heaps of dust—but we loved
    them so!
  There are trinkets and tresses of hair;

There are fragments of song that nobody
    sings,
  And a part of an infant's prayer;
There's a lute unswept, and a harp without
    strings;
There are broken vows and pieces of rings,
  And the garments that she used to wear.

There are hands that are waved when the
    fairy shore
  By the mirage is lifted in air;
And we sometimes hear through the turbu-
    lent roar
Sweet voices we heard in the days gone
    before,
  When the wind down the river is fair.

Oh, remembered for aye be the blessed isle,
  All the day of our life until night;
When the evening comes with its beautiful
    smile,
And our eyes are closing to slumber a while,
  May that "Greenwood" of soul be in sight!
                              —Benjamin F. Taylor.

Marriage and Divorce

Licenses

Marriage licenses are required in all the states and territories except Alaska, New Jersey (required for non-residents), New Mexico and South Carolina. California and New York require prospective bride and groom to appear and be examined under oath. The legal age at which marriage may be contracted without consent of parents in most of the states having laws on the subject is 21 years for men; in California, Delaware, Idaho and North Dakota, 18 years; in Tennessee, 16 years. For women the legal age is 21 years in Florida, Iowa, Kentucky, Louisiana, Minnesota, Montana, Nebraska, North Carolina, Pennsylvania, Rhode Island, South Carolina, Kansas, South Dakota, Utah, Virginia, West Virginia, Wisconsin and Wyoming; 18 years in all other states having laws on the subject, except Delaware, District of Columbia, Idaho, Maryland, New York and Tennessee, where it is 16 years, and California and North Dakota, 15 years.

Prohibited Marriages

Marriage is prohibited and punishable between whites and persons of negro descent in Alabama, Arizona, Arkansas, California, Colorado, Delaware, Florida, Georgia, Idaho, Indiana, Kentucky, Louisiana, Maryland, Mississippi, Missouri, Nebraska, North Carolina, Oklahoma, Oregon, South Carolina, Tennessee, Texas, Utah, Virginia and West Virginia. Marriages between whites and Indians are void in Arizona, North Carolina, Oregon and South Carolina and between whites and Chinese in Arizona, California, Mississippi, Oregon and Utah. Marriage between first cousins is forbidden in Alaska, Arizona, Arkansas, Illinois, Indiana, Kansas, Missouri, Nevada, New Hampshire, North Dakota, Ohio, Oklahoma, Oregon, Pennsylvania, North Dakota, Washington and Wyoming. In some of these states such marriage is declared void. Marriage with step relatives of near degree is forbidden in all states except Florida, Hawaii, Iowa, Kentucky, Minnesota, New York, Tennessee and Wisconsin. The marriage of an epileptic imbecile or feeble-minded woman under 45 years of age is prohibited in Connecticut and Minnesota. The marriage of lunatics is void in the District of Columbia, Kentucky, Maine, Massachusetts and Nebraska; also of persons having sexual diseases in Michigan.

Standard Time

A standard of time was established by mutual agreement in 1883, primarily for the convenience of the railroads, by which trains are run and local time regulated. According to this system, the United States, extending from 65° to 125° west longitude, is divided into four time sections, each 15° of longitude, exactly equivalent to one hour, commencing with the 75th meridian. The first (Eastern) section includes all territory between the Atlantic coast and an irregular line drawn from Detroit to Charleston, S C, the latter being its most southern point. The second (Central) section includes all the territory between the last-named line and an irregular line from Bismarck, N D, to the mouth of the Rio Grande. The third (Mountain) includes all territory between the last-named line and nearly the western borders of Idaho, Utah and Arizona. The fourth (Pacific) section covers the rest of the country to the Pacific coast.

Standard time is uniform inside each of these sections, and the time of each section differs from that next to it by exactly one hour. Thus at 12 noon in New York city (Eastern time), the time at Chicago (Central time) is 11 o'clock a m; at Denver (Mountain time), 10 o'clock a m, and at San Francisco (Pacific time), 9 o'clock a m. Standard time is 16 minutes slower at Boston than true local time, 4 minutes slower at New York, 8 minutes faster at Washington, 19 minutes faster at Charleston, 28 minutes slower at Detroit, 18 minutes faster at Kansas City, 10 minutes slower at Chicago, 1 minute faster at St Louis, 28 minutes faster at Salt Lake City, and 10 minutes faster at San Francisco.

Catholic Bible

The Douay version of the Bible is used in the Roman Catholic church. This translation was authorized by the pope. The Old Testament was published by the English college at Douay in France in 1609, and the New Testament at Rheims in 1582. The text of the Douay Bible is copiously explained by notes of Roman Catholic divines, and is a translation of the Latin Vulgate.

All That's Necessary

"Why in the world did you elect that dumb man as your representative?"

"That's all right. He may not be able to talk, but he can make motions."

Birds in Summer

How pleasant the life of a bird must be,
Flitting about in each leafy tree;
In the leafy trees so broad and tall,
Like a green and beautiful palace hall,
With its airy chambers light and boon,
That open to sun and stars and moon;
That open to the bright blue sky,
And the frolicsome winds as they wander by.

They have left their nests on the forest
   bough;
Those homes of delight they need not now:
And the young and the old they wander out,
And traverse their green world round about;
And hark! at the top of this leafy hall,
How one to the other in love they call!
"Come up! come up!" they seem to say,
Where the topmost twigs in the breezes
   sway.

"Come up! come up! for the world is fair
Where the merry leaves dance in the sum-
   mer air."
And the birds below give back the cry,
"We come, we come to the branches high."
How pleasant the lives of the birds must be,
Living in love in a leafy tree!
And away through the air what joy to go,
And to look on the green, bright earth
   below!

What joy it must be, like a living breeze,
To flutter about 'mid the flowering trees;
Lightly to soar, and to see beneath
The wastes of the blossoming purple heath,
And the yellow furze, like fields of gold,
That gladdened some fairy region old!
On the mountain tops, on the billowy sea,
On the leafy stems of a forest tree,
How pleasant the life of a bird must be!
                              —Mary Howitt.

You may break, you may shatter the vase
   if you will,
But the scent of the roses will hang
   round it still.
                              —Moore.

Here hath been dawning
   Another blue day;
Think, wilt thou let it
   Slip useless away?
                              —Carlyle.

Glossary of Words Used in the Agricultural Sciences

A Guide to the Meaning of Terms that Everyone Ought to Know About, but with Reference to Farming and Kindred Interests

ABDOMEN—The part of an insect lying behind the thorax.

ABOMASUM—The fourth stomach of ruminants.

ACID—A chemical name given to many sour substances. Vinegar and lemon juice owe their sour taste to the acid in them.

AD LIBITUM—At pleasure. In case of feeding farm animals, all they will eat of any particular feeding stuff.

ADULT—A person, animal or plant grown to full size and strength.

ALBUMINOIDS—The more complex forms of protein. They are usually insoluble in water or may be rendered so by heat.

ALIMENTARY TRACT OR CANAL—The duct comprising the stomach, intestines, etc, by which food (aliment) is conveyed through the body and the useless parts evacuated.

AMIDES—A class of chemical compounds formed by substituting acid for nitrogen atoms in ammonia.

AMMONIA—(*ammonium*)—A compound of nitrogen readily usable as a plant food. It is one of the products of decay.

ANNUAL—A plant that bears seed during the first year of its existence and then dies.

ANTHER—The part of a stamen that bears the pollen.

ATMOSPHERIC NITROGEN—Nitrogen in the air. Great quantities of this valuable plant food are in the air; but, strange to say, most plants cannot use it directly from the air, but must take it in other forms, as nitrates, etc. The legumes are an exception, as they can use atmospheric nitrogen.

ASH—The portion of a feeding stuff which remains after it has been burned.

ASSIMILATE—The conversion of digested nutrients into the fluid or solid substances of the body.

AVAILABLE PLANT FOOD—Food in such condition that plants can use it.

BACTERIA—A name applied to a number of kinds of very small living beings, some beneficial, some harmful, some disease-producing.

BALANCED RATION—A combination of farm foods containing the various nutrients in such proportion and amount as will nurture the animal for 24 hours, with the least waste of nutrients.

BIENNIAL—A plant that produces seed during the second year of its existence and then dies.

BLIGHT—A diseased condition in plants in which the whole or a part of a plant withers or dries up.

BLUESTONE—A chemical; copper sulphate. It is used to kill fungi, etc.

BOLUS—A rounded mass; portion of food ready to be swallowed at one time.

BORDEAUX MIXTURE—A mixture invented in Bordeaux, France, to destroy disease-producing fungi.

BUD—(noun)—An undeveloped branch.

BUD—(verb)—To inset a bud from the scion upon stock to insure better fruit.

BUD VARIATION—Occasionally one bud on a plant will produce a branch differing in some ways from the rest of the branches; this is bud variation. The shoot that is produced by bud variation is called a sport.

CALORIE—The amount of heat required to raise the temperature of 1 kilogram of water 1° Centigrade (or 1 pound of water 4° Fahrenheit).

CALYX—The outermost row of leaves in a flower.

CANON—The shank bone above the fetlock in the fore and hind legs of a horse.

CARBOHYDRATES—A group of nutrients rich in carbon and containing oxygen and hydrogen in the proportion in which they form water. The carbohydrates do not contain nitrogen.

CARBOLIC ACID—A chemical often used to kill or prevent the growth of germs, bacteria, fungi, etc.

CARBON—A chemical element. Charcoal is nearly pure carbon.

CARBON DISULPHIDE—A chemical used to kill insects.

CARBONIC-ACID-GAS—A gas consisting of carbon and oxygen. It is produced by breathing, and whenever carbon is burned. The source of carbon in plants.

(*To Page 92*)

The Old Swimmin'-Hole

Oh! the old swimmin'-hole! whare the crick
   so still and deep
Looked like a baby river that was layin'
   half asleep,
And the gurgle of the worter round the drift
   jest below
Sounded like the laugh of something we
   onc't ust to know
Before we could remember anything but the
   eyes
Of the angels lookin' out as we left Paradise;
But the merry days of youth is beyond our
   control,
And it's hard to part forever with the old
   swimmin'-hole.

Oh! the old swimmin'-hole! In the long,
   lazy days
When the hum-drum of school made so
   many run-a-ways,
How pleasant was the jurney down the
   old dusty lane.
Whare the tracks of our bare feet was all
   printed so plain.
You could tell by the dent of the heel and
   the sole
They was lots of fun on hands at the old
   swimmin'-hole.

Thare the bullrushes growed, and the cat-
   tails so tall,
And the sunshine and shadder fell over it
   all;
And it mottled the worter with amber and
   gold
Till the glad lilies rocked in the ripples that
   rolled;
And the snake-feeder's four gauzy wings
   fluttered by
Like the ghost of a daisy dropped out of
   the sky.

Oh! the old swimmin'-hole! When I last
   saw the place,
The scenes was all changed like the change
   in my face;
The bridge of the railroad now crosses the
   spot
Whare the old divin'-log lays sunk and fer-
   got.
And I stray down the banks whare the trees
   ust to be—
But never again will their shade shelter me!
And I wish in my sorrow I could strip to
   the soul,
And dive off in my grave like the old swimmin'-hole.

—James Whitcomb Riley.

Glossary

(From Page 90)

CARNIVOROUS—A term applied to animals that feed chiefly on flesh.

CASEIN—The protein substance of milk which is coagulated by rennet or acids.

CASTOR-OIL BEAN—The seed of Ricinus communis.

CATHARTIC—A medicine that acts as a purge.

CELLULOSE—The cell tissue of plants. The lint of cotton and wool pulp are almost pure cellulose.

CEREAL—The name given to grasses that are raised for the food contained in their seeds, such as corn, wheat, rice.

CHYLE—A milky fluid found in the lacteals, consisting of digested but unassimilated nutrients in solution, and the digested fatty matter of the food in a state of emulsion.

CONCENTRATED—When applied to food the word means that it contains much feeding value in small bulk.

CORN FODDER, or FODDER CORN—Stalks of corn, either green or dried, which are grown for forage and from which the ears or nubbins, if they carry any, have not been removed.

CORN STOVER—(See Stover).

CROSS-POLLINATION—The pollination of a flower by pollen brought from a flower on some other plant.

CRUDE FIBER—The framework forming the walls of the cells of plants. It is composed of cellulose and lignin, the latter being the more woody portion.

CRUDE PROTEIN—(See Protein).

CURCULIO—A kind of beetle or weevil.

DIASTASE—The ferment found in seeds while germinating, especially in malting barley, by aid of which starch is converted into glucose.

DIGESTIBLE MATTER—The part of feeding stuffs brought into solution or semi-solution by the digestive fluids.

DIGESTIBLE NUTRIENTS—The portion of any food constituent that is digested by animals.

DIGESTION COEFFICIENT—The percentage of any particular nutrient of a feeding stuff which is found to be digestible.

DIGESTIVE TRACT—(See Alimentary Tract).

DRY MATTER—The portion of a feeding stuff remaining after the water or moisture contained therein has been driven off by heat.

ENSILAGE—Green foods preserved in a silo.

EMULSION—A milklike mixture of a liquid and a solid, or of two liquids in which one of the constituents, generally fat or oil, is present in suspension in an exceedingly fine mechanical condition.

ETHER EXTRACT—That which is dissolved from a water-free feeding stuff by means of ether. It is often termed "fat" by agricultural writers.

EVAPORATE—To pass off in vapor, as a fluid often does; to change from a solid or liquid state into vapor, usually by heat.

EXCREMENT—The indigestible or refuse matter of farm foods voided by animals.

FERMENTATION—A chemical change produced by bacteria, yeast, etc. A common example of fermentation is the change of cider into vinegar.

FERTILIZING CONSTITUENTS—The nitrogen and mineral components of feeding stuffs. Generally the term applies only to nitrogen, phosphoric acid and potash, since these are most apt to be lacking in the soil or present in insufficient quantities.

FETLOCK—The long-haired cushion on the back side of a horse's leg, just above the hoof.

FIBER—Any fine, slender thread or threadlike substance, as the rootlets of plants or the lint of cotton.

FOOT-TON—The work performed in raising a weight of one ton to a height of one foot.

FORMALIN—A 40% solution of a chemical known as formaldehyde. Formalin is used to kill fungi, bacteria, etc.

FUNGICIDE—A substance used to kill or prevent the growth of fungi; for example, bordeaux mixture or copper sulphate.

FUNGOUS—Belonging to or caused by fungi.

FUNGUS—(plural Fungi)—A low kind of plant life lacking in green color. Molds and toadstools are examples.

GERM—That from which anything springs. The term is often applied to any very small organism or living thing, particularly if it causes great effects, such as disease, fermentation, etc.

GLOBULE—A small particle of matter shaped like a globe.

GLUCOSE—A kind of sugar very common in plants. The sugar from grapes, honey, etc, is glucose. That from the sugar cane is not.

GLUTEN—A vegetable form of protein found in cereals.

(To Page 94

The Birth of the Opal

The Sunbeam loved the Moonbeam,
  And followed her low and high,
But the Moonbeam fled and hid her head,
  She was so shy—so shy.

The Sunbeam wooed with passion·
  , Ah, he was a lover bold!
And his heart was afire with mad desire
  For the Moonbeam pale and cold.

She fled like a dream before him,
  Her hair was a shining sheen,
And, oh, that Fate would annihilate
  The space that lay between

Just as the day lay panting
  In the arms of the twilight dim,

The Sunbeam caught the one he sought
  And drew her close to him.

But out of his warm arms, startled
  And stirred by Love's first shock,
She sprang afraid, like a trembling maid,
  And hid in the niche of a rock.

And the Sunbeam followed and found her,
  And led her to Love's own feast;
And they were wed on that rocky bed,
  And the dying day was their priest.

And lo! the beautiful Opal—
  That rare and wondrous gem—
Where the moon and sun blend into one,
  Is the child that was born to them
          —Ella Wheeler Wilcox.

A Song of Living

I would live—
Not go through life content, as cattle are
With field and stream, hemmed in with gate
    and bar,
To peaceful commonplaces always chained,
Without one thought of all the unattained.
Give me the rapture of enthralling strife,
The nerves that thrill, the blood with pas-
    sion rife,

The dreams that reach to things beyond the
    real,
The love that fashions its own fair ideal.
Not ease nor fame nor what men call suc-
    cess
I ask, but that each day I may possess
A will to do, a soul undaunted still,
A heart where life's most poignant pulses
    thrill.

Glossary

(From Page 92)

GRAFT—To place a living branch or stem on another living stem so that it may grow there. It insures the growth of the desired kind of plant.

GRAIN EQUIVALENT—The term used to designate the comparative value of grain and less concentrated feeding stuffs, such as milk, whey, roots, etc.

GYPSUM—Land plaster.

HEREDITY—The resemblance of offspring to parent.

HIBERNATING—To pass the winter in a torrid or inactive state in close quarters.

HOCK—The joint in the hind leg of quadrupeds between the leg and the shank. It corresponds to the ankle in man.

HUMUS—The portion of the soil caused by the decay of animal or vegetable matter.

HYBRID—The result of breeding two different kinds of plants together.

HYDROGEN—A chemical element. It is present in water and in all living things.

INOCULATE—To give a disease by inserting the germ that causes it in a healthy being.

INTESTINE—The lower part of the alimentary canal.

KAINIT—Salts of potash used in making fertilizers.

KILOGRAM—A metric weight equal to 2.2+ pounds.

LACTEALS—Minute tubes which take the chyle from the alimentary canal.

LEGUME—A plant belonging to the family of the pea, clover and bean; that is, having a flower of similar structure.

LICHEN—A kind of flowerless plant that grows on stones, trees, boards, etc.

LITER—A metric measure of capacity equaling 1.05+ quarts.

LOAM—An earthy mixture of clay and sand with organic matter.

LYMPH—The colorless fluid found in the lymphatics of the animal body.

LYMPHATICS—The veinlike vessels that convey lymph.

LYMPHATIC SYSTEM—The system of lymph vessels which collect and convey the lymph.

MAINTENANCE RATION—An allowance of food sufficient to maintain a resting animal—neither gaining nor losing in weight.

MASTICATE—To crush or grind food with the teeth.

MEMBRANE—A thin layer or fold of animal or vegetable matter.

MILDEW—A cobwebby growth of fungi on diseased or decaying things.

MINERAL MATTER—(See Ash).

MULCH—A covering of straw, leaves, or like substance over the roots of plants to protect them from heat, drouth, etc, and to preserve moisture.

NECTAR—A sweetish substance in blossoms of flowers from which bees make honey.

NITRATE—A readily usable form of nitrogen. The most common nitrate is saltpeter.

NITROGEN—A chemical element; one of the most important and most expensive plant foods. It exists in fertilizers, in ammonia, in nitrates and in organic matter.

NODULE—A little knot or bump.

NUTRIENT—Any substance which nourishes or promotes growth.

NUTRITIVE RATIO—The proportion of digestible protein to digestible carbohydrates and ether extract in a ration, the percentage of ether extract being multiplied by 2.4 and added to the carbohydrates.

OIL MEAL—As understood by American farmers, this term applies only to linseed-oil cake reduced to meal by grinding.

OMASUM—The third stomach of ruminants.

OMNIVEROUS—Eating or living upon animal or vegetable food indiscriminately.

ORGANIC MATTER—Substances made through the growth of plants or animals.

OVARY—The particular part of the pistil that bears the immature seed.

OXYGEN—A gas present in the air and necessary to breathing.

PEPSIN—The digestive ferment found in the stomach of animals.

PERENNIAL—Living through several years. All trees are perennial.

PERIOD OF GESTATION—The length of time of carrying the young; from conception to birth.

PERIOD OF LACTATION—The time during which the animal suckles her young; with dairy cows, the period from calving to drying off.

PHOSPHORIC ACID—An important plant food occurring in bones and rock phosphates.

PISTIL—The part of the blossom that contains the immature seeds.

(To Page 96)

King Corn

With a tassel for a scepter
  And a wild bloom for a crown,
The king of autumn harvest
  Brings his golden store to town;
The wigwams of the fodder
  On the garnered hillsides glow,
With smoke wreaths curling upward
  From the lips of long ago:
      King Corn, King Corn is coming,
        And his ranks are on the hill;
      A tassel for a scepter
        And the meal bag at the mill!

Down all the Maryland valleys
  And on her mountain heights
King Corn is in his glory
  Of his harvest home delights;
The savers of the fodder
  Tie the bundles, pull the blades;
The rabbit's in the hollows
  And the partridge in the glades:
      King Corn, King Corn is coming,
        He's walled the fields with gold;
      The Shorthorns in the barnyard
        And the Southdowns in the fold!

The sunlight weaves its shadows
  Round the shock that stands in line;
The mist is on the meadows
  And the grapes are ripe for wine;
The cider press is going
  And across the fallow sweet

They sing the song of sowing
  As they cast the winter wheat:
      King Corn, King Corn is coming,
        From the valleys, from the dells,
      With music of the mill wheels
        And the jingle of the bells!

The golden harvests brighten
  On the sweet old mountain slopes;
The voices in the valleys
  Lift the song of harvest hopes;
The pumpkin in the corn rows
  Drinks the sunshine as it dies,
And falls to sleep with dreaming
  Of the hand that makes the pies:
      King Corn, King Corn is coming,
        And the cornstalk fiddles hum;
      And the squirrels are shinkapinning,
        And the Bob White sounds his drum!

The land is filled with beauty
  For the king that comes this way;
The earth and sky have married
  And they've brought forth lovely day;
Like castanets of silver
  On the limbs of fairy queen,
The cricket plinks his banjo
  In the lanes of living green:
      King Corn, King Corn is coming,
        And it's howdy do, my dear!
      A tassel for a scepter
        And his crown a golden ear!
          —Folger McKinney.

Glossary

(*From Page 94*)

POLLEN—The powdery substance borne by the stamen of the flower. It is necessary to seed production.

POLLINATION—The act of carrying pollen from stamens to pistils. It is usually done by the wind or insects.

POTASH—An important part of plant foods. The chief source of potash is kainit, muriate of potash, sulphate of potash, wood ashes and cotton-hull ashes.

PROPAGATE—To cause plants or animals to increase in numbers.

PROTEIN—A term used to characterize the constituents of feeding stuffs which contain nitrogen. The organic part of the bones, muscles, tendons, internal organs, skin, etc, of the animal body are formed from the protein nutrients of feeding stuffs. Wheat gluten and white of egg are examples of protein. On the average, 16% of protein in compounds is nitrogen, the other elements being the same as in carbohydrates and fat. The protein compounds in feeding stuffs can be divided into albuminoids and amides, which see. The terms "nitrogenous compound" and "nitrogenous substance" have the same meaning as protein.

PROTOPLASM—The jelly-like or granular substance of living plant cells.

PRUNING—Trimming or cutting parts that are not needed or that are injurious.

PUPA—An insect in the stage of its life that comes just before the adult condition.

RATION—A fixed daily allowance of food for an animal.

RENNET—The ferment found in the lining of the rennet stomach of young mammals.

RESPIRATION APPARATUS—An apparatus for determining the waste matter thrown off by the lungs of an animal.

RETICULUM, or HONEYCOMB—The second stomach of ruminants.

ROUGHAGE—The coarse portion of a ration, including such feeding stuffs as hay, corn fodder, silage, roots, etc.

RUMEN, or PAUNCH—The first stomach of ruminants.

RUMINANT—An animal that chews the cud.

SALIVA—The secretion of the salivary glands of the mouth, the office of which is to moisten the food and through its ferment, ptyalin, partially digest the starchy components of the food.

SCION—A shoot, sprout or branch taken to graft or bud upon another plant.

SEEDLING—A young plant just from the seed.

SEPAL—One of the leaves in the calyx.

SILAGE—A succulent forage preserved in the silo.

SILO—An air-tight structure used for the preservation of forage in a succulent condition.

SOILING—The system of feeding farm animals in a barn or enclosure with fresh grass or green fodders, as corn, rye, oats, etc.

SMUT—A disease of plants, particularly of cereals, which causes the plant or some part of it to become a powdery mass.

SPORE—A small body formed by a fungus to reproduce the fungus. It serves the same use as seeds do for flowering plants.

STAMEN—The part of the flower that bears the pollen.

STERILIZE—To destroy all the germs or spores in or on anything. Sterilizing is often done by heat or chemicals.

STIGMA—The part of the pistil that receives the pollen.

STOCK—The stem or main part of a tree or plant. In grafting or budding the scion is inserted upon the stock.

STOVER—The dry stalks of corn from which the ears have been removed.

SUBSOIL—The soil under the topsoil.

SUCCULENT FEED—Feed containing much water, as grass, silage, roots.

TAPROOT—The main root of a plant, which runs directly down into the earth to a considerable depth without dividing.

THORAX—The middle part of the body of an insect. The thorax lies between the abdomen and the head.

TILLAGE—The act of preparing land for seed, and keeping the ground in a proper state for the growth of crops.

TRANSPLANT—A plant grown in a bed with a view to being removed to other soil.

TUBERCLE—A small, wartlike growth on the roots of legumes.

UDDER—The milk vessel of a cow.

VARIETY—A particular kind. For example, the Winesap, Bonum, Esop, etc, are different varieties of apples.

VILLI—Minute hairlike projections on the inside of the intestines, through which the larger portion of the digested nutrients is absorbed.

VIRGIN SOIL—A soil which has never been cultivated.

(*To Page 98*)

The House by the Side of the Road

There are hermit souls that live withdrawn
 In the place of their self-content;
There are souls, like stars, that dwell
 apart,
 In a fellowless firmament;
There are pioneer souls that blaze their
 paths
 Where the highways never ran—
But let me live by the side of the road
 And be a friend to man.

Let me live in a house by the side of the
 road,
 Where the race of men go by—
The men who are good and the men who
 are bad,
 As good and as bad as I,
I would not sit in the scorner's seat,
 Or hurl the cynic's ban—
Let me live in a house by the side of the
 road
 And be a friend to man.

I see from my house by the side of the
 road,
 By the side of the highway of life,
The men who press with the ardor of hope,
 The men who are faint with the strife.
But I turn not away from their smiles nor
 their tears—

Both parts of an infinite plan—
Let me live in my house by the side of the
 road
 And be a friend to man.

I know there are brook-gladden meadows
 ahead
 And mountains of wearisome height;
That the road passes on through the long
 afternoon
 And stretches away to the night.
But still I rejoice when the travelers re-
 joice,
 And weep with the strangers that moan.
Nor live in my house by the side of the
 road
 Like a man who dwells alone.

Let me live in my house by the side of the
 road
 Where the race of men go by—
They are good, they are bad, they are weak,
 they are strong,
 Wise, foolish—so am I.
Then why should I sit in the corner's seat,
 Or hurl the cynic's ban?
Let me live in my house by the side of the
 road
 And be a friend to man.
 —Sam Walter Foss.

Glossary
(*From Page 96*)

WEATHERING—The action of moisture, air, frost, etc. upon rocks.

WEED—A plant out of place. A wheat plant in a rose bed or a rose in the wheat field would be regarded as a weed, as would any plant growing in a place in which it is not wanted.

WITHERS—The ridge between the shoulder bones of a horse, at the base of the neck.

YEAST—A preparation containing the yeast plant used to make bread rise, etc.

How to Secure a Patent

Application for a patent must be made in writing to the Commissioner of Patents, Washington. The applicant must also file in the Patent Office a written description of the invention, and of the process of making and using it, in such full and exact terms as to enable any person skilled in the art or science to which it appertains to make, construct, compound and use the same; and in case of a machine, he must explain the principle thereof, and the best mode in which he has contemplated applying that principle, so as to distinguish it from other inventions, and partially point out and distinctly claim the part, improvement or combination which he claims as his invention or discovery. The specification and claim must be signed by the inventor and attested by two witnesses.

FEES

Fees must be paid in advance, and are as follows: On filing each original application for a patent, $15. On issuing each original patent, $20. In design cases: For three years and six months, $10; for seven years, $15; for fourteen years, $30. On filing each caveat, $10. On every application for the reissue of a patent, $30. On each disclaimer, $10. For certified copies of patents and other papers in manuscript, 10 cents per 100 words and 25 cents for the certificate; for certified copies of printed patents, 80 cents. For uncertified printed copies of specifications and drawings of patents, for single copies, or any number of unclassified copies, 5 cents each; for copies by subclasses, 3 cents each; by classes, 2 cents each. For recording every assignment, agreement, power of attorney or other paper of 300 words, or under, $1; of over 300 words or under 1,000, $2; or over 1,000 words, $3. For copies of drawings, the reasonable cost of making them. The Patent Office is prepared to furnish photographic copies of any drawing, foreign or domestic, in the possession of the office, in sizes and at rates as follows: Large size, 10x15 inches, 25 cents; medium size, 7x11 inches, 15 cents. Fee for examining and registering trade-mark $10, which includes certificate. Stamps cannot be accepted by the Patent Office in payment of fee. Stamps and stamped envelopes should not be sent to the office for replies to letters, as stamps are not required on mail matter from the Patent Office.

Birth-Months for Happiness

Astrologers say that if one would be happy when married it will be necessary to consult the birth-month of the future husband or wife. By deductions taken from the planets a table has been made of those months which are not in accord.

January cannot agree with July.

February cannot agree with August.

March cannot agree with September.

April cannot agree with October.

May cannot agree with November.

June cannot agree with December.

For the encouragement of those who need to be cheered there is another table of correct affinities.

Those born in	will agree with those born in
Jan	March and May
Feb	April and June
March	May and July
April	June and Aug
May	July and Sept
June	Aug and Oct
July	Sept and Nov
Aug	Oct and Dec
Sept	Nov and Jan
Oct	Dec and Feb
Nov	Jan and March
Dec	Feb and April

Other combinations may prove satisfactory, if conditions of age, health, religion, temperament, habits and prosperity are mutually congenial.

Advice to Mothers

Don't expect your boy to be noiseless; if he is, hunt for him at once.

Don't keep drumming at your boy to be as good as his father; it may set him to watching the old man.

Don't try to force your children to show off before company; it will be embarrassing to the company, to the children and to yourself.

The Road to Yesterday

If I could only find the road,
   The road to yesterday,
I'd ease my heart of many a load
   That burdens it today,
Recall the words so harsh, unkind,
Kiss clean the stabs I made when blind,
Plant love for hate, if I could find
   The road to yesterday.

   The road to yesterday,
   The road to yesterday,
Unlock, O blessed angel guide,

My night of sleep and open wide
The gates that intervene and hide
   The road to yesterday!

Repentant turn and walk again
   The road to yesterday!
Rewrite the page with cleaner pen
   And wipe out yesterday.
With wiser heart I would retrace
The stains of sin and wrong efface;
My tortured soul seeks means of grace
   To re-live yesterday.
                  —W. N. Hull.

L'Envoi

When Earth's last picture is painted and
   the tubes are twisted and dried,
When the oldest colors have faded, and the
   youngest critic has died,
We shall rest, and, faith, we shall need it—
   lie down for an aeon or two,
Till the Master of All Good Workmen shall
   put us to work anew!

And those that were good shall be happy;
   they shall sit in a golden chair;
They shall splash at a ten-league canvas
   with brushes of comets' hair;

They shall find real saints to draw from—
   Magdalene, Peter and Paul;
They shall work for an age at a sitting
   and never be tired at all!

And only the Master shall praise us, and
   only the Master shall blame;
And no one shall work for money, and no
   one shall work for fame,
But each for the joy of working, and each,
   in his separate star,
Shall draw the Thing as he sees It for the
   God of Things as They Are!
                —Rudyard Kipling.

China's Constitution

A 9-years' program has been made out through which China is to become a constitutional monarchy. In the main the constitution follows that of Japan. The steps toward self-government under the constitution promulgated by the empress dowager and the emperor are to be taken as follows:

1908—Local self-government.

1909—Election of provincial assemblymen; issuing of school books.

1910—Provincial assemblies open.

1911—Local self-government continued; rules on imperial taxation; extension of schools.

1912—Completion of general arrangement of urban self-government.

1913—Police registration; courts; criminal code promulgated.

1914—Rural self-government established.

1915—Imperial household expenses fixed; public accounting enforced; police system complete.

1916—Promulgation of full constitution; appointment of a premier.

The fact that China has granted this constitution by the imperial will is reiterated again and again. It is set forth that the imperial government under the constitution shall not be criticised, on the principle that the sacred majesty of the sovereign may not be offended against, and that the leaders of the political parties are to be appointed by the throne.

The document declares: " Officers and people who keep within the law will have freedom of speech, of the press and of assemblies. They shall not be liable to arrest or restrictions or punishments except as prescribed by law. They shall not be disturbed without cause in their possession of property, nor interfered with in their dwellings, and they have the obligation to pay taxes and render military service and the duty of obedience to the law of the land.

" Members of parliament shall not speak disrespectfully of the court nor slander others. Violators of this law will be punished. In the 43d year of Kuang Hau, or 1917, China will be, by following this plan, a parliamentary country, like Japan or Russia."

The estimated wealth of the four leading nations is: United States, 116 billions of dollars; Great Britain, 62 billions; Germany, 59 billions; France, 42 billions.

British Taxation of Wealth

Approximately 40 per cent of the entire receipts of the British government from taxes are derived from two sources—death duties and the income tax. In both cases certain exemptions are allowed. There is a differentiation of the income tax on earned and unearned incomes, so that earned incomes under $10,000, for instance, pay 18 cents, while unearned incomes pay 24 cents in the pound sterling. By the revision last year of the death duties millionaires' estates are more heavily taxed than before, so that above $750,000 the rate is 7 per cent, above $1,250,000 8 per cent, above $2,500,000 9 per cent and so on until estates of $15,000,000 or over pay 10 per cent on the first $5,000,000 and 15 per cent on every other $5,000,000.

According to the report issued by the government for the fiscal year ending March 31, 1908, the gross amount that passed at death during the year was over $1,575,000,000, which produced in death duties about $72,500,000. In Great Britain immense fortunes are less common than in this country. In the year 1906, which was notable for the number of millionaires that died, of 12 estates probated only 2 reached $15,000,000 and the entire 12 barely amounted to $100,000,000. In this country 2 estates, recently appraised, that of Russell Sage at $64,000,000 and that of William B. Leeds at $40,000,000, together exceeded that sum, and neither was subject to a Federal tax.

The British income tax produced $159,000,000. At the same time only 20 persons had incomes assessed at more than $250,000, 241 persons between $50,000 and $250,000, and 517 persons between $25,000 and $50,000. For the year 1906-7 the gross income was nearly $4,720,000,000 and the income on which the tax was collected $3,200,000,000, the difference between the 2 sums showing in a way the extent to which small incomes are favored.

These 2 taxes on wealth which produce 2-5 of the British government receipts from taxation are virtually unknown in this country save where an occasional state imposes a small inheritance tax.

All nations have equal rights on the high seas. The sea which washes the coast of a nation to the extent of 3 miles is now deemed to be part of that nation, but international law recognizes the right of a nation to exercise jurisdiction anywhere on the high seas for the purpose of enforcing its own municipal regulations.

ECLIPSES FOR 1909.

Standard Time.

There will be four Eclipses this year, two of the Sun and two of the Moon.

I.   A Total Eclipse of the Moon June 3, visible in New England, the Moon rising partly eclipsed; the beginning generally visible in South America, Africa, Europe, and the southwestern part of Asia; and the ending generally visible in Africa, central and western Europe, South America, and in North America except in the northwestern portion.

Begins 6 h. 43.4 m., P. M. Middle 8 h. 28.8 m., P. M. Ends 10 h. 14.3 m., P. M.

II.   A Central Eclipse of the Sun June 17.   The eclipse will be visible throughout the greater part of North America, in northern and eastern Asia, northeastern Russia, a portion of the Pacific Ocean, and about the North Pole.   The eclipse will be total except for a fraction of a minute at the beginning and end when it will be slightly annular.   The line of totality runs from the southern part of Greenland to the northern part of Asia, and very close to the North Pole.   It will be visible in New England as a partial eclipse, the Sun setting eclipsed.

Begins 6 h 54.5m., P. M. Ends after sunset.

III.   A Total Eclipse of the Moon Nov. 27; visible in New England, the beginning generally visible in North and South America and northeastern Asia; and the ending generally visible in North America, the northwestern part of South America, eastern and northern Asia; and in Australia.

Begins 2 h. 11.0 m., A. M. Ends 5 h. 38.2 m., A. M.

IV.   A Partial Eclipse of the Sun Dec 12; invisible in New England, but visible in New Zealand, the southern part of Australia, the South Pacific and South Indian Oceans and about the South Pole.

THE TWELVE SIGNS OF ZODIAC.

THE RAM. Aries. ♈ HEAD & FACE.

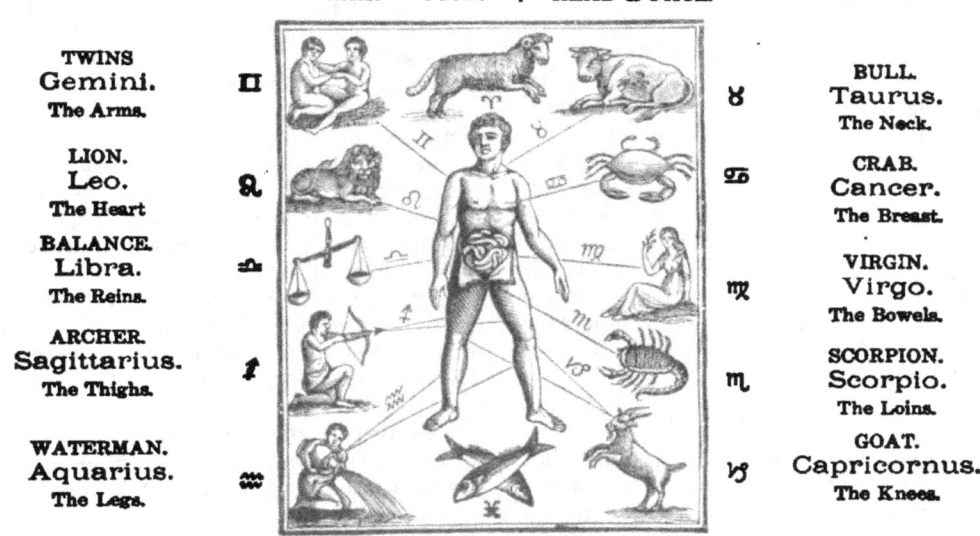

TWINS
Gemini.
The Arms.

LION.
Leo.
The Heart

BALANCE.
Libra.
The Reins.

ARCHER.
Sagittarius.
The Thighs.

WATERMAN.
Aquarius.
The Legs.

BULL.
Taurus.
The Neck.

CRAB.
Cancer.
The Breast.

VIRGIN.
Virgo.
The Bowels.

SCORPION.
Scorpio.
The Loins.

GOAT.
Capricornus.
The Knees.

FISHES. Pisces. ♓ THE FEET.

THE SEASONS.

		Eastern Time
Vernal Equinox (Spring begins)– – – – – – –	March	21 d. 1 h. A. M.
Summer Solstice (Summer begins) – – – – –	June	21 d. 9 h. P. M.
Autumn Equinox (Autumn begins)– – – – –	September	23 d. 0 h. P. M.
Winter Solstice (Winter begins) – – – – – –	December	22 d. 6 h. A. M

Some of the Many Wonders Revealed to us in Astronomy

What they Mean to the People on this Earth—Important Facts and Figures Verified by Prof David Todd, Amherst College

The Earth's Motions

The earth has five motions, briefly described as follows, in the order of their discovery:

1. It revolves on its own axis once in 24 sidereal hours, hence causing day and night. In this movement, its surface (at our latitude) travels about 13 miles every minute. Who discovered this rotation is not known. It probably antedates Pythagoras, about 600 B C.

2. The earth shoots in its elliptical orbit of 585 million miles around the sun every 365¼ days (hence one year), traveling nearly 19 miles a second. Aristarchus of Alexandria divined this fact 18 centuries before Copernicus established it.

3. There is a large change in direction of the earth's axis, called precession of the equinoxes, by which it will point 47° away from the North Star of the present day, but it will take nearly 13,000 years to get there.

4. The earth is journeying with the sun and all the other planets toward Vega, the bright star in Lyra, at a speed of about 12 miles to every second. Sir William Herschel first discovered this motion more than a hundred years ago.

5. There is also a minute change in the earth's axis, called variation of latitude. The pole describes an irregular circuit around its mean position in the course of about 16 months, but does not vary from the mean center more than 30 feet during this period.

The Size of the Planets

Earth is one of the 8 major planets that revolve around the sun. Its circumference is about 25,000 miles at the equator.

These planets (and the diameter of each in miles) are: Mercury, 3,000; Venus, 7,700; Earth, 7,920; Mars, 4,200; Jupiter, 87,000; Saturn, 73,000; Uranus, 32,000; Neptune, 32,000.

Four largest asteroids: Ceres, about 500; Pallas, 300; Juno, 120; Vesta, 240.

The diameter of the sun is about 865,000 miles. All the rest of the solar system put together is about 1-700 of the sun.

About 650 smaller bodies (between Mars and Jupiter) called Asteroids, from a few miles to several hundred in diameter; several comets, and much cosmic dust, also travel around the sun.

These 8 planets have in all 26 moons. Earth has but one moon. Its attraction and that of the sun cause the tides. The other moons are: Mars, 2; Jupiter, 8; Saturn, 10; Uranus, 4; Neptune, 1.

Solar Distances

Our moon is about 240,000 miles from earth. The distance from earth to sun is about 93,000,000 miles.

The distance from each planet to the sun (each in millions of miles) is: Mercury, 36; Venus, 67; Earth, 93; Mars, 141; Jupiter, 483; Saturn, 886; Uranus, 1,780; Neptune, 2,790.

The outermost planet, Neptune, being 2,790 millions of miles from the sun, twice that distance, or 5,580 millions of miles, may be assumed as the diameter of earth's solar system, so far as planets at present known are concerned. A few of the comets recede to more than double the distance of Neptune.

Such distances are beyond the mind's grasp. Used only for star distances as the light-year, therefore, a vastly larger and simpler unit of measurement is taken—the velocity of light.

Light travels at the rate of 186,300 miles per second. It requires 8 1-3 minutes for the sun's light to reach the earth. The diameter of the sun is, say, 5 light-seconds.

The origin of the sun's heat is due to its contraction. It is doubtless growing smaller by a few feet each year; and it will take centuries of observation to find out just how much.

Movements of the Polar System

The planets are held in place in space by the attraction of the sun according to the law of gravitation.

The sun revolves on its axis once in 25¼ days. It is shooting through space at the rate of about 12 miles per second, dragging its planets with it.

Our solar system is thus traveling toward the constellation Lyra at the rate of 375,000,000 miles annually.

But don't be scared. There's no danger of sun or earth hitting its neighbors. The nearest known solar system to ours is that of Alpha Centauri. It's so far away that its light consumes over 4 years in reaching earth.

The Stars are Suns

The stars in the heavens, other than the planets of our own solar system, are suns, each probably with several planets revolving about it.

Some of these suns are vastly bigger than our sun. Canopus may be possibly 100,000 times the bulk of our sun!

These suns all seem to be flying through space, apparently helter-skelter, but probably in obedience to some as yet unknown law.

But there's room enough and to spare. For the diameter of so much of the universe as can be registered on photographic plates in the most powerful telescopes, with a good margin for imagination, is 3,000 light-years.

Then what is beyond that?

Whether our Mars is inhabited may never be known. Very possibly it is; but reason indicates that there may be planets in other systems that are habitable, and doubtless are inhabited now.

The universe seems directed by law, down to life itself and activities of man.

Protection for Oil

Mineral oil and oil products are on the free list, but are subject to an exception that gives them the highest rate of protection enjoyed by the products of any trust. The exception is that when another country imposes a duty on oil and oil products shipped into the United States, the United States shall impose an equal rate of duty on these products coming from that country. The United States has only one competitor in the oil business, and that is Russia. Russia imposes a duty of from 150 to 250% on oil and oil products. Hence, to all intents and purposes, oil produced in the United States is protected by duties ranging from 150 to 250%, for oil in competitive quantities could come only from Russia.

The President's Cabinet

The United States cabinet is a body of men, nine in number, constituting the official advisors of the president, and charged with the administration of the executive departments of the government. They are appointed by the president, with the advice and consent of the senate. Arranged in the order of succession for the presidency, declared by chapter 4, acts of the Forty-ninth congress, first session, they follow: Secretary of state, established in 1780; secretary of the treasury, established in 1789; secretary of war, established in 1789; attorney-general, established in 1789—although the office of attorney-general was created in 1789, the attorney-general did not become a member of the cabinet until 1814; postmaster-general, established in 1829; secretary of the navy, established in 1789; secretary of the interior, established in 1849; secretary of agriculture, established in 1889; secretary of commerce and labor, established in 1903. Each department has its official seal for public documents. The salary of each member of the cabinet is $12,000.

Ex-Governor N. J. Bachelder

The Master of the National Grange is Ex-Governor N. J. Bachelder of New Hampshire, one of the wisest leaders that agriculture has ever had. Governor Bachelder was educated in the grange, serving his state as master for many years and the State Board of Agriculture as secretary for nearly a score of years. He is one of the great agricultural leaders of this country.

Weights and Measures Compiled in Many Useful Tables and Rules

The More Familiar Tables of Arithmetic Supplemented by the Less Familiar Ones of Special Value to the Farmer

Table of Solids

128 solid feet (4x4x8) make 1 cord.

40 solid feet of round timber make 1 ton.

50 solid feet of hewn timber make 1 ton.

1 11-45 solid feet of shelled corn make 1 bushel.

6 2-9 solid feet of shelled corn make 1 barrel.

2 22-45 solid feet of ear corn make 1 bushel.

12 4-9 solid feet of ear corn make 1 barrel.

27⅞ solid inches make 1 wine pint.

231 solid inches make 1 wine gallon.

282 solid inches make 1 beer gallon.

268 4-5 solid inches make 1 gallon, dry measure.

1828 solid inches make 1 bushel unslaked lime, coal or coke.

A bucket or other cylindrical vessel 7 inches in diameter and 6 inches deep holds 1 gallon, wine measure, and a similar vessel 7 inches in diameter and 7 1-3 inches deep holds 1 gallon, beer measure.

Box Measure

A box 24 inches by 16 inches square, and 28 inches deep, will contain a barrel, or 10,752 cubic inches.

A box 24 inches by 16 inches square, and 14 inches deep, will contain ½ barrel, or 5376 cubic inches.

A box 16 inches by 16.8 inches square, and 8 inches deep, will contain a bushel, or 2150.4 cubic inches.

A box 14x14x13¾ inches in the clear holds 1 bushel.

A box 14x7x13¾ inches in the clear holds ½ bushel.

A box 12 inches by 11.2 inches square, and 8 inches deep, will contain ½ bushel, or 1075.2 cubic inches.

A box 7x7x13¾ inches in the clear holds 1 peck.

A box 8 inches by 8.4 inches square, and 8 inches deep, will contain 1 peck, or 537.6 cubic inches.

A box 8 inches by 8 inches square, and 4.2 inches deep, will contain ½ peck, or 268.8 cubic inches.

A box 7 inches by 4 inches square, and 4.8 inches deep, will contain ½ gallon, or 134.4 cubic inches.

A box 4 inches by 4 inches square, and 4.2 inches deep, will contain 1 quart, or 67.2 cubic inches.

The measures all come within a small fraction of a cubic inch of being perfectly accurate; as near, indeed, as any measures of capacity have ever yet been made for common use. The difficulty of making them with absolute exactness has never yet been overcome.

Weights

Apothecaries' weight. 20 grains=1 scruple, 3 scruples=1 dram, 8 drams= 1 ounce, 12 ounces=1 pound.

Avoirdupois weight (short ton). 27 11-32 grains=1 dram, 16 drams=1 ounce, 16 ounces=1 pound, 25 pounds=1 quarter, 4 quarters=1 hundredweight, 20 hundredweights=1 ton (2000 pounds).

Avoirdupois weight (long ton). 27 11-32 grains=1 dram, 16 drams=1 ounce, 16 ounces=1 pound, 112 pounds=1 hundredweight, 20 hundredweights=1 ton (2240 pounds).

Troy weight. 24 grains=1 pennyweight, 20 pennyweights=1 ounce, 12 ounces=1 pound.

Diamond weight. 4 grains=1 carat, 16 parts=1 grain=0.8 Troy grain, carat =3.2 Troy grains.

Iron, lead, etc. 14 pounds=1 stone, 21½ stone=1 pig, 8 pigs=1 fother.

Measures

Dry measure. 2 pints=1 quart, 8 quarts=1 peck, 4 pecks=1 bushel.

Liquid measure. 4 gills=1 pint, 2 pints=1 quart, 4 quarts=1 gallon, 31½ gallons=1 barrel, 2 barrels=1 hogshead.

Fluid measure. The minim=0.95 grain, 60 minims=1 fluid drachm, 8 fluid drachms =1 fluid ounce (455.69 grains) or 480 minims.

Long measure. 12 inches=1 foot, 3 feet=1 yard, 5½ yards=1 rod or pole, 40 rods=1 furlong, 8 furlongs=1 statute mile, 3 miles=1 league.

Nautical measure. 6 feet=1 fathom, 608 fathoms=1 cable length, 7½ cable lengths=1 mile, 5280 feet=1 statute mile, 6080.27 feet=1 nautical mile.

Square measure. 144 square inches= 1 square foot, 9 square feet=1 square yard, 30¼ square yards=1 square rod or perch, 40 square rods=1 rood, 4 roods=1 acre, 640 acres=1 square mile, 36 square miles (6 miles square)=1 township.

Cubic measure. 1728 cubic inches=1 cubic foot, 27 cubic feet=1 cubic yard.

Circular measure. 60 seconds=1 minute, 60 minutes=1 degree, 30 degrees=1 sign, 12 signs=1 circle.

Time measure. 60 seconds=1 minute, 60 minutes=1 hour, 24 hours=1 day, 7 days=1 week, 4 weeks=1 lunar month, 365 days=1 year, 366 days=1 leap year.

Measure of number. 12 units=1 dozen, 12 dozen=1 gross, 20 units=1 score.

12 gross (144 dozen) make 1 great gross.

20 units make one score.

The commercial weights and measures of the United States are the avoirdupois pound (7000 grains)=16 ounces of 437.5 grains each. The wine gallon (231 cubic inches)=4 quarts, or 8 pints of 16 fluid ounces to each pint.

Various miles. The distance called a mile varies greatly in different countries. Its length in yards is as follows: Norway 12,182, Sweden 11,660, Hungary 9139, Switzerland 8548, Austria 8297, Prussia 8238, Poland 8100, Italy 2025, England and the United States 1760, Spain 1522, Netherlands 1094. The nautical mile is 1-60th the length of a degree at the equator, or 2025 yards.

Cloth measure. 2¼ inches=1 nail, 4 nails=1 quarter, 4 quarters=1 yard.

Chain measure. 7.92 inches=1 link, 25 links=1 rod, 100 links=1 chain, 80 chains=1 mile, 10 square chains=1 acre.

Paper measure. 24 sheets=1 quire, 20 quires=1 ream, 2 reams=1 bundle, 5 bundles=1 bale.

Area. One acre contains 160 square rods, 4840 square yards, 43,560 square feet. One rod contains 30¼ square yards, 272¼ square feet. One square yard contains 9 square feet. The side of a square must measure as follows to contain:

	FEET	RODS	PACES
10 acres	660.00	40.00	
1 acre	280.71	12.65	64
½ acre	147.58	8.95	45
1-3 acre	120.50	7.30	37
¼ acre	104.38	6.32	32
⅛ acre	73.79	4.47	22½

To double the length of the side makes four times the area of the field.

Various units. A cubit is 4 hands and a half, or 1 foot and a half. A yard is 36 inches or 2 cubits. A square yard is 9 square feet. A cubical yard is 27 cubical feet. An ell is 1 yard and a quarter, or 45 inches. A geometrical space is 5 feet. A fathom is 6 feet, or 2 yards. A square is 100 square feet. A pace is 3 feet. A palm is 3 inches. A hand is 4 inches. A span is 6 inches. A bible cubit is 21.8 inches.

Liquids. 1 gallon oil weighs 9.32 pounds avoirdupois, 1 gallon distilled water 10.32 pounds, 1 gallon proof spirits 9.08 pounds.

Miscellaneous Table

A book composed of sheets folded into 2 leaves is a folio.

A book composed of sheets folded into 4 leaves is a quarto.

A book composed of sheets folded into 8 leaves is an octavo (8vo).

A book composed of sheets folded into 12 leaves is a duodecimo (12mo).

A book composed of sheets folded into 16 leaves is a 16mo.

56 pounds of butter make 1 firkin.

100 pounds of fish make 1 quintal.

196 pounds of flour make 1 barrel.

200 pounds of beef, pork, shad or salmon make 1 barrel.

24 sheets of paper make 1 quire.

20 quires make 1 ream.

2 reams make 1 bundle.

5 bundles make 1 bale.

3 barleycorns make 1 inch.

18 inches make 1 cubit.

22 inches make one sacred cubit.

9 gallons make 1 English firkin.

2 firkins make 1 kilderkin.

2 kilderkins make 1 barrel.

25 pounds make 1 keg (powder).

100 pounds make 1 cental (grain measure).

280 pounds make 1 barrel of salt.

31½ gallons make 1 barrel (wine measure).

42 gallons make 1 tierce (wine measure).

63 gallons make 1 hogshead (wine measure).

84 gallons make 1 puncheon (wine measure).

126 gallons make 1 pipe (wine measure).

252 gallons make 1 tun (wine measure).

8 bushels of wheat (70 pounds each) make 1 quarter (European measure).

14 pounds make 1 stone.

24¾ cubic feet (masonry) make 1 perch.

100 square feet (carpentry) make 1 square.

1760 yards (5280 feet) make 1 statute mile.

2028.63 yards (6085.9 feet) make 1 nautical mile.

3 miles make 1 league.

69 1-6 statute miles make 1 degree (of latitude).

60 geographical miles make 1 degree (of latitude).

360 degrees make a circle.

640 acres make 1 square mile.

36 square miles make 1 township.

60 pairs of shoes make 1 case.

4 inches make 1 hand (measuring horses).

Legal Weights of the Bushel

States and Territories	Barley	Buckwheat	Corn meal	Potatoes, sweet	Onions	Turnips	Beets	Apples	Dried apples	Dried peaches	Castor beans	Flaxseed	Hungarian grass seed
United States	48	48	48	—	—	—	—	—	—	—	50	56	—
Alabama	47	—	48	55	—	55	—	—	24	33	—	—	—
Alaska	—	—	—	—	—	—	—	—	—	—	—	—	—
Arizona	45	—	—	—	—	—	—	—	—	—	—	—	—
Arkansas	48	52	48	50	57	57	—	50	24	33	—	56	—
California	50	40	—	—	—	—	—	—	—	—	—	—	—
Colorado	48	52	50	—	57	—	—	—	—	—	—	—	—
Connecticut	48	48	50	54	52	50	60	48	25	33	—	55	—
Delaware	—	—	48	—	—	—	—	—	—	—	—	—	—
Dist. of Columbia	—	—	—	—	—	—	—	—	—	—	—	—	—
Florida	48	—	48	60	56	54	—	48	24	33	48	—	—
Georgia	47	52	48	55	57	55	—	—	24	33	—	56	—
Hawaii	48	—	—	—	—	—	—	—	—	—	—	—	—
Idaho	48	42	—	—	—	—	—	45	28	28	—	56	—
Illinois	48	52	48	50	57	55	—	—	24	33	46	56	—
Indiana	48	50	50	55	48	55	—	—	25	33	46	—	—
Indian Territory	—	—	—	—	—	—	—	—	—	—	—	—	—
Iowa	48	52	—	46	57	—	—	48	24	33	46	46	50
Kansas	48	50	50	50	57	55	—	48	24	33	46	56	50
Kentucky	47	56	50	55	57	60	—	—	24	39	45	56	50
Louisiana	—	—	—	—	—	—	—	—	—	—	—	—	—
Maine	48	48	50	—	52	50	60	44	—	—	—	—	—
Maryland	—	—	—	—	—	—	—	—	—	—	—	—	—
Massachusetts	48	48	50	54	52	—	—	48	25	33	—	55	—
Michigan	48	48	50	56	54	48	—	48	22	28	46	56	50
Minnesota	45	50	—	55	52	52	50	50	28	28	—	—	48
Mississippi	48	48	48	60	57	55	—	—	26	33	64	56	50
Missouri	48	52	50	56	57	42	—	48	24	33	46	56	48
Montana	48	52	50	—	57	—	50	45	—	—	—	56	50
Nebraska	48	52	50	50	57	55	—	—	24	33	46	56	50
Nevada	—	—	—	—	—	—	—	—	—	—	—	—	—
New Hampshire	—	—	50	—	—	—	—	—	—	—	—	—	—
New Jersey	48	50	—	54	57	—	—	50	25	33	—	55	—
New Mexico	—	—	—	—	—	—	—	—	—	—	—	—	—
New York	48	48	50	54	57	—	—	48	25	33	—	55	—
North Carolina	48	50	48	—	—	—	—	—	—	—	—	—	—
North Dakota	48	42	—	46	52	60	60	—	—	—	—	56	—
Ohio	48	50	—	50	55	60	56	50	24	33	—	56	50
Oklahoma	48	42	—	46	52	60	60	—	—	—	—	56	—
Oregon	46	42	—	—	—	—	—	45	28	28	—	—	—
Pennsylvania	47	48	—	—	50	—	—	—	—	—	—	—	—
Philippines	—	—	—	—	—	—	—	—	—	—	—	—	—
Porto Rico	—	—	—	—	—	—	—	—	—	—	—	—	—
Rhode Island	48	48	50	54	50	50	50	48	25	33	46	56	50
Samoa	—	—	—	—	—	—	—	—	—	—	—	—	—
South Carolina	—	—	48	—	—	—	—	—	—	—	—	—	—
South Dakota	48	42	—	46	52	60	60	—	—	—	—	56	—
Tennessee	48	50	48	50	56	50	50	50	24	26	46	56	48
Texas	48	42	—	55	57	55	—	45	28	28	—	56	48
Utah	—	—	—	—	—	—	—	—	—	—	—	—	—
Vermont	48	48	—	—	52	60	60	46	—	—	—	—	—
Virginia	48	52	50	56	57	55	—	45	28	32	—	56	48
Washington	48	42	—	—	—	—	—	45	28	28	—	56	—
West Virginia	48	52	—	—	—	—	—	—	25	33	—	56	—
Wisconsin	48	50	50	54	57	42	50	50	25	33	—	56	48
Wyoming	—	—	—	—	—	—	—	—	—	—	—	—	—

Metric System

WEIGHT

10 milligrams=1 centigram.
10 centigrams=1 decigram.
10 decigrams=1 gram.
10 grams=1 decagram.
10 decagrams=1 hectogram.
10 hectograms=1 kilogram.
10 kilograms=1 myriagram.
10 myriagrams=1 quintal.
10 quintals=1 ton (metric).

LINEAR MEASURE

10 millimeters=1 centimeter.
10 centimeters=1 decimeter.

10 decimeters=1 meter.
10 meters=1 decameter.
10 decameters=1 hectometer.
10 hectometers=1 kilo.

CUBIC AND CAPACITY MEASURE

10 millimeters=1 centiliter.
10 centiliters=1 deciliter.
10 deciliters=1 liter.
10 liters=1 decaliter.
10 decaliters=1 hectoliter.
10 hectoliters=1 kiloliter.

EQUIVALENTS

1 acre=.0407 hectare.
1 bushel=35.24 liters.
1 centimeter=.3937 inch.
1 cubic foot=.023 cubic meter.
1 cubic inch=16.39 cubic centimeters.
1 cubic meter=35.31 cubic feet.
1 cubic yard=.7645 cubic meter.
1 foot=30.48 centimeters.
1 gallon=3.785 liters.
1 grain=.0648 gram.
1 gram=15.43 grains.
1 hectare=2.471 acres.
1 inch=25.40 millimeters.
1 kilogram=2.205 pounds.
1 kilometer=.6214 mile.
1 liter=.9081 quart (dry).
1 liter=1.057 quart (liquid).
1 yard=.9144 meter.
1 meter=3.281 feet.
1 mile=1.609 kilometers.
1 millimeter=0.3937 inch.
1 ounce (avoirdupois)=28.35 grams.
1 ounce (Troy)=31.10 grams.
1 pint=.4732 liter.
1 pound=.4536 kilogram.
1 quart (dry)=1.101 liters.
1 quart (liquid)=.9464 liter.
1 square centimeter=.1550 square inch.
1 square foot=.0929 square meter.
1 square inch=6.452 square centimeters.
1 square meter=1.196 square yards.
1 square meter=10.76 square feet.
1 square yard=.8361 square meter.
1 ton (2000 pounds)=.9072 metric ton.
1 ton (2240 pounds)=1.017 metric tons.
1 ton (metric)=.9842 ton (2240 pounds).

The surface units in the metric system are the linear units squared, and for land measures 100 square meters are called the "ar" (for area).

100 ars=1 hectare.

Guide for Estimates and Computation

**Rules and Facts for Short Cuts in Figures
Quick Aids to the Practical
Farmer**

Weight and Measure

To find the number of tons of hay in a mow or stack, multiply together the length, breadth and depth in feet, and divide the product by 510 if the hay is not well settled or by 460 if the hay is well packed.

A solid cubic foot of anthracite coal weighs about 93 pounds. When broken for use it weighs about 54 pounds. Bituminous coal when broken up for use weighs about 50 pounds. Rule: Multiply the length in feet by the height in feet, and again by the breadth in feet, and this result by 54 for anthracite coal, or by 50 for bituminous coal, and the result will equal the number of pounds. To find the number of tons, divide by 2,000.

To ascertain the weight of cattle, measure the girth close behind the shoulder, and the length from the fore part of the shoulder blade along the back to the bone at the tail, which is in a vertical line with the buttock, both in feet. Multiply the square of the girth, expressed in feet, by five times the length, and divide the product by 21; the quotient is the weight, nearly, of the four quarters, in Imperial stones of 14 pounds avoirdupois. In very fat cattle, the quarters will be about 1-20th more, while in very lean ones, there will be about 1-20th less.

To find the bushels of apples, potatoes, shelled corn, etc, in the bins, divide the cubic contents in inches by 2747.7 (the cubic inches in a heaped bushel). If corn is in the ear, deduct one-third from the result.

Tank and Barrel Measurement

To find the contents of a round tank: Multiply the square of the diameter in feet by the depth in feet, and multiply this result by 6, and you have the approximate contents of the tank in gallons. (For exact results multiply the product by 5⅞ instead of 6.) If the tank is larger at the bottom than at the top, find the average diameter by measuring the middle part of the tank half way between the top and bottom.

To find capacity of barrels: Add the head and bung diameters in inches, and divide by 2 for the mean diameter. Then multiply the average diameter by itself in inches, and again by the height in inches, then multiply by 8; cut off the right-hand figure, and you have the number of cubic inches. Divide by 277¼ and you have the number of gallons. A barrel is estimated usually at 31½ gallons; the hogshead at 63 gallons.

To find the contents of a watering-trough: Multiply the height in feet by the length in feet, and the product by the width in feet, and divide the result by 4, and you will have the contents in barrels of 31½ gallons each. For exact results multiply the length in inches by the height in inches, by the width in inches, and divide the result by 231, and you will have the contents in gallons.

A Table for Circular Tanks, 1 foot in Depth

5 feet in diameter holds...	4½	barrels
6 feet in diameter holds...	6¾	"
7 feet in diameter holds...	9	"
8 feet in diameter holds...	12	"
9 feet in diameter holds...	15	"
10 feet in diameter holds...	19½	"

To find the contents of a tank by the table, multiply the contents of one foot in depth by the number of feet deep.

To Measure Wells or Cisterns

Square the diameter in inches, multiply by the decimal .7854, and the product by the depth of the well or cistern in inches. The result will be the full capacity of the well in cubic inches. If the actual quantity of water be sought, multiply by the depth of water in inches, and in either case divide by 231 for the number of gallons.

Circular Cisterns, 1 foot in Depth, Computed

Diameter in inches	Contents in gallons
12	5.875
15	9.18
16	10.44
18	13.218
20	16.32
21	18.

For any greater depth than 1 foot, multiply by the number of feet and fractions of a foot.

Other Rules

To compute the contents of circular cisterns: Multiply the square of the diameter in feet by the depth in feet, and that product by 373-4000 for the contents in hogsheads, or by 373-2000 for barrels, by 47-8 for the contents in gallons.

Square cisterns: Multiply the width in feet by the length in feet, and that by the depth in feet, and that again by 19-100 for hogsheads, or 19-80 for barrels, or 7 48-100 for gallons.

Another and simpler method is to multiply together the length, width and depth, in inches, and divide by 231, which will give the contents in gallons.

Cask gauging: To measure the contents of cylindrical vessels, multiply the square of the diameter in inches by 34, and that by the height in inches, and point off four figures. The result will be the contents or capacity, in wine gallons and decimals of a gallon. For beer gallons multiply by the height of the liquid instead of the height of the cask, to ascertain actual contents. In ascertaining the diameter, measure the diameter at the bung and at the head, add together, and divide by 2 for the mean diameter.

Facts for Builders

100 square feet of surface, 4 inches to the weather, requires about 1000 shingles.

1000 shingles require of shingle nails about 5 pounds.

70 yards of surface will require about 1000 laths.

100 square yards of plaster will require 16 bushels sand, 8 bushels lime, 1 bushel hair.

1000 laths will require lath nails 11 pounds.

100 cubic feet of wall will require 1 cord stone, 3 bushels lime and 1 cubic yard of sand.

One-fifth more siding is required than surface measure, to allow for lap.

Number of Shingles Required for a Roof

Rule—Multiply the length of the ridge pole by twice the length of one rafter, and, if the shingles are to be exposed 4½ inches to the weather, multiply by 8, and if exposed 5 inches to the weather, multiply by 7 1-5, and you have the number.

Shingles are 16 inches long, and average about 4 inches wide. They are put up in bundles of 250 each.

One bundle 16-inch shingles will cover 30 square feet.

One bundle 18-inch shingles will cover 33 square feet.

When laid 5 inches to the weather, 5 pounds 4-penny or 3¾ pounds 3-penny nails will lay 1000 shingles.

Number of Laths for a Room

Laths are 4 feet long and 1½ inches wide, and 16 laths are generally estimated to the square yard.

Rule—Find the number of square yards in the room and multiply by 16, and the result will equal the number of laths necessary to cover the room.

To find the number of square yards in a ceiling or wall, multiply the length by the width or height (in feet) and divide the product by 9; the result will be the square yards.

Stonework, Brickwork and Plastering

STONEWORK

A cord of stone, 3 bushels of lime and a cubic yard of sand will make 100 cubic feet of wall.

One cubic foot of stonework weighs from 130 to 175 pounds.

BRICKWORK

Five courses of brick will make 1 foot in height on a chimney.

One cubic foot of brickwork, with common mortar, weighs from 100 to 110 pounds.

A cask of lime will make mortar sufficient for 1,000 bricks.

PLASTERING

Six bushels of lime, 40 cubic feet of sand (there are about $1\frac{1}{4}$ cubic feet in a bushel) and $1\frac{1}{2}$ bushels of hair will plaster 100 square yards with two coats of mortar.

Brick in a Wall or Building

A brick is 8 inches long, 4 inches wide and 2 inches thick, and contains 64 cubic inches. Twenty-seven bricks make 1 cubic foot of wall, without mortar, and it takes from 2 to 22 bricks, according to the amount of mortar used, to make a cubic foot of wall with mortar.

Rule—Multiply the length of the wall in feet by the height in feet, and that by its thickness in feet, and then multiply that result by 20, and the product will be the number of bricks in the wall.

For a wall 8 inches thick multiply the length in feet by the height in feet, and that result by 15, and the product will equal the number of bricks.

When doors and windows occur in the wall, multiply their height, width and thickness together and deduct the amount from the solid contents of the wall before multiplying by 20 or 13, as the case may be

Short Method of Estimating Stonework

Rule—Multiply the length in feet by the height in feet, and that by the thickness in feet, and divide this result by 22

and the quotient will be the number of perches of stone in the wall.

In a perch of stone there are $24\frac{3}{4}$ cubic feet, but $2\frac{3}{4}$ cubic feet are generally allowed for the mortar filling.

Cords of Stone to Build Cellar and Barn Walls

Rule—Multiply the length, height and thickness together in feet, and divide the result by 100.

There are 128 cubic feet in a cord, but the mortar and sand make it necessary to use but 100 cubic feet of stone.

Rules for Measuring Land

To find the number of acres in a rectangular piece of land: Multiply the length in rods by the breadth in rods and divide by 160.

Triangular pieces, when the triangle is right-angled: Multiply the width by the length and divide by 2. If the triangle is without a right angle, a perpendicular has to be found. Multiply the base in rods by the perpendicular height in rods and divide by 2, and you have the area in square rods.

To find the area of a quadrangular piece of land when only two of the opposite sides are parallel: Add the two parallel sides together and divide by 2, and you have the average length. Then multiply the width in rods by the length in rods and divide by 160, and you have the number of acres.

To Lay Off Small Lots of Land

Farmers and gardeners often find it necessary to lay off small portions of land for the purpose of experimenting with different crops, fertilizers, etc. To such the following rules will be helpful: One acre contains 160 square rods, or 4,840 square yards, or 43,560 square feet. To measure off one acre it will take 208 7-10 feet each way; one-half acre, $147\frac{1}{2}$ feet each way; one-third acre, $120\frac{1}{2}$ feet each way; one-fourth acre, $104\frac{3}{8}$ feet each way; one-eighth acre, $73\frac{3}{4}$ feet each way.

To measure town lots: Multiply the length in feet by the width in feet and divide the result by 43,560 and you will have the fractional part of an acre in the lot.

To find the number of acres in a given number of square rods: Remove the decimal point two places to the left in the number of square rods, divide by 8 and multiply by 5, and you have the number of acres.

Some Measurement Rules

To find the circumference of a circle multiply the diameter by 3.1416.

To find the diameter of a circle multiply the circumference by .31831.

To find the area of a circle multiply the square of the diameter by .7854.

To find the area of a triangle multiply the base by ½ of the perpendicular height.

To find the surface of a ball multiply the square of the diameter by 3.1416.

To find the cubic inches in a ball multiply the cube of the diameter by .5236.

Doubling the diameter of a pipe increases its capacity about four times.

A cubic foot of water contains 7½ gallons, and weighs 62½ pounds.

Quick Method for Calculating Interest

This is probably the shortest and simplest method known. Multiply the principal by the number of days, and

For 4 per cent, divide by 90
For 5 per cent, divide by 72
For 6 per cent, divide by 60
For 7 per cent, divide by 52
For 8 per cent, divide by 45
For 9 per cent, divide by 40
For 10 per cent, divide by 36
For 12 per cent, divide by 30

BANKER'S METHOD

To find the interest on any sum at 6 per cent for any number of days: Remove the decimal point two places to the left, and you have the interest for 60 days. When the time is more or less than 60 days, first find the interest for 60 days, and from that to the time required.

For 120 days, multiply by 2.
For 90 days, add ½ of itself.
For 75 days, add ¼ of itself.
For 30 days, divide by 2.
For 15 days, divide by 4.
For 3 days, divide by 20.

Wheat Harvest Calendar

January—Australia, New Zealand, Chile, Argentine Republic.

February and March—Upper Egypt, India.

April—Lower Egypt, India, Syria, Cyprus, Persia, Asia Minor, Mexico, Cuba.

May—Texas, Algeria, China, Japan, Morocco.

June—California, Oregon, Mississippi, Alabama, Georgia, North Carolina, South Carolina, Tennessee, Virginia, Kentucky, Kansas, Arkansas, Utah, Colorado, Missouri, Turkey, Greece, Italy, Spain, Portugal, France.

July—New England, New York, Pennsylvania, Ohio, Indiana, Michigan, Illinois, Iowa, Wisconsin, Southern Minnesota, Nebraska, Upper Canada, Roumania, Bulgaria, Austria, Hungary, South of Russia, Germany, Switzerland, South of England.

August—Central and Northern Minnesota, Dakotas, Manitoba, Lower Canada, British Columbia, Belgium, Holland, Great Britain, Denmark, Poland, Central Russia.

September and October—Scotland, Sweden, Norway, North of Russia.

November—Peru, South Africa.

December—Burmah, New South Wales.

A lot of people go through the world picking up pins, because they never see anything better than pins to pick up.

Dr W. A. Henry

For nearly 25 years Dr Henry was Dean of the College of Agriculture and Director of the Agricultural Experiment Station of Wisconsin. During this period of service he organized and developed the most conspicuous, helpful and practical college of agriculture in the entire world. The short-course movement, including all winter schools of agriculture, have come about as the result of the initiative given by Dr Henry. The contributions given to animal industry fields and to all problems of feeding have been larger from Dr Henry than from any contemporary.

Value of Rare Coins

A good idea of the value of premium coins of the United States can be obtained from the following list. The two prices quoted for each coin show the difference in value between a piece in good condition and in fine, "uncirculated" condition. When badly mutilated or worn the same coins will as a general rule bring about one-tenth of the first price given.

SILVER DOLLARS

	Good	Fine		Good	Fine
1794....	$20.00	$50.00	1803....	$1.20	$2.00
1795....	1.25	2.00	1804....	250.00	400.00
1796....	1.50	3.00	1836....	3.00	7.00
1797....	1.50	3.00	1838....	35.00	50.00
1798....	1.15	1.60	1839....	20.00	30.00
1799....	1.15	1.75	1851....	20.00	30.00
1800....	1.15	1.75	1852....	20.00	30.00
1801....	1.25	2.00	1858....	15.00	20.00
1802....	1.20	2.00			

HALF DOLLARS

1794....	$2.00	$4.00	1836, milled		
1795....	.65	1.00	edge...	$1.00	$2.00
1796....	25.00	50.00	1838, O under		
1797....	20.00	35.00	bust...	5.00	15.00
1801....	2.00	4.00	1851....	10.00	2.00
1802....	2.50	4.00	1852....	1.00	2.00
1815....	3.00	5.00			

QUARTER DOLLARS

1796....	$1.00	$2.50	1824....	$0.50	$1.50
1804....	1.25	3.00	1827....	25.00	50.00
1807....	.40	1.00	1853, no rays		
1823....	25.00	50.00	or arrows	2.00	3.50

20-CENT PIECES

1877....	$1.00	$2.00	1878....	$1.00	$2.00

DIMES

1796....	$1.00	$2.00	1804....	$3.00	$10.00
1797....	2.00	5.00	1805....	.25	.75
1798....	1.00	3.00	1807....	.25	.75
1800....	2.00	4.00	1809....	.50	1.50
1801....	1.50	3.00	1811....	.50	1.50
1802....	2.00	4.00	1822....	1.00	3.00
1803....	1.00	3.00	1846....	.40	.75

HALF DIMES

1794....	$1.50	$3.00	1802....	$20.00	$50.00
1795....	.40	1.00	1803....	1.00	2.00
1796....	1.25	2.00	1805....	2.00	5.00
1797....	.75	1.50	1838, no		
1800....	.40	1.00	stars...	.15	.50
1801....	1.25	2.00	1846....	1.00	3.00

SILVER 3-CENT PIECES

1863 to 1873, inclusive........... 15c to 25c each
1877, three and five cent nickels.. 50c to $1.00 each

NICKEL 1-CENT PIECES

1856, flying eagle............... $1.50 to $3.00
1873, bronze two cent........... 50c to 1.00

LARGE COPPER CENTS

(Badly worn or mutilated specimens not wanted.)

From			From		
1793....	$1.00 to	$8.00	1807....	$0.10 to	$0.50
1794....	.25 to	1.00	1808....	.20 to	.75
1795....	.25 to	1.00	1809....	.50 to	1.50
1796....	.25 to	1.00	1810....	.10 to	.25
1797....	.10 to	.50	1811....	.25 to	1.00
1798....	.05 to	.20	1812....	.05 to	.25

From			From		
1799....	$5.00 to	$25.00	1813....	$0.20 to	$0.50
1800....	.15 to	.40	1814....	.05 to	.25
1801....	.20 to	.50	1821....	.10 to	.40
1802....	.10 to	.40	1823....	.25 to	2.00
1803....	.05 to	.25	1824....	.10 to	.40
1804....	4.00 to	12.00	1825....	.05 to	.25
1805....	.20 to	.60	1857....	.10 to	.40
1806....	.25 to	1.00			

HALF CENTS

From			From		
1793....	$0.50 to	$1.50	1831....	$3.00 to	$5.00
1794....	.10 to	.50	1836....	3.00 to	5.00
1795....	.10 to	.50	1840 to 1848		
1796....	5.00 to	30.00		3.00 to	5.00
1897....	.10 to	.40	1849, small		
1802....	.40 to	1.00	date...	3.00 to	5.00
1810....	.10 to	.40	1852....	3.00 to	5.00
1811....	.25 to	1.00			

Legal Tender

Gold coins, silver dollars and United States notes or greenbacks are full legal tender for any amount. Gold and silver certificates and national banknotes are not legal tender. Silver coins below a dollar are legal tender for $5 or less. Nickels and pennies are legal tender for sums of 25 cents or less.

State Flowers

Adopted in most instances by vote of the public school pupils of the respective states:

Alabama	Golden Rod	Montana	Bitter Root
Arkansas	Apple Blossom	Nebraska	Golden Rod
California	Eschscholtzia	New York	Rose
	(California Poppy)	North Dakota	Wild Rose
Colorado	Columbine	Ohio	Scarlet Carnation
Delaware	Peach Blossom	Oregon	Oregon Grape
Idaho	Syringa	Pennsylvania	
Illinois	Rose		Golden Rod
Indiana	Corn	Rhode Island	Violet
Iowa	Wild Rose	South Dakota	Pasque
Kansas	Sunflower		(Anemone)
Kentucky	Golden Rod	Texas	Blue Bonnet
Louisiana	Magnolia	Utah	Sego Lily
Maryland	Golden Rod	Vermont	Red Clover
Michigan	Apple Blossom	Washington	
Minnesota	Moccasin		Rhododendron
Mississippi	Magnolia	West Virginia	
Missouri	Golden Rod		Rhododendron

The four oldest colleges in the United States are: Harvard, founded at Cambridge, Mass, in 1636; William and Mary college, founded at Williamsburg, Va, in 1693; Yale (which was first known as the Collegiate School of Connecticut), founded in 1701; and Brown university, founded at Providence, R I, in 1765.

Largest Cities of the World

	Year	Population
London.................	1906	7,113,561
New York.............	1905	4,014,304
Paris..................	1901	2,714,068
Chicago...............	1906	3,049,185
Berlin................	1905	2,033,900
Tokio.................	1903	1,818,655
Vienna...............	1901	1,674,957
Canton...............	Est.	1,600,000
Peking...............	Est.	1,600,000
Philadelphia..........	1900	1,441,735
St. Petersburg........	1897	1,373,390
Calcutta.............	1901	1,125,400
Constantinople.......	Est.	1,125,000
Moscow..............	1897	1,092,360
Buenos Ayres.........	1905	1,000,250

Postal Rates

Domestic

First class.—Letters, postal cards, and matter wholly or partly in writing, whether sealed or unsealed (except manuscript copy accompanying proof-sheets of the same), and all matter sealed or otherwise closed against inspection.

Rate.—Two cents per ounce or fraction thereof. Postal cards, 1 cent each. On "drop" letters, 2 cents per ounce or fraction thereof, when mailed at letter-carrier office; and 1 cent per ounce or fraction thereof at other offices.

Second class.—Newspapers and publications issued at stated intervals as often as four times a year, bearing a date of issue and numbered consecutively, issued from a known office of publication, and formed of printed sheets, without board, cloth, leather, or other substantial binding. Such publications must be originated and published for the dissemination of information of a public character, or devoted to literature, the sciences, art, or some special industry. They must have a legitimate list of subscribers, and must not be designed primarily for advertising purposes, or for free circulation at nominal rates.

Rate.—One cent per pound or fraction thereof when sent by publisher thereof and from office of publication including sample copies, or when sent from news agency to actual subscribers or other news agents.

One cent for each 4 ounces or fraction thereof on newspapers and periodical publications of second class, when sent by other than publisher or news agent. One cent each on newspapers (excepting weeklies) and periodicals not exceeding 2 ounces in weight, when deposited in letter-carrier office for delivery by carrier, 2 cents each on periodicals weighing more than 2 ounces.

One cent per pound on newspapers, other than weeklies, and periodicals when deposited by publisher or news agent in letter-carrier office for general or box delivery; 1 cent for 4 ounces or fraction thereof when deposited by other than publishers or news agents for general or box delivery. One cent per pound or fraction thereof on weekly newspapers deposited by publisher or news agent in letter-carrier office for letter or box delivery, or delivery by carrier; free when one copy is sent to each actual subscriber residing in county where same are printed, in whole or in part, and published; but at rate of 1 cent per pound when delivered at letter-carrier office, or distributed by carriers.

Third class.—Books, circulars and pamphlets, and matter wholly in print (not included in second class), proof-sheets, corrected proof-sheets and manuscript copy accompanying the same.

"Printed matter" is the reproduction upon paper, by any process, except that of handwriting, of any words, letters, characters, figures, or images, or of any combination thereof, not having the character of an actual and personal correspondence.

A "circular" is a printed letter, which, according to internal evidence, is being sent in identical terms to several persons. It is permissible to write, in circulars, the date, the name of the person addressed, or of the sender, and to correct mere typographical errors.

Seeds, bulbs, roots, scions and plants are also mailable at the rate of third-class postage, such as samples of wheat or other grain in its natural condition, seedling potatoes, beans, peas, acorns, etc. Cut flowers and botanical specimens go as fourth class.

Rate.—One cent for each 2 ounces or fraction thereof.

Fourth class.—Merchandise; namely, all matter not embraced in the other three classes, and which is not in its form or nature liable to destroy, deface, or otherwise damage the contents of the mail bag, or harm the person of any one engaged in the postal service, and not above the weight provided by law. Includes artificial flowers, cut flowers, dried plants, botanical and geological specimens, samples of flour or other manufactured grain for food purposes, blank address tags or labels, queen bees when properly packed, dried fruit.

Rate.—One cent per ounce or fraction thereof.

Foreign Postage

To Canada, Newfoundland and Mexico the rates are similar to the United States domestic postage, except on second-class matter to Canada. The latter is now 1 cent for each 4 ounces, or fraction thereof, in bulk for publishers, which is the same as the rate for single periodicals mailed by the general public. Letters cost 2 cents per ounce; merchandise not exceeding 4 pounds 6 ounces, 1 cent per ounce.

To Great Britain and Ireland and Germany the letter rate is 2 cents per ounce or fraction thereof.

In the Universal Postal Union, which includes nearly all the countries of the world, rates are as follows: Letters, 1 ounce or fraction thereof, 5 cents; each succeeding ounce or fraction thereof, 3 cents; postal cards, each, 2 cents; newspapers and other printed matter per 2 ounces, 1 cent; samples of merchandise, same as printed matter, except that lowest rate is 2 cents.

Unsealed packages of mailable merchandise may be sent by parcels post to most countries in the postal union at the rate of 12 cents for not exceeding 1 pound in weight, and 12 cents for each additional pound or fraction thereof. The limit of weight is 4 pounds 6 ounces, to certain parts of Mexico and to all of Germany, Norway, Hong Kong, Japan, Belgium, Great Britain and Ireland, Australia, Denmark, Sweden and China. In other countries the maximum weight allowed is 11 pounds.

Registration fee on foreign mail, 8 cents.

Foreign money orders cost about the same as domestic post-office money orders, except that the minimum fee is 8 cents in some countries and in others 10 cents. Domestic rates apply to money orders for Canada, Cuba, Newfoundland, Jamaica and several other less important places.

International response coupons, exchangeable in any country in the postal union for the equivalent of the United States 5-cent stamp, are purchasable at postoffices for 6 cents each. The purpose of these coupons is to enable persons to send return postage to foreign countries, or small sums without the expense of buying a money order.

How the President is Elected

Most of us are accustomed to speak of the election of president as having occurred on election day in November. As a matter of fact, the real election does not occur until February. At the November election electors are chosen in the several states; these electors meet in their respective state capitals on the second Monday in January and cast their votes for president. These votes are sealed and sent to the president of the United States senate.

On the second Wednesday in February the senate and the house of representatives meet in joint session. The president of the senate presides and the votes are then counted officially. If no candidate receives a majority of the electoral votes the election is taken to the house of representatives where the members voting by states, each state having one vote, ballot for president. The three candidates having the largest number of votes, if there are that number of candidates, must be the ones voted for.

The qualifications of the president are fixed by the constitution. They are citizenship acquired by birth in the United States, and the age of 35 years. The president is inaugurated on March 4, following his election. The qualifications for vice-president are the same as those for president, and he is elected and inaugurated with the president.

The man who attends strictly to his own business seldom has a headache the next morning.

Dr Thomas Forsythe Hunt

One of the most conspicuous leaders of agricultural thought is Dean Hunt of Pennsylvania, who is also Director of the Pennsylvania Agricultural Experiment Station. Dr Hunt is one of the clearest thinkers in the land, and has given direction to the agricultural college movement. It was he who conceived the idea of a graduate school in agriculture that has been so helpful in training teachers and investigators; and it was he who performed the great service of developing, systematizing and outlining agricultural education as it concerns the college course.

Some Big Railroad Projects Under Way

The River Tunnels and Terminal Stations at New York—The Line to Key West

New York city is the scene of railroad improvements of unprecedented magnitude. The Pennsylvania railroad is spending $125,000,000 for river tunnels minus will occupy 40 acres in the heart of New York city. Underground tracks will be put in to the extent of 24 acres of space. The new terminal will have more than 64 acres. There will be 39 tracks on the upper level to be used by express trains, and 15 tracks on the lower level for local trains. The New York Central had spent about $20,000,000 on its electrical equipment alone up to the end of 1908. The extension now being pushed

PENNSYLVANIA RAILROAD TUNNELS AT NEW YORK

and a great railroad station, which, with the trackage leading into it, occupies 8 city blocks. The New York Central system is spending $70,000,000 for improvements at its New York terminus. The Grand Central station will be extended and rebuilt and the trackage area will be increased 17 acres, so that the total area of the New York Central ter- to Croton will cost as much more. In the latter part of 1908 there were 35 electric locomotives running on the terminal tracks that cost $30,000 and 125 motor cars costing $15,000 each. The New York, New Haven and Hartford railroad also operates over the New York Central tracks at the New York city terminal 40 electric locomotives.

-

One of the most daring conceptions in railroad finance was a project of the Pennsylvania railroad to expend $125,000,000 upon the New York city terminus, while at the same time expending millions in improvements upon other sections of its road, including, for instance, a new union station at Washington, and a massive and costly bridge across the Susquehanna river. The magnificent new terminal station will be connected east and west with tunnels that lead from the island of Manhattan to the New Jersey shore and to Long Island. We present an illustration of a cross-section view of the river tunnels. The tunnels are completed and work on the terminus is being pushed as rapidly as possible.

The Pennsylvania has adopted plans for a bridge 3 miles long, from Port Morris, a suburb of New York, to Long Island. It will span Hell Gate ship channel 1,000 feet in the clear. The train floor will hang from the crown of the arch, which latter will be 300 feet above the water, and the floor 140 feet.

From the standpoint of money involved, the Florida East Coast railroad ranks third among railroads of the United States engaged in great construction projects. It has already spent $15,000,000 on its extension from Miami to Key West. This extension of 156 miles savors more of marine engineering than of railroad construction, as practically the entire road is a series of viaducts connecting the numerous little islands and reefs. The first of the viaducts begins at Long Key and extends 2 miles with 184 concrete arches. Another viaduct crosses Moser Key channel 7,800 feet, and the span across Bahia Honda channel is 14,950 feet long. This construction work costs about $100,000 per mile. It is expected that when the Panama canal is opened the railroad having its terminal at Key West will control a strategic point for interoceanic commerce and will be in strong competition with the ports of New Orleans and Galveston. It is proposed to maintain regular ferry service for trains between Key West and Havana, Cuba.

American Canals

Showing the cost and date of construction, length and navigable depth of the principal canals of the United States used for commercial purposes.

Name	Cost of construction	When completed	Length miles	Depth feet
Albemarle and Chesapeake (Va. and N. C.)	$1,641,363	1860	41	7½
Augusta (Ga.)	1,500,000	1847	9	11
Black River (N. Y.)	3,581,954	1849	35	4
Cayuga and Seneca (N. Y.)	2,232,632	1832	25	7
Champlain (N. Y.)	4,044,000	1822	81	6
Chesapeake and Delaware (Md. and Del.)	3,730,230	1829	14	9
Chesapeake and Ohio (Md. and D. C.)	11,230,327	1850	181	6
Companys (La.)	90,000	1847	22	6
Delaware and Raritan (N. J.)	4,888,749	1818	66	7
Delaware Division (Pa.)	2,433,350	1830	60	6
Des Moines Rapids (Iowa)	4,582,009	1877	7½	5
Dismal Swamps (Va. and N. C.)	2,800,000	1822	22	6
Erie (N. Y.)	52,540,800	1826	387	7
Galveston and Brazos (Tex.)	340,000	1851	38	3½
Hocking (Ohio)	975,481	1843	42	4
Illinois and Michigan (Ill.)	7,357,787	1848	102	6
Illinois and Mississippi (Ill.)	7,350,000	1895	75	7
Lehigh Coal and Navigation Co. (Pa.)	4,455,000	1821	108	6
Louisville and Portland (Ky.)	5,578,631	1872	2½	..
Miami and Erie (Ohio)	8,062,680	1835	274	5½
Morris (Pa. and N. J.)	6,000,000	1836	103	5
Muscle Shoals and Elk R Shoals (Tenn.)	3,156,919	1889	16	6
Ogeechee (Ga.)	407,810	1840	3	3
Ohio (Ohio)	4,695,201	1835	317	4
Oswego (N Y.)	5,239,526	1828	38	7
Pennsylvania (Pa.)	7,731,750	1839	193	6
Portage Lake and Lake Superior (Mich.)	529,892	1873	25	15
Santa Fe (Fla.)	70,000	1880	10	5
Sault Ste. Marie (ship canal) (Mich.)	4,000,000	1895	3	18
Schuylkill Navigation Co. (Pa.)	12,461,600	1826	108	6¼
Sturgeon Bay and Lake Michigan (Wis.)	99,661	1881	1½	15
St. Mary's Falls (Mich.)	7,909,667	1896	1½	21
Susquehanna and Tidewater (Pa. and Md.)	4,931,345	1840	45	5½
Walhonding (Ohio)	607,269	1843	25	4
Welland (ship canal) (Ont.)	23,736,353		26½	14

Foreign Canals

Name	Length miles	Depth feet	Bottom width feet	Cost
Suez, Mediterranean and Red Seas	90	31	108	$100,000,000
Cronstadt, St. Petersburg	16	20½	...	10,000,000
Corinth, Corinth and Ægina Gulfs	4	26½	72	10,000,000
Manchester Ship, Manchester and Liverpool	35½	26	120	75,000,000
Kaiser Wilhelm, Baltic and North Seas	61	29½	72	40,000,000
Elbe and Trove	41	10	72	6,000,000

The Use of Concrete for Farm Structures

A Plain and Practical Article on How to Mix Concrete and How to Make and Use the Forms

The Composition of Concrete

Concrete consists of a mixture of broken stone or gravel, clean sand and Portland cement. The proportions of stone, sand and cement vary according to the uses to which the concrete is to be put. The strength varies almost directly with the proportion of cement used, increasing as the proportion of the cement increases. The ideal mixture contains just enough sand and cement to fill the spaces between the larger pieces of stone or gravel, making an absolutely solid substance. Broken stone with sharp corners is preferable to round gravel, and the hard stones are much to be preferred. Soft sandstone or limestone, slate, etc, should be avoided. Cinders may be used, but only in places where great strength is unnecessary.

Good Materials Important

The sand used with concrete should be clean and coarse. To test whether it is clean or not rub some between the hands and note if the hands are discolored or not. If they are, it contains too much clay or loam. Another test is to drop a handful into a pail of clear water and if the **water is clear enough** to see the sand at

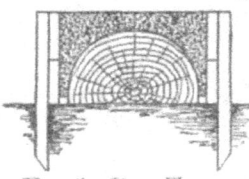

FIG 1. PIG TROUGH

the bottom in two minutes it is clean enough for concrete work. Coarse sand should have grains ranging from 1-32 up to ¼ of an inch in diameter. Fine sand can be used, but a larger proportion of cement is needed.

Proportions of the Mixture

For very heavy work, such as foundation for machines, surfaces of floors, tanks and water-tight work, a " rich " mixture is desirable. The table presented below gives the amounts of the three materials needed to make each kind of a mixture. For ordinary thin walls of buildings, sidewalks, etc, a " medium " mixture is desirable, while for heavy walls, piers and abutments, an " ordinary " mixture will do. For solid work, where there is but little strain except as pres-

sure, such as foundations for floors, etc, a " lean " mixture will be sufficient.

The measuring of the various constituents should be done carefully and accurately. For small jobs an ordinary pail will be all right, and the mixing can be done with a shovel on an ordinary platform. First mix the sand and cement together, then add the stone and shovel it over again and finally put on the water until the entire mixture is just soft enough to be mushy and not stand the weight of a man when placed in the form. In large jobs a mixing machine will save expense. A sprinkling pot or a hose should be used to add the water, as throwing it on with a pail will often wash away the cement.

MATERIALS FOR ONE CUBIC YARD

	Proportions	Bbls Cement	Bbls Sand	Bbls Stone
Rich	1:2:4	1.57	3.14	6.28
Medium	1:2½:5	1.29	3.23	6.45
Ordinary	1:3:6	1.10	3.30	6.60
Lean	1:4:8	0.85	3.40	6.80

Forms for Movable Work

Concrete is especially adapted to making small structures for use about the farm, such as fence posts, feed troughs, water troughs, tile, and foundation stones for various purposes. For the forms use a green timber which will not absorb much of the moisture from the concrete. If a smooth surface is desired, plane off the timber used to make the forms. For ordinary work grease the inside of the forms with soap, linseed oil or axle grease before putting the concrete into the forms. Greasing may be avoided if the forms are thoroughly wet just before the concrete is placed in them. An "ordinary" mixture is most desirable for small, movable work. For finishing and painting the surface a mixture of one part cement and three parts clean sand will be best. As shown in Fig 1 a form placed upon the ground between stakes with the half of a round log for inside mold is excellent to make a trough for feeding live stock. After this has been taken from the molds it should be painted with a mixture of cement and water to make the surface smooth. No special reinforcement is necessary in small jobs of this sort.

Solid Concrete Work

For foundations of buildings, floors, etc, the ground should be excavated enough to allow the concrete to extend below the frost line where foundation is exposed to the water. Otherwise freezing may heave and crack the wall. For ordinary straight walls a form such as shown in Fig 2 is easily constructed out of ordinary 1-in timber. For a wall of this sort the "ordinary" mixture is sufficiently strong, and where it is made quite thick without an air space the "lean" mixture will be sufficient. When a wall higher than can be inclosed in forms braced from the ground is desired, the two forms attached by bolts which pass through the concrete are most desirable.

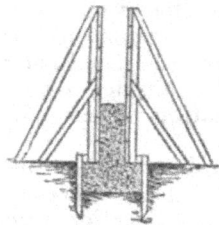

FIG 2. PLAIN FORM

The bolts should be greased before the concrete is put in place so that they can be easily removed when it has set. For such work as floors the foundation should be laid only on soils where there is no danger of frost getting in below the floor. Drainage should be provided for, and the dirt tamped hard before a coarse mixture of broken stone is placed upon it and then the concrete laid upon this. It will be found most desirable to lay the floor in blocks from 6 to 12 ft wide, with joints to allow for expansion, due to change of temperature. A surface coat about 1 in thick of one part cement and one and one-half parts sand will give the best finish. In stables this should be made rough by scratching it with a broom so as to prevent slipping. Fig 3 shows a very satisfactory arrangement for a stable floor of this character.

Reinforced Concrete

For all structures where concrete is used in comparatively thin layers and exposed to severe changes of temperature or great pressure it should be reinforced or strengthened by the addition of metal laid in the concrete at the time it is molded. For this purpose all such materials as wire, iron or steel bars, old buggy tires, rake teeth, etc, are useful and may be worked in with satisfactory results. In some cases a heavy, galvanized wire netting such as used on poultry yards will give sufficient strength. Where the second floor or roof of the building are constructed of concrete, specially placed reinforcements are necessary, which require oversight by an architect to be most satisfactory. The building of the Orange Judd Company in which the eastern editions of the American Agriculturist weeklies are printed, which is eight stories high, is built entirely of reinforced concrete, heavy twisted steel being used for this purpose.

Concrete Blocks

A form of concrete building which is very popular and adaptable to many uses is the concrete block which is molded before it is built into place. There are many types of molds on the market for this purpose and any ingenious farmer

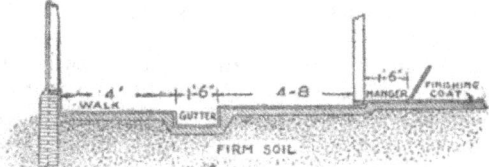

FIG 3. SECTION OF A GOOD CEMENT FLOOR

can make a mold of his own for making these blocks. They are usually constructed with an air space within, which is useful in walls, because it prevents moisture condensing upon the inside of the walls or buildings.

Concrete Plaster

A very popular method of using concrete is as a plaster for so-called stucco work. A wooden frame is erected for the building and covered with 1-in sheathing, over which is put one or two layers of roofing paper. Thin strips ½ in thick are nailed 1 ft apart, and to this wire lathing or fine galvanized wire netting is fastened. A rough coat of stucco is then put on and pressed through the laths and left with a rough surface. After this is allowed to set a finishing coat ½ in to 1 in thick is added and left smooth or rough as desired.

The Kansas Emigrant's Song

We cross the prairies as of old
   The pilgrims crossed the sea,
To make the West, as they the East,
   The homestead of the free.

CHORUS

The homestead of the free, my boys,
   The homestead of the free!
To make the West, as they the East,
   The homestead of the free.

We go to rear a wall of men
   On Freedom's southern line,
And plant beside the cotton tree,
   The rugged Northern pine!

We're flowing from our native hills,
   As our free rivers flow;
The blessing of our mother-land
   Is on us as we go.

We go to plant her common schools
   On distant prairie swells,
And give the Sabbaths of the wild
   The music of her bells.
          —Whittier.

Denatured Alcohol Law and Regulations

How the Measure was Modified to Benefit Farmers—The Chief Provisions

The amended law relating to denatured alcohol, passed by the 59th congress, went into effect September 1, 1907. The purpose of the amendments was to further reduce the cost of denatured alcohol, and place its benefits within the reach of the farmers as well as the large distillers. It enables those who wish to produce alcohol on a small scale to distill it in suitably locked stills, and to. have it denatured without the expense of a bonded warehouse, which was necessary under the original law of 1906. The provisions allowing transportation of denatured alcohol in tank cars should materially reduce its cost to the consumer. In substance, the amended law makes the following provisions:

Domestic alcohol, when suitably denatured, may be withdrawn from bond tax free, and used in the manufacture of certain definite chemical substances where alcohol is changed into some other chemical substance and does not appear in the finished product as alcohol.

The provisions of the denatured alcohol law are extended to apply to rum.

The commissioner of internal revenue, with the approval of the secretary of the treasury, may authorize the establishment of central denaturing bonded warehouses other than those at distilleries. To these alcohol of the required standard may be transported without payment of internal revenue tax and in these warehouses the alcohol may be stored and denatured.

The establishment, operation and custody of such warehouses shall be under regulations and upon the execution of bonds such as may be prescribed by the commissioner of internal revenue.

Alcohol of the required proof may be drawn off for denaturing from cisterns of a distillery, for transfer by pipes direct to any denaturing bonded warehouses on the distillery premises, or to storage tanks in such warehouses. The denatured alcohol may be transported in the same manner and by means of packages, tanks, or tank cars, on execution of bonds and under regulations prescribed by the commissioner of internal revenue. Alcohol to be denatured may be transferred to central denaturing plants in such packages, tanks and tank cars as come under the regulations of the commissioner.

The section of the new law that chiefly interests the farmers states that distilleries producing alcohol from any substance for denaturing only, and having a daily producing capacity of not over 100 gallons, may use cisterns or tanks of such size and construction as may be deemed expedient in lieu of distillery bonded warehouses. The commissioner of internal revenue will prescribe regulations as to the manner and process of denaturing on the premises where the alcohol is produced and of transportation of such alcohol.

Farmers who contemplate going into the denatured alcohol business, either on their own account or in co-operation with their neighbors, should write the commissioner of internal revenue, Washington, D C, asking for full instructions in the matter, also for circulars regarding the process of denaturing alcohol. This will enable the producer to avoid conflicting with the federal law. Full details with reference to denatured alcohol will be found in the book, Alcohol, sent by Orange Judd Company, for which the price is $1.

Denatured Alcohol Making

Alcohol Making Materials

Alcohol for denaturing is made by the same process and from the same materials as ordinary alcohol. Potatoes have always been a chief source, and grains are also frequently used. Beets are an important alcohol making vegetable, and practically all fruits and vegetables which contain sugar can be used to some extent. By the best methods 220 lbs wheat will make 7 gals pure alcohol; a similar quantity of rye 6 gals, corn 5½ gals, and potatoes vary in value for alcohol making, but yield more alcohol than grains.

The Process

Three steps are essential in the making of alcohol from any vegetable product; preparation of material, fermentation and distillation. The first step often requires some special machinery, such as grinders, crushers or steaming apparatus. The fermentation is usually accomplished in large vats, which are comparatively inexpensive. The distilling requires a still, which may be simple or complex, according to the purity requirements of the product. Where grains are used, such as wheat, rye, barley, oats, buckwheat, corn or rice, they are usually ground to a coarse flour before they can be prepared.

The crushed grain is then steeped in a vat until it has swollen. It must then be mashed with yeast. Then an infusion is made and this is put through the fermenting process and finally distilled.

Potatoes must be crushed and steamed, run through a crusher, fermented with malt and yeast and then distilled. When beets are used they must be prepared by rasping, pressing out the juice and fermenting this, or by treating them similar to potatoes. They may sometimes be distilled direct.

Why Regulations Are Strict

As the process first involves the making of pure alcohol before it is denatured, all government regulations applying to the production of alcohol must be enforced, in view of the fact that the revenue to the government each year amounts to many millions of dollars. It must safeguard its income by exercising a careful oversight over even the smallest stills. The revised government regulations, while seemingly very complicated, are really simple, being designed to accomplish one thing. i e, that the revenue officer may know absolutely the amount of alcohol that has been produced. Where the business is sufficiently large, an internal revenue officer will be on hand when alcohol is being distilled and denatured. If the business is small, especially constructed storage tanks, etc, securely locked. will be provided, so that the officer may occasionally visit the plant and determine the amount of alcohol produced. In certain cases the law provides that if the collector of revenues is satisfied as to the character of the distiller, he may allow the still to be operated in his absence.

Apparatus Necessary

Elaborate or costly apparatus is not needed to make alcohol, as has been demonstrated by the "moonshiners" in the southern mountains. Crushing grinders, presses and vats are already owned on almost every farm where cider is made. The still and the holders of the alcohol must conform with the law. The essential feature of a still is a boiling vessel with a tube leading from the closed top of the vessel, passing as a coil through cold water, forming what is known as the "worm" wherein the spirituous vapor is condensed into a liquid. The alcohol is separated from water by boiling at a certain temperature, as it becomes a vapor at a lower temperature than water.

While small stills have not been sold in this country to any extent, American manufacturers of alcohol making apparatus are taking up the making of small outfits. The success of small alcohol distillers in other countries has proved beyond question that the American farmer can master this process and be able to make alcohol on a paying basis from farm by-products.

The making of denatured alcohol offers great possibility for co-operative effort among farmers, as a centralized plant equipped to use the waste products of several farms would doubtless be a paying proposition.

No men living are more worthy to be trusted than those who toil up from poverty; none less inclined to take or touch aught which they have not earned honestly. [Lincoln.

PROF CHAS E. THORNE

An experiment station director who has been long in the service is Prof Chas E. Thorne, Director of the Ohio Experiment Station. Prof Thorne is a born executive and has organized for Ohio one of the best experiment stations in the entire land, and his investigations of maintenance of fertility and the rational use of farm manures have supplied farmers with positive information of the utmost value to them. If an estimate might be made of the value of Director Thorne's work, it would run away up into the millions.

How to Improve Our Farm Poultry

Selection of the Best Layers from Well-Bred Stock—Use of the Trap Nest

Never has interest been so widespread in the improvement of poultry as today. The common scrub hen, puddle duck, farm goose and turkey are far less popular than they ever have been before. This is because so much attention has been directed toward the improvement of poultry, not merely from the standpoint of the fancier but from the standpoint of the practical farmer who must have a profitable strain of fowls in order to make poultry keeping worth his while.

Probably no one thing ever has had so much to do with the improvement of chickens, more particularly, than the desire to secure a plentiful supply of eggs in winter when prices are high. Two means have been widely adopted to reach this end: First, the fowls have been reared early in the season so as to be fully developed by the time cold weather arrives; and second, the trap nest has been used to discover when hens are the layers and which ones are the drones. To be sure, many poultrymen believe they can distinguish good layers from poor ones by the form, activity and other qualities of the individual hen, but this means of judging is only relative; it is not absolute. Similarly, the early rearing and the trap-nest methods are also relative, but they are more reliable and definite in the hands of the novice than the judging by appearances. A combination of these three is of course the most desirable means of securing a flock of good layers, but few people appreciate well enough just what is meant in a word description of form to apply this method, so the trap nest and the early hatching practices are most reliable for the majority.

It is needless to say that well-bred stock should be given the preference when selecting for layers, but where this does not seem feasible, the poultryman may start with just the flock he has. The first thing to do then is to confine the fowls in yards where they will be obliged to use the trap nest. Each hen should bear a numbered leg band and records should be made of her achievement in laying. Then the hen that lays the largest number of eggs when prices are

highest should be given the preference as a possible mother of future flocks. Of course, it may seem desirable to have merely the largest total number taken into consideration, especially where eggs are sold at advanced prices for hatching, but for the home flock, where the home table or the market is the object, the hen that lays during the cold months is most likely to be the most desirable.

In the latitude of New York, when one is breeding the American or English classes, the hatching should be done in March or April; when breeding Mediterranean varieties the hatching may be done as late as the first week in May. During the whole growing season the chicks should be supplied with everything necessary to produce sturdy fowls which will begin to lay at five or six months old. But it is not desirable to force these fowls into laying. They should be allowed to become well-developed, vigorous pullets rather than precocious egg machines. If so treated, they are much more likely to turn out satisfactory continuous layers than if forced.

Of course, among any flock, no matter how well its ancestors may have behaved at the egg basket, there will be found drones. These must be eliminated in the same way as before; namely, by means of a trap nest, and only those layers that have made good records should be kept the following season for breeding purposes. Not only so, but roosters raised from such good layers should be the only ones permitted in the flock. A few years of this practice has greatly increased the egg yield in thousands of flocks all over the country. It is just as applicable to the so-called meat breeds (the American, English and Asiatic classes as a whole) as to the egg laying breeds which nearly all belong to the Mediterranean class; and as a rule it will be found most satisfactory with these breeds because the large size and well-formed carcass of these classes are much more highly appreciated in the market than are the less fleshy and smaller Mediterraneans.

There is no difficulty in raising chickens as early in the season as indicated, especially where one uses incubators and brooders, but many farmers hesitate to apply the trap-nest method because they fear it will require too much time in releasing the hens and the keeping of records. The simplest method to obviate this difficulty is to have trap nests which open into two different yards, one yard containing the flock, the other being empty. If three yards are placed side by side, larger flocks may be kept in the two outside yards and the hens allowed to escape from the trap nest into the middle yard.

Of course, in such cases, the trap nest should be self-setting, so that when the hen escapes, the door will open for the next hen to come in. Then in the evening, when the fowls retire to the roost, records can be made of all the fowls that have laid and these individuals turned back into the original flocks. When careful breeding is desired, this plan will not be quite so desirable as where the confining trap nests are employed, but for ordinary farm purposes it is most convenient and satisfactory. Anyone can thus improve his flock and develop his own strain of good laying fowls, not only to his profit but to his satisfaction.

Aids Jewish Farmers

The Jewish Agricultural and Industrial Aid Society was organized in 1900. Its headquarters are 174 Second avenue, New York. The president is Cyrus L. Sulzberger, the secretary, Percy C. Straus, treasurer, Eugene Meyer, and general manager, Leonard G. Robinson.

The society assists and encourages Jewish immigrants to become farmers; helps them to find suitable farms and grants loans, on easy terms and at low rate of interest, toward the purchase of the farms and for their equipment; buys desirable moderate-priced farms and sells them to worthy immigrants at cost and on easy terms; loans money to those already on farms requiring financial assistance; publishes The Jewish Farmer, a monthly agricultural paper in Yiddish, for the benefit of those farmers who, on account of their unfamiliarity with English, cannot take advantage of the government bulletins and agricultural publications; maintains a farm labor bureau, finding employment on farms for Jewish immigrants without cost to employer or employe; renders free advice in agricultural matters to farmers or those intending to become farmers and aids them in various ways to improve their condition. It has aided nearly 1,000 Jewish farmers in 22 states and in Canada with loans aggregating over $500,000.

Some men are unable to see the straight and narrow path because a big round dollar shuts off their view.

The Farmer's Garden Through the Season

Preparing the Soil, Selecting the Seed, Planting and Cultivating—Profitable Crops to Raise

Every year the number of farm gardens worthy of the name increases. This is because a larger number of farmers each year awake to the fact that it is more satisfactory to cultivate the farm garden according to modern methods than to practice the old-fashioned ways. A piece of land close to the house is selected, and instead of being laid off in the short rows for hand cultivation, the rows are made long so the horse may be used. This gradually reduces the amount of work that must be done by hand and thus increases the amount of vegetables that can be secured. By judicious arrangement of the rows, everything can be grown that anybody will need for the whole year through.

The perennial crops, such as asparagus and rhubarb, are placed preferably at the far side where they can be out of the way during the greater part of the season. Next to these come the plants that require a long season to mature. Then the short-season crops nearer the house. All these latter crops should be planned for double cropping, so that when the early ones are off, a second crop of something else may be put in the ground at once. Another way to secure the same end is practiced with such crops as are transplanted. Cabbage, tomatoes and peppers may be placed between plants in rows of such crops as lettuce, radish, early onions and so on. Many combinations of these two plants may be made so as to compel a comparatively small area to produce two or three times as much yield as the old methods required with their greatly enlarged space and extra work.

The prime requisite, however, in having a good garden is to have first-class seed; without this there is sure to be disappointment. The ordinary garden seed purchased at the local store is too often poor, either because of age, bad selection or improper care. Criticism is widely applied to the seed sent out by congress. It pays, therefore, to spend a little more money to secure a first-class sample of seed from a reliable seedsman whose reputation is at stake than to indulge in the penny-wise-pound-foolish policy indicated.

The land selected for the garden should be a well-drained loam, preferably light .

than heavy. It should be well supplied with humus and be liberally dressed with barnyard manure. The humus may be added annually by sowing crimson clover at the last cultivation for the season and turning this under the following spring. The soil should be prepared very thoroughly by deep plowing and harrowing, then some of the handwheel tools may be used for sowing and cultivating.

As an adjunct to the farm garden a hotbed and a coldframe are highly desirable. Such plants as tomatoes, peppers and other rather tender vegetables may be started in the hotbed and the more hardy ones, such as cabbage and cauliflower, in the coldframe. This method of raising plants is far superior to the growing of plants in the house, because the plants can be hardened off and thus better prepared to meet the conditions of field culture when placed in the garden than those plants raised in the house.

During the season the surface should be kept free from weeds and loose by frequent cultivation. Once in two weeks and after each rain, which forms a crust, is often enough. By practicing such methods farmers all over the country will recognize the fact that the garden is one of the greatest sources of pleasure and profit to the household they have on their farms. Not only does the garden supply the table with great variety of household products, but it reduces the cost of living very markedly, principally because it supplies a welcome substitute for meat.

From the time the first radishes, lettuce and spring onions come in until they come in again the following year it is quite possible to have a continuous supply of vegetables. No one who has ever had a good asparagus bed and plenty of rhubarb will be willing to do without these two plants. They appear just when the appetite is craving something fresh, after the long winter diet of cabbage, potatoes, turnips, beets and other root crops. Quickly following and accompanying these two come the first early spring crops of lettuce, onions and radishes. Where salads are enjoyed, nothing is nicer than mustard and cress, which can be obtained in three weeks or less, from time of sowing, and may be served as salads or as greens.

Garden peas and beans are among the most enjoyed of all vegetables and can be so easily raised that it is a marvel farmers ever buy them or fall back on field peas and beans. Cucumbers, squash, tomatoes, pumpkins and a score of other vegetables fill out the season so well that there need not be a day when something new and fresh is not upon the table. And where these vegetables are grown at home they, as a rule, can be had in so much better condition than when purchased that it is strange there are still any people who patronize the hucksters. All that is needed is a little pioneer determination, a little application of the principles mentioned and a desire to enjoy the best products of the soil.

Marking Cattle

It is often a difficult matter to be certain of the identity of individual animals. We illustrate the system largely used by breeders of Aberdeen-Angus cattle, and many other breeders are coming to adopt the same or a similar system.

The notches in the ears of the animals represent numbers as follows:

A notch in bottom of left ear equals 1, two notches equal 2.

A notch in top of left ear equals 3, two notches equal 6.

PERMANENT MARKING

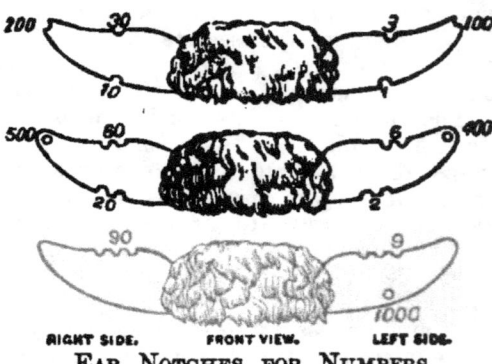

RIGHT SIDE.   FRONT VIEW.   LEFT SIDE.

EAR NOTCHES FOR NUMBERS

A notch in bottom of right ear equals 10, two notches equal 20.

A notch in top of left ear equals 3, two notches equal 60, three notches equal 90.

A notch in end of left ear equals 100.

A notch in end of right ear equals 200.

A hole in end of left ear equals 400.

A hole in end of right ear equals 500.

A hole in bottom of left ear equals 1,000.

Numbers can thus be made from 1 to 1,000.

Examples:—Notches made as follows represent 217:

1 notch in bottom of left ear equals.. 1
2 notches in top of left ear equal.... 6
1 notch in bottom of right ear equals 10
1 notch in end of right ear equals..200

Total equals No. 217

How Man May Conquer the Soil That Our Harvests May be Great

The Science and Practice of Successful Tillage—Plowing and Cultivation —A Common-Sense Talk by the Man Who Knows How

I—The Role That Tillage Plays

Nature's tools are not meant for fast-working man: too much, now, is required of the producing power of lands for modern men to depend upon these ancient, these earliest forms of tillage. Nor are meant for our use today the ancient forms designed by the early man: the crooked stick has been displaced; the wooden plow and the wooden harrow have disappeared; so, also, all ancient, out-of-date tools and implements for every purpose have been replaced by kinds more suited to the needs and the demands of present-day requirements.

II—Draining the Land

A wise man once was asked: "What is the most valuable discovery in agriculture?" He answered: "Drainage."

In draining the land, we are concerned, for the most part, with the surplus water and its removal. For drainage acts thus: it removes the gravitational water—the kind that often injures plants, the kind that drowns the roots—and it increases the quantity of capillary water—the kind useful to plants, the kind that draws into solution the needed plant food salts, and secures them for roots and stems and leaves and for all the growing tissues of the plant.

Deepening the Soil

It is perfectly evident to any thinking man that a soil that is well drained is a more habitable place for plant roots than one filled with standing water. We do not need to theorize about this proposition. You need only to observe, as you pass along any highway, to see how slight is vegetation, and how sickly are cultivated crops on lands not drained. A soil that is constantly saturated with water will not permit a good growth of crops. The essential conditions for growth are wanting. It is understood readily that where a tile drain, or, in fact, any sort of substitute, when constructed and placed three or four feet below the surface of the ground, the water level is naturally lowered to a point on a level with the bottom of the drain. Drainage, therefore, provides a large pasture ground for plant roots and a deep one, also, as a consequence, for all time to come.

Perfectly drained soils, drained to a depth of three or four feet, show plant roots throughout this body limit. It stands to reason that such a root-foraging ground is more desirable than a shallow one, made so by a high water table near the surface of the ground. And here are the reasons: there is more room for the roots; there is more plant food to be secured; there is more warmth in the soil; there is more air to be used; hence, there is a more comfortable home for the roots provided in drained land.

Air Gets Into the Soil

It has been pointed out that both air and oxygen are essential for good root development, as well as for high crop production. But air and oxygen are excluded from the soil when water fills up all of the air spaces in the soil. Drainage removes this water, and, hence, increases the air content of the soil. Air goes just as deeply into the soil as the water table allows, and as it goes down, it leaves all along its way its helpful gifts—scores of beneficial influences that stand for better crops.

And still two other things: It supplies the roots with oxygen, and it breaks down complex substances, fitting them for the call that other plants will soon make.

Manure is Made More Effective

Vegetable matter and other humus-forming materials are of no value in the soil until they are thoroughly decomposed and destroyed. Hence, it follows that good results, from the use of manure, will be obtained in the highest degree only when the rotting influences of the soil are best; for undrained soils do their work in this respect very unsatisfactorily. The drained soil makes the best use of manures. There is in this connection another point to be considered:

Useful bacteria find favorable development only in the presence of an abundance of oxygen; they find enjoyable the work of breaking down compounds, and of building up nitrates only when air is furnished abundantly, and when the soil is open and warm and sweet.

But they do not like wet soils and in them they do but little work. On the other hand, much work is done by the nitrogen-feeding bacteria; the evil-doers which release soil-nitrogen, and send it back again into the atmosphere.

The Soil is Made Warmer

Wet soils are always cold. And since warmth is necessary for both germination and for active growth of plants, it follows that wet soils never can equal well-drained ones when it comes to high production. It is out of the question to expect otherwise.

The Season is Lengthened

Wet soils often become dry in the summer, when hot and dry weather is the rule, but for a period so short no paying crop can result. Nature is too slow here and your work will be unsatisfactory and unprofitable, if you depend upon her, alone, to drain areas naturally wet. Better not do it. Use tiles instead. The first noticeable difference, after drainage is done, is the lengthening of the growing season; work may be begun earlier in the spring and it may be extended far into the autumn. This means that stagnant water is removed from the land, both in the spring and in the fall, to the advantage of your work, your crop and yourself. You can handle, often, well-drained soils three to five weeks earlier than like soils in undrained condition.

Tillage is More Easily Done

Soils are injured frequently (if not ruined for the time being) by tillage operations, if done when the land is wet. It is perfectly evident, therefore, that a soil undrained, either naturally or artificially, makes all tillage operations a burden, and the work a drag.

Plowing is often done later in the season, consequently often unsatisfactorily, and the crop suffers, for it is planted in haste; and, as a result, it is hindered throughout its period of growth. Drained lands are easily tilled and easily cultivated. They permit all tillage operations to be done easily and satisfactorily, and at a time when most urgently needed or demanded.

Less Danger of Drouth

One of the proved facts that scientific investigation has shown is this: a soil contains more available moisture after drainage than before. The explanation of this seeming inconsistency lies in the fact that the physical condition of undrained soil is being improved; the soil is made loose and mellow; the soil grains are more open; and the interspaces admit and hold air—the capillary water is more freely introduced, when demanded, and more readily handed out to the roots, as they call for it.

When you open and mellow and fine the soil you increase the moisture content of the soil. When this is done, a larger store of water is secured in the soil, down to a considerable depth, all of which will be available, when dry weather arrives, furnishing what wet soils fail always in doing.

Washing is Prevented

If a soil is saturated with water, the only means of escape for rains is by means of surface washing, a most injurious operation for the soil. For this reason: surface washing picks up dissolved plant food, and fine particles of soil, and carries both to lower depths. I have seen vast areas, gullied, and ridged, and torn, and made so by surface washing of the land.

Often plowing is done poorly; so shallow that rains never find their way down into the soil, carrying as they go their good effects to the lower depths; but, on the other hand, they flee along the hillsides, carrying away treasures that are valuable and sadly needed, just where they are—taking them from the place where most able to do good. Drainage provides, therefore, both an entrance into, and, at the same time, an exit out of the soil, for all water that falls as rain.

III—Saving Water by Cultivation

The work of the farmer is to induce water to enter the soil both in summer and winter. But it is more than this. He must save it, once it is secured. And now we come back to our original proposition: cultivation checks the water loss. Until you grasp this idea, until you come to a full realization of its force and importance, you will never be able to compel your soils to expend their fullest powers toward the production of maximum crops.

The principle of moisture-saving, briefly stated, is this: Water is carried from the water storehouse of the lower depths of the soil by capillarity. It rises in the soil from soil particle to soil particle, just as oil creeps up in the lamp-wick. It moves sidewise and diagonally and upward; it goes in the direction of the hardest pull.

But always in the end, unless prevented by some obstacle—a dry mulch so acts—it finds the surface of the soil, at which point it passes into vapor and leaps into the atmosphere.

While frequent stirring of the soil, during the growing season, and, especially in the time of drouth, tends to produce better crops than if this work is neglected, still, it is a wiser practice to begin the work of moisture conservation before the drouth period sets in. Hence, you must have much water admitted to the soil. You must keep it and preserve it until it is in greatest demand.

Keeping the surface crust broken and loose and mellow is the first step; it takes the water in. Conserving it after it is stored is the second step; it holds for the plants.

Mulch-Making Makes It Effective

In periods of abundant rainfall, it matters little to you whether you stir your soil a half inch or 4 inches deep; for the time being, you are not concerned with the moisture content. But during dry weather you ought to be careful; you ought to interrupt and break the capillary tubes that connect the surface and the immediate lower region, so that the escaping water may be kept within the smallest limits.

Now, as to the depth of the mulch: an inch is good if it is even and level and completely separated from the tubes below; if it provides an effective blanket over the surface of the soil. Even with so slight a mulch, the operation is beneficial.

Experiments on this subject indicate the following:

1. That the water content of the soil is increased to a very appreciable extent when the soil is evenly and uniformly stirred.

2. That the water content is increased in proportion to the depth and effectiveness of the mulch.

3. That the water content increases less rapidly as the depth of cultivation is increased beyond 3 inches.

4. That the water content is greater when cultivation is provided in a form of mulch, than by ridge culture or broken tillage.

IV—Plowing

One of the most expensive things a man can do is to move dirt.

No tool has ever been invented that moves so great quantities of soil for so little money as the plow. No farm implement is more in use, nor is any more essential, yet Prof Roberts declares that in America plowing is the least understood and the most imperfectly performed operation in connection with our preparation of land for crops.

We know how to plow, but how few of us really know when and why to plow. The only reasons why people used to plow were to get crops in and to kill the weeds. It is no wonder to me that at one time people hated to plow. With primitive tools it was hard work; and it, too, was slow work.

The First Plow Was a Sharpened Stick

The first plow was the sharpened stick. But man is lazy; he soon abandons this most primitive of all forms of tillage, selects a forked stick, ties it to the horns of a bull and makes the animal do most of the work.

Thousands of plows have been invented since this early type, but there is no change in the principle. The motive power has changed; the long end of the forked stick has been succeeded by a beam of finished wood or steel; the short end has been metamorphosed into a chilled steel point and moldboard; the rough hand knot has been supplanted by curving handles or a driver's seat; but for all these improvements there has been no radical change in the tool.

The Work the Modern Plow Should Do

Now, what should this modern plow, evolved from the crooked stick, do for the land? How shall a man know when he has a good plow, how shall he know when he is doing capital work? In the first place, the effective plow turns the land; the furrow slice is laid entirely over, or it is set up well on edge. In either case, it must cover manure, trash or green crops.

In the second place, the plow should go deep into the ground. This for two reasons: deep plowing, as stated heretofore, enables the soil to drink in and to hold more water against the time of drouth. Deep plowing gives plant roots a wider pasture.

In the third place, the effective plow must pulverize the furrow slice turned out. Turning the land is not enough; the soil must be broken, fined and mellowed. We get these results by means of the sharp, bold curve that is given the moldboard. A plow that does not thoroughly pulverize the soil is a poor plow. It may make a handsome furrow, cover the ground well, plunge far into the ground, and still do poor plowing; unless it leaves the soil in so friable a condition that the other tillage tools can easily and economically do their part, it has fallen short of its duty.

Aim to get a furrow slice that is set well on edge, with a snap as it comes from the moldboard. This is the sort that the harrow uses best for completing the bed for seeds.

In addition to plowing, in order to get a pulverized, deep, warm, moisture-holding plant bed, we must plow with a view to bettering the physical condition of the land. Hence, we should aim to get deep and uniform plowing done in every field.

V—The Tools of Preparation

The harrow follows the plow. You need this tool in connection with the roller to complete the pulverization of the soil begun by the plow. Both of these tools mellow the soil and push the particles nearer to one another. You have observed that the cloudy spots found in a fertile field make a poor harvest. In these places the bad condition of the soil excludes moisture and pens in plant food, hence this lack of fruitfulness.

The harrow and roller will correct this trouble. You cannot be too painstaking, when it comes to harrowing. A field may look well after a harrow has gone over it, but this does not necessarily mean that the work has been done well. For this reason you should always examine carefully to see whether the soil has been uniformly pulverized, and the particles pressed in close continuity.

While few men catch the spirit of plowing, a still less number catch the spirit of harrowing. The harrow is the tool to complete disintegration and pulverization. It should go 3, 4 or even 5 inches into the soil. The harrow teeth should go down well below the surface, and work among clods and lumps; they should either break all clods and lumps, or bring them to the surface where they can be ground and crumbled by subsequent tillage.

A field is never well harrowed until the interstices between the coarser particles are filled with a sifting in of the finer particles. When this has been ac-complished, the seeds have a perfect chance to sprout and grow; the soil is well fitted to take care of its water supply.

One kind of harrow is not enough: it will not do for all seasons, nor for all soils. Here are the things the harrow must do: it must smooth, cut, level, spade, pulverize and compact. No harrow can do all these, hence you will need different kinds to do all the work involved in harrowing well and effectively.

The fine-tooth smoothing harrow should have a place on every farm. It levels and disintegrates, and it comes in handily for intertillage; does splendidly on corn and cotton land after planting is done.

The spring-tooth harrow should be had, as it comes in nicely where you have leveling and smoothing to do, or where a heavy rain has compacted the soil too hard for seeding purposes.

In addition to these, you should have a disk or cutaway harrow. You will find it very valuable—in fact, indispensable—for many kinds of farm work. Such a harrow takes the place of the plow in seeding wheat or rye after corn or cotton or cowpeas or potatoes. The rolling disk cuts and turns and pulverizes, and thus does the work of a plow, although it does not go so deeply. But since you want a compact soil, excepting the top, this becomes the very implement for your work. Fields that have just been disked and then crossed with a spring-tooth harrow are usually left smooth and mellow, and in fine condition for the seeding tools.

The disk harrow is an excellent tool to use immediately after the harvest as a means of opening the soils to catch summer rains, and of conserving the moisture already present in the soil. In our dryer sections, especially in semi-arid regions, this plan of soil treatment is coming into practice, and is to be commended. If you find clods on top, the wooden drag or roller will be the next implement to use. The wooden drag grinds the clods and lumps; and is also a good implement for leveling purposes. The roller is primarily a crushing and compacting tool. While it is effective as a crusher, it drives the dry, hard clods into the surface soil.

You will find the roller most desirable during dry seasons for compacting the soil so that capillarity may be restored, and moisture from the great reservoir down in the soil be drawn up to the seeds and roots. This, however, may prove harmful, for you may induce a too rapid evaporation, and thus destroy your reserve supply. You can minimize this harmful effect by using a smoothing harrow a few days after the roller. Fortunate is the farmer who becomes a believer in the practice of making the soil firm for all seeds and plants.

Compacting Soil Increases Water Content

The market gardener has led the way here. Just note how he uses his feet to accomplish this very purpose; he knows that it pays—in fact, that it is clearly necessary to make firm soil about the newly planted plants or seeds. This is important in dry, hot weather in loose, poor soils. The farm roller and gardener's feet both accomplish the same work. Both compact the soil; start the water in the soil below on an upward course that brings it to the seeds and roots that need it.

When this has been accomplished, a smoothing harrow, with its little teeth, should be run over the land so as to break off the tops of the capillary tubes; and thus make a mulch of the top soil and check the evaporation of the soil water.

The cultivating tools in the tillage of the soil answer three purposes: They kill weeds; they provide a much-needed mulch, especially in dry climates, so as to preserve moisture; and they release plant food. The old, one-shovel plow is fast giving way to the shallow cultivator with several shovels. We no longer expect to use intercultural tools for preparing the soil for root development; we do that now at seeding time.

VI—Cultivation

How deep shall we cultivate? That question has been answered with quite a good deal of certainty. At least a half hundred carefully planned and executed experiments have, by their results, answered in favor of shallow cultivation. Since then we have heard much about this new idea in cultivating the soil. But we are in danger of going to the other extreme. Our fathers "plowed" corn; they cultivated too deep. Some of us, perhaps, cultivate too shallow; we get in trouble with weeds; and because of our thin mulch, let the water get away from the soil.

In sections where there is much rain, the shallow extreme may do; but where moisture is demanded—in the north, where the ground is frozen for so many months; in the semi-arid regions, where the supply is generally limited—a deeper mulch and a more effective mulch is to be preferred. Four inches, perhaps, is too much and 1 inch is too little. A better depth is from 2 to 3 inches; better for weed destruction and good enough for mulch making.

Level Culture a Most Important Point

You will find farmers who still ridge their crops; they "hill" the crop that it may not be blown over by winds, nor pulled down by storms and rain. But have you ever noticed that near-by crops, although given level culture, are no more troubled by storms and wind than the hilled and ridged crops? Often not so much, is the true situation.

Hilling and ridging the crop is advisable for just one reason: to drain the land. With proper drainage and seed-bed preparation, there is no occasion for either of these expensive practices.

Level culture, since it exposes a smaller area to sun and wind than ridge culture, actually protects, with greater efficiency, the water stores in the soil. Bedding the land is often advisable with some soils (although it increases the cost of planting), for the reason it secures a small amount of drainage and a greater warmth to the soil.

When to Cultivate

You must be in sympathy with the spirit of cultivation if you would get the best results. You must do it at the time when the soil is in the best condition to profit by the work. Just after a rain, the word goes out. But use your judgment here, else you may cultivate too early after the rain and "puddle" your land. When the next rain comes, the crust caused by the cultivation may be so hard and stiff the rain may slip away before it can secure entrance through the stubborn top.

Here is the better plan: just wait until the soil is slightly dried; enough so that when it is stirred it will not settle and connect with the capillary tubes below, thus defeating the very object you set about to secure. In times when you are depending upon cultivation for water preservation it will be worth your while to watch the mulch, to see if it is still an effective blanket or if the connection with the capillary tubes below is beginning to take place. If the latter be so, it is high time that you repeat the cultivating work.

Water Saving Means Early Work

Water saving falls into two means— the catching and holding of it. You first must get water into the soil, and then you can use it; provided, of course, you do not let it escape before it is needed. Too many tillers of the soil fail to understand that the most important principle at stake in water saving is to till and cultivate in such a manner that there is free access of water into the soil. Then it can be preserved by cultivation and mulches throughout the season. But failures in supplying water, although effective culture—mulch making—is given during the growing season, are certain to happen if no water is in the soil to be conserved. If you would have water for plants for the time when they shall need it, if you would have soil water for them for later use, make no mistake about first getting it in the soil, and the rest of the work will be easy.

Just Bear in Mind these Suggestions

1. Getting ready for crops—opening soils and catching water—is of more importance than after cultivation.

2. Get water deep into the soil and you will have bigger stores of supply.

3. Cultivate after every rain, not when the soil is really wet, but before it becomes very dry.

4. Make your mulch deep enough—3 inches is none too deep in dry regions.

5. Open the soil early in the spring with a disk if you have not fall plowed or winter tilled.

6. Stir unused summer lands frequently, so as to let water in and to keep it in for the next crop.

7. Lands frozen up for long periods —as in the New England territory— are as needful of water saving as those of the semi-arid or dry farming districts.

VII—Applying Manure

Manure should be applied as fast as it is made, unless some good provision can be had for its protection and preservation against loss by fermentation or by leaching. The covered barnyard and the manure pit have come into use and popularity with recent years, doing much in the way of saving manure against loss. These provisions are good only for certain seasons of the year when it is impracticable to get out on the fields with the spreader, so as to make direct application of the manure to the land.

Broadly speaking, the sooner the manure can be got into the soil the better, for these reasons: The organic matter is still intact and the plant food is preserved. The rotting of manure means a waste of organic matter. Such rotting should be allowed to take place within the soil. As the manure rots, so will the soil rot; so will the compounds containing plant food rot and thereby furnish available plant food.

We want a lot of organic matter in the soil, for the reason that organic matter is the basis of humus supply; and hence it regulates the water content of the soil and the activity of bacteria, whose work is so intimately connected with the growth of crops.

That manure materially decreases in bulk and plant food value is shown in an experiment recorded by Prof Roberts. Starting with 4,000 pounds of manure, the amount decreased to 1,730 pounds; because of poor preservation, 60% of the nitrogen escaped into the air, 75% of the potassium and 40% of the phosphorus leached away in rain water—in all a loss so great that no farm can stand it even for a short time.

When to Apply

When this pile of manure is considered from the standpoint of its money value, we find that at the beginning it was worth $5.48; but after being exposed for five months, the plant food value was only $2.03—scarcely one-third its original value. Surely no farmer can afford to follow any method so wasteful as this. What method do you follow? Just bear this in mind: If you haul manure to the field and spread thinly over the soil as fast as it is made, say each day or once a week, you will not only save all the plant food it contains, but you will give the soil all the benefit of the action of fermentation on the soil.

When we consider that at least half the entire amount of manure made on our farms is as carelessly handled, we can realize in short order the enormous loss that annually takes place; a loss in real value as large as the entire crop of American wheat or cotton is worth. Just take this direction and you will find an explanation for the depletion of so many lands; you will find the real cause of so much poor farming and of lessened yields; you will find, in a large measure, the true meaning of abandoned farms; you will find the gist of all the troubles that infect the soil, the farm and the farmer.

These evil results may be eliminated— at least reduced to a minimum—if the manure be applied direct to the fields.

Where to Apply Manure

You ought not to apply manure on lands containing a large amount of nitrates. There is too great danger that these will be broken up by the decay of the manure; hence, manure should go to the fields where the supply of nitrates is at its lowest point; during the fall, just after crops have fructified; during winter, when nitrification is slow or inactive; in the spring, when the supply of nitrates is still low. It is unwise, perhaps, in the summer, when the nitrate supply is unused and still large, to apply manure to cultivated lands; the risk is too much, for these nitrates are likely to be lost. There is no objection, however, in sending manure at all times, and especially during the winter and spring months, to grass or pasture or mowing fields. A clover sod that is to be planted to corn in the spring is an ideal place for a thin, even, and uniform covering of manure.

How Much Manure to Apply

As a general rule, it is more scientific to apply small amounts of manure frequently than to apply large amounts at longer intervals. This is the best fixed rule that we are able to give. This point has been tested at several stations. In New Hampshire 4 tons per acre applied each year for three years furnished an increase of 31% more of corn than a single application of 9 tons per acre at one time. Commenting on an experiment of a similar nature at the Ohio Station, Director Thorne says: "We have compared the value of manure applied at the rate of 4 and 8 tons per acre. The result has been that the increase per ton of manure has been more than 25% greater when used at the smaller rate, although the increase per acre has been larger when used at the larger rate; hence, when manure is scarce, it is better to apply it in smaller quantities, so as to cover all the land in crop, rather than to spread it over part of the land only and leave part unmanured."

VIII—Computing Fertilizer Value

In a commercial way, nitrogen is about three times as costly as phosphorus or potassium. The cost of the fertilizing element varies from year to year, but as a rule, nitrogen is worth 15 cents per pound and phosphorus and potassium each 5 cents per pound. In computing relative values, bear in mind that 1% means 1 pound in a hundred, or 20 pounds in a ton.

It is also a good plan in computing the value of a fertilizer to use the lowest figure representing the percentage, since that more nearly represents the true value.

Sliding figures are used more to deceive the purchaser than to help him or to give him a larger quantity at the cost of a smaller amount.

In order to show the process of computing the value of a fertilizer, let us take a problem for the purpose of finding the plant food value of a ton of fertilizer. Here is the problem:

What is the money value of the plant food in a fertilizer containing 1.95% of ammonia, 7 to 8% of phosphoric acid, and from 2 to 2.75% of potash—the commercial value being $30 per ton?

Process: First, reduce the ammonia to nitrogen, since it is the real element of plant food. Ammonia sounds larger, and hence is used in the fertilizer formulæ. Remember that ammonia is not nitrogen. It is only fourteen-seventeenths nitrogen, the other three-seventeenths being hydrogen, which has no value whatever as a fertilizer.

So to get the real amount of nitrogen in the ammonia we shall have to divide the ammonia percentage by 1.214, so as to get the percentage of nitrogen.

Just do it this way: $1.95 \div 1.214 = 1.60$; the nitrogen percentage. We will then multiply each of the several percentages (use only the smallest figures) by 20, so as to obtain the number of pounds in a ton, and then multiply this product by the value per pound, and we have the value on the basis of a ton.

The following shows the process:

Nitrogen 2 x 20 = 40 lbs. at 15 cents =	$6.00
Phosphorus 9 x 20 = 180 lbs. at 5 cents =	9.00
Potassium 2 x 20 = 40 lbs. at 5.4 cents =	2.16
Value of plant food in a ton	**$17.16**

So here is all there is to this estimate. When several fertilizers are available, just make the calculation in this way, and you can then determine in which fertilizer you get the largest quantity of plant food for the least money. For the purpose of comparison, we will take another fertilizer that sells for $29 per ton, just one dollar less; its analysis is: Nitrogen, 2%, phosphoric acid, 9%, potash, 2%.

With a first glance the average farmer might think the first fertilizer, since it sells for a dollar a ton more, is, therefore, a better fertilizer; but let us see, calculating as we did before:

Nitrogen 1.60 x 20 = 32 lbs. at 15 cents =	$4.80
Phosphorus 7 x 20 = 140 lbs. at 5 cents =	7.00
Potassium 2 x 20 = 40 lbs. at 5.4 cents =	2.16
Value of plant food in a ton	**$13.96**

Now you have your comparison: If you take the first fertilizer, you get in each ton $13.96 worth of plant food.

which costs you $30; if, on the other hand, you purchase the second, you get $17.16 worth of plant food for $29. The difference between the value of the plant food and the selling price is due. to the cost of manufacture, profits, agent's commission, etc. In the case of the first this difference is $16.04, while in the second it is but $11.84; a clear saving of $4.20 on each ton, and the latter is equal to the former in every sense of the word.

IX—Rotation of Crops

"No branch of husbandry requires more sagacity and skill than a proper rotation of crops, so as to keep the ground always in heart, and yet to draw from it the greatest possible profit." So wrote Lord Kames a great many years ago. And with every form of scientific investigation, with all improvements in agriculture—improved soils, better bred plants, more perfected tools of tillage, cultivation and harvesting—there has come into use no method that contributes quite so much as a wise, well-systematized scheme of crop rotation to the maintenance of fertility and to the production of maximum and profitable crops.

It is Nature's plan; she favors giving crops fresh lands to grow in. Note the forest: When trees are cut, new and different kinds grow in place of those removed. Note the grass: Timothy and clover may grow abundantly, but in the end Bermuda (in the south) and blue grass drive both away. Note the cultivated crop: Corn does better after clover or alfalfa, wheat after corn or potatoes, cotton after cowpeas or grass, than either crop after its own kind.

A soil is severely injured when a cultivated crop like corn or cotton is grown on it year after year. Even wheat or oats, timothy or cowpeas (when cultivated) bring about the same ill effect. The humus is burned out, the soil hardens and deadens, the elements of plant food, especially needed for these special crops, become scant. Hence the soil loses its power to successfully produce the constant crop. You can correct this trouble, to a great extent, by a change of crops.

A few principles that enter into the scheme of crop succession are:

Plants place their roots differently in the soil.

All plants exhaust the soil.

Plants do not exhaust the soil in the same manner.

All plants do not exhaust soils equally.

Some plants add nitrogen to the soil.

Some plants act favorably to weed growth, while others do not.

Plants grown constantly on the same land favor the spread of insects and diseases.

X—Grow Legumes Constantly

Nothing helps old, worn-out soils more than the legumes. They give nitrogen and humus, and they open the subsoil to air and water. Clover and cowpeas come first. Either one or the other will grow in your climate and fit into your work. Take the cowpea, for instance. It is an admirable plant for a depleted soil. Though poor tillage be provided, though the soil be hard and dead, the cowpea will respond with a luxurious crop. Look into the soil and you will find the evidences of the little fairies that did the work—the bacteria and their tubercle homes—gathering nitrogen for the plant and leaving what was unused in the soil for the following crop. But this first cowpea crop may not be what you like to see; it may lack vigor and aggressiveness. But just wait, and the next year repeat the work—use the same crop over again. Now you will see a difference, for the bacteria have increased sufficiently to meet all the demands. Now you get your reward! Now you become a friend of the cowpea! And the same is true of all other legumes—of the clovers, of the soy bean, of the vetches, of the alfalfa.

You should use these legumes in every kind of rotation—a legume every year if possible, and cowpeas in every crop of corn, using the last cultivation of the crop as the seeding time for the cowpeas. This is a practical way to do this, so practical that thousands of farmers in all parts of the country have adopted it. The author harvested 36 bushels of corn and 12 bushels of cowpeas from a field a few years before abandoned and forsaken because of its worn-out condition, which has been restored to high productive powers in just the way herein described.

Let Green Manures Help

Some soils are so completely devoid of humus it often is best to center the first effort in humus supply to them. This may be done by the use of green manures. You may have to pick your crop, for the reason the soil is so poor it may refuse to do much. You had better use the cowpea for this purpose. It seldom will fail. Use a bushel of seed per acre, applying them broadcast. When mature, plow under, turning the soil an inch or two deeper than the previous preparation. The following season either disk the land or replow and sow the second crop of cowpeas, using the same quantity per acre and seeding broadcast.

This second crop will tempt you greatly; you will be inclined to harvest it as hay. But it will pay you to remain firm to your original resolution; let it mature and be plowed into the soil, where it is needed for the nitrogen it holds and for the humus locked in its rich tissues.

The old soils deficient in fertility it will pay you to assist. Help the soil and crop through an application of fertilizers. Something like this will do: Mix acid phosphate and kainit together—1,500 pounds of the former and 500 pounds of the latter. Of this mixture use from 200 to 400 pounds per acre, depending upon the productive power of the soil. With this treatment given, your old soils will soon be on the way to recovery; they soon will be available for all sorts of crops.

Rotate Crops on the Old Land

Now, do not neglect crop rotation. Remember that this neglect in the past was one of the reasons why your soils became worn out and exhausted—one reason why they became "run down." Surely you do not want this to happen a second time. Crop rotation will largely help in preventing such a condition. It matters not what money crops you grow, give your soil a change. Introduce legume crops frequently and constantly. They will keep nitrogen and humus in the soil; they will keep the soil mellow and friable; they will open the subsoil to other roots; and they will save the land.

Woodman, Spare That Tree

Woodman, spare that tree!
 Touch not a single bough!
In youth it sheltered me,
 And I'll protect it now.
'T was my forefather's hand
 That placed it near his cot;
There, woodman, let it stand,
 Thy ax shall harm it not.

When but an idle boy
 I sought its grateful shade;
In all their gushing joy
 Here, too, my sister played.
My mother kissed me here;
 My father pressed my hand—
Forgive this foolish tear,
 But let that old oak stand.

My heartstrings around thee cling,
 Close as thy bark, old friend!
Here shall the wild bird sing,
 And still thy branches bend.
Old tree! the storm still brave!
 And, woodman, leave the spot;
While I've a hand to save
 Thy ax shall harm it not.
 —George Pope Morris.

What Lime May Do

This picture shows the effect of lime upon a growth of clover. The two piles of clover were from equal areas, each treated with the same quantity of a complete fertilizer. The picture shows the second crop. The crop on the unlimed ground was so small that it had to be

THE EFFECT OF LIMING ON CLOVER

cut with a jackknife and was placed in a straw hat (b) or it would hardly have been visible upon the ground at all. On the ground that was limed the stand was good (a) and the crop was cut in the usual manner. This demonstration occurred at the agricultural experiment station at Kingston, R I, under the supervision of H. J. Wheeler, the station director. At Slocum, R I, the growth was increased 100 times by liming, and at Moosup Valley, R I, in the case of mangelwurzels, it was raised from 6,072 pounds per acre to 41,760 pounds per acre. It represents a gain of nearly 18 tons per acre. This latter experiment was conducted on an old pasture where no chemicals or other fertilizer had been applied for years, and a like amount of commercial fertilizer had been used to each plat. In this instance the ill effects could not be due to any residue of any previous fertilizing; the effect was due solely to other causes.

Easter

The method by which Easter day is now determined is that of the first Sunday after the paschal full moon (fourteenth day of the calendar moon, or the full moon which happens upon or next after March 21). If the full moon happens on a Sunday, then Easter day is the first Sunday following.

The church of San Miguel in Santa Fe, N M, is the oldest church in the United States. It was erected in 1545, 20 years before the founding of St Augustine, Fla, and 53 years after the landing of Columbus.

Brief and Practical Farm Arithmetic

Rules Applied to Milk and Milk Products —Feeding and Feed Values

I—Dairy Problems

Much dairy work is based upon arithmetical rules and principles. Calculations incidental to dairy practice on the farm or in the creamery can be readily made by referring to one of the rules following:

FINDING WEIGHT OF ANY CONSTITUENT

To find the weight of any constituent in milk or milk products, when the weight of the milk or its product and the per cent of the constituent are known, multiply the weight by the number indicating per cent of the constituent and divide the result by 100. Example: How many pounds of fat in 675 pounds milk testing 4.6% of fat?

675x4.6÷100=31.05, the number of pounds fat.

EXAMPLES FOR PRACTICE

1. How many pounds of fat in 2,000 pounds cheese containing 35% of fat?
2. How much fat is there in 5,000 pounds skim milk testing .15% of fat?

FINDING PER CENT OF ANY CONSTITUENT

Rule—To find the per cent of any constituent in milk, etc, when the weight of the milk, etc, and the weight of the constituent are known, multiply the weight of the constituent by 100 and divide the result by the weight of the milk, etc. Example: What is the per cent of fat in 675 pounds milk containing 31.05 pounds fat? 31.05x100÷100=4.6%.

EXAMPLES FOR PRACTICE

1. What is the per cent of fat in 120 pounds butter containing 96 pounds fat?
2. What is the per cent of water in 600 pounds cheese containing 210 pounds water?

FINDING THE YIELD OF BUTTER

Rule—To find the yield of butter when the per cent of fat in milk and the weight of milk are known, find the number of pounds of fat in milk by Rule 1 and multiply this result by 1.17 or 1 1-6. Example: How much butter is made from 1,000 pounds milk containing 4% of fat? Applying Rule 1, 1000x4÷100= 40 pounds fat in milk; and 40x1.17= 46.8 pounds butter yield.

CHANGING POUNDS OF MILK INTO QUARTS

Rule—Divide the number of pounds milk by 2.15. Example: How many quarts of milk in 100 pounds? 100÷2.15 =46.5 quarts.

CHANGING QUARTS OF MILK INTO POUNDS

Rule—Multiply the number of quarts by 2.15. Example: How many pounds in 40 quarts of milk? 40x2.15=86 pounds.

CHANGING DEGREES FAHRENHEIT INTO DEGREES CENTIGRADE

Rule—From the degree F substract 32 and multiply the result by 5-9. Example: 162° F.=(162-32)x5-9=72° C.

CHANGING DEGREES CENTIGRADE INTO DEGREES FAHRENHEIT

Rule—Multiply the degrees C by 9-5 and add 32. Example: 72° C=(72x 9-5)x32=162° F.

FINDING AMOUNT OF CREAM

Rule—To find the amount of cream produced for 100 pounds milk when the per cent of fat in milk and in cream is known, divide the per cent of fat in milk by the per cent of fat in cream and multiply the result by 100. Example: How many pounds of cream containing 25% of fat are produced from 100 pounds of milk containing 5% of fat? 5÷25=.2. .2x100=20, number of pounds of cream with 25% of fat.

II—Feeding Problems

Animals depend upon plants directly or indirectly for food. Plants get their food from the soil and from the air. Soils get their producing power from the decay of earth, animal and vegetable life.

WHAT PLANTS CONTAIN

Growing and mature plants contain (1) water, (2) ash, (3) protein, (4) crude fiber, (5) starch and sugar (nitrogen-free extract) and (6) oil or fat.

FRESH PASTURE GRASS CONTAINS

	Per cent
Water	75.3
Ash	2.5
Protein	4.0
Crude Fiber	5.9
Nitrogen-free extract	11.4
Fat	.9

DIGESTIBLE NUTRIENTS IN FEEDING STUFF

Food eaten is not wholly assimilated in the animal body; just a part of it.

This percentage-expressing amount assimilated is known as coefficient of digestibility. Hence to know the exact amount of every feed that is digested, it becomes necessary to determine this for each feed and each constituent in it.

In 100 pounds of wheat bran there are 15.4 pounds of protein; 79% (coefficient of digestibility) of which, when eaten, is digested. How many pounds of protein are digested?

$$15.4 \times .79 = 12.17$$

The composition and coefficient of digestibility of corn stover are as follows:

	Per cent Composition	Coefficient Digestibility
Protein	3.8	45
Crude Fiber	19.7	67
Nitrogen free extract (starch)	31.5	61
Fat	1.1	62

How many pounds of digestible protein, carbohydrates (fiber and starch) and fat in 100 pounds?

In 100 pounds of cottonseed meal there are 42.3 pounds of protein; of this 37.2 are digested. What is the coefficient of digestibility?

The number of pounds of digestible protein in a ton of clover hay is 136. When clover hay contains 12.3% of protein, what is the coefficient of digestibility?

Wheat bran contains 9% crude fiber, of which 22% is digestible and 53.9% of nitrogen-free extract, of which 69% is digestible. What is the percentage digestible of each constituent?

In 600 pounds of wheat bran what is the total digestible quantity of each of these constituents?

THE IMPORTANT THREE COMPOUNDS

Ash usually is contained in feeds in sufficient quantities to supply all needs of the body; water is supplied through drink; leaving protein, carbohydrates (nitrogen-free extract and fiber) and fat as the important three constituents to be considered in the feeding of animals.

WHAT THESE DO

1. Protein (the muscle maker) is used for the formation of muscle, bone, organs of the body, blood, skin and also for milk.

2. Carbohydrates (heat and fat producers) are used as heat, fat and energy.

HEAT VALUE

Careful calculation has shown that 1 pound of fat will produce 2.4 times as much heat as 1 pound of carbohydrates.

NUTRITIVE RATIO

Food may be said to do two things: (1) it furnishes the protein for growth, daily waste, blood, etc; (2) it furnishes the materials for fat, heat and energy.

The ratio between digestible protein and digestible heat and fat-producing elements (carbohydrates and fat) for any food or combination of foods is called the nutritive ratio.

FEEDING STANDARDS

These are guides for supplying protein, carbohydrates and fat in correct proportion and in sufficient quantities for any class of animals to do their special work. A dairy cow fed "fattening feeds" will grow fat, but lose in milk flow. Another cow giving a large quantity of milk requires more protein in her food, if she is expected to continue to produce a heavy daily yield of milk. The horse calls for a different daily ration than the beef cow; the pig than the plow horse.

STANDARDS

On basis of 1000 pounds live weight.

Kind of animal	Dry matter	Digestible Nutrients		
		Protein	Carbohydrates	Fat
Ox at rest	18.	.7	8.	.1
Calves	30.	2.5	15.	.5
Dairy Cow:				
11 lbs. milk daily	25.	1.6	10.	.3
22 lbs. milk daily	30.	3.5	13.	.5
Horses:				
Light work	20.	1.5	9.5	.4
Moderate work	24.	2.	11.	.6
Pigs	36.	4.5	25.	.7

IMPORTANT TRUTH

The feeding standard is simply a guide that is to be used in feeding animals. Based on 1,000 pounds live weight, it is to be proportionately less for small animals and more for large ones of each class. It is not to be regarded as a feeding recipe but to be varied as circumstances indicate.

COMPOUNDING OF RATIONS

Every animal uses food for five different and distinct purposes:

1. To replace the waste of all parts of the body.

2. To produce heat to keep the body warm.

3. To make growth or increase the body in muscle, fat, flesh, bone, etc.

4. To produce energy so that work may be done.

5. To produce hair, wool, milk, etc.

To produce these requires food, protein, carbohydrates and fat. To feed these nutrients in the quantity and proportion they should be fed for one or more of the above purposes gives rise to the selection and compounding of feeding rations.

VARIETY NECESSARY

Animals, to do their work efficiently, must be properly fed. A feeding stuff is valuable only in the proportion to the quantity of digestible nutrients it contains. Fresh pasture grass is the only feed of any consequence that is balanced. For this reason rational feeding calls for a variety of food that all nutrients may be furnished in correct proportions and quantities. A ration properly compounded may mean not only better work, or more milk or more rapid development, but it may mean a saving in cost as well.

COST OF DIGESTIBLE NUTRIENTS

That the daily ration fed an animal should be furnished as cheaply as possible there is no question. Since feeding stuffs vary in cost or value, as well as in amount of digestible nutrients, it follows that in compounding rations market prices of feeds should always be considered. The farm is a factory for producing carbohydrates and fat. As much protein as possible should be grown also. Usually, however, this cannot be done sufficiently to supply every need; hence, it must be purchased else best results may not follow.

WHAT PROTEIN COSTS

It is possible for the farmer to purchase corn, oats, gluten meal, cottonseed meal, wheat bran and numerous other grains or feeding stuffs containing protein. In what form shall he purchase it?

Corn contains 7.9% digestible protein. When corn sells for $20 per ton, what is the cost of a pound of digestible protein?

7.9x20=158 pounds in a ton.

158 pounds cost $20.

1 pound costs 12.6 cents.

Cottonseed meal contains 37.2% of digestible protein. When it sells for $26 per ton, what is the cost of each pound of digestible protein?

When a pound of digestible protein costs 12.6 cents in corn and 3.4 cents in cottonseed meal, how many more times cheaper is it in the latter feeding stuff?

When gluten meal sells at $25 per ton a pound of digestible protein costs 4.9 cents.

JUDGMENT MUST BE EXERCISED

in the selection of a concentrated feeding stuff. As a rule, protein is the nutrient that is wanted and for it the purchase

is especially made. Even if the cost for a pound of nutrients in bran and cottonseed meal, for instance, were the same, cottonseed meal would be preferable, since it contains nearly three times as much protein as bran. Since protein is the most difficult nutrient to obtain, and hence the most valuable, that feeding stuff supplying it most abundantly, and cheaply, is always to be selected rather than the one where the opposite is true.

PURCHASE OF HAY

What has been said in reference to concentrates is also true with roughage feeds. Select those that furnish nutrients at least cost and at the same time contain the highest quantity of protein.

Carefully consider the protein content in the sale or purchase of hays just as you should do with the concentrates. At the same price per ton, cowpea hay and alfalfa hay are to be preferred to timothy because of the greater amount of protein and total digestible nutrients they contain. It is wise farming not only to grow all hay and roughage material for feed, but also to grow as well those kinds that are rich in protein—the costly and most important nutrient.

Speak Gently to the Erring

Speak gently to the erring.
  Ye know not all the power
With which the dark temptation came
  In some unguarded hour;
Ye may not know how earnestly
  They struggled, or how well,
Until the hour of darkness came,
  And sadly thus they fell.

Speak gently of the erring,
  Oh! do not thou forget,
However darkly stained by sin,
  He is thy brother yet.
Heir of the selfsame heritage,
  Child of the selfsame God,
He hath but stumbled in the path
  Thou hast in weakness trod.

Speak gently to the erring,
  For is it not enough
That innocence and peace are gone
  Without thy censure rough?
It surely is a weary lot,
  That sin-crushed heart to bear;
And they who share a happier fate,
  Their chidings well may spare.

Speak kindly of the erring—
  Thou yet may'st lead him back,
With holy words and tones of love,
  From misery's thorny track;
Forget not thou hast often sinned,
  And sinful yet must be;
Deal kindly with the erring one,
  As God has dealt with thee!
          —Mary Elizabeth Lee.

New times demand new measures and new
    men;
The world advances, and in time out-
    grows
The laws that in our father's days were
    best.

Breeding Guide for Farm Animals and Fowls

Period of Gestation

The average duration of the period of gestation in domestic animals is as follows:

Ass	363 days	Sow	116 days
Mare	340 "	Dog	63 "
Cow	284 "	Cat	50 "
Sheep	152 "	Rabbit	30 "
Goat	149 "	Guinea Pig	21 "
Goose	30 "	Guinea Hen	26 "
Turkey	29 "	Hen	21 "
Duck	29 "	Pigeon	18 "

Range of Variation

Mare	.	.	295 days to 370 days.
Cow	.	.	265 " " 300 "
Ewe	.	.	145 " " 154 "
Sow	.	.	110 " " 118 "
Goose	.	.	27 " " 33 "
Turkey	.	.	26 " " 30 "
Duck	.	.	26 " " 32 "
Peahen	.	.	28 " " 30 "
Guinea Hen	.	.	25 " " 26 "
Pigeon	.	.	16 " " 20 "

Wolff's Animal Gestation Calendar

Date of Serving	Mares 340 Days	Cows 284 Days	Ewes 152 Days	Sows 116 Days	Date of Serving	Mares 340 Days	Cows 284 Days	Ewes 152 Days	Sows 116 Days
1 Jan.	6 Dec.	11 Oct.	1 June	26 April	5 July	9 June	14 April	3 Dec.	28 Oct.
6 "	11 "	16 "	6 "	1 May	10 "	14 "	19 "	8 "	3 Nov.
11 "	16 "	21 "	11 "	6 "	15 "	19 "	24 "	13 "	8 "
16 "	21 "	26 "	16 "	11 "	20 "	24 "	29 "	18 "	14 "
21 "	26 "	31 "	21 "	16 "	25 "	29 "	4 May	23 "	18 "
26 "	31 "	5 Nov.	26 "	21 "	30 "	4 July	9 "	28 "	22 "
31 "	5 Jan.	10 "	1 July	26 "	4 Aug.	9 "	14 "	2 Jan.	28 "
5 Feb.	10 "	15 "	6 "	31 "	9 "	14 "	19 "	7 "	3 Dec.
10 "	15 "	20 "	11 "	5 June	14 "	19 "	24 "	12 "	8 "
15 "	20 "	25 "	16 "	10 "	19 "	24 "	29 "	17 "	13 "
20 "	25 "	30 "	21 "	15 "	24 "	29 "	3 June	22 "	18 "
25 "	30 "	5 Dec.	26 "	20 "	29 "	3 Aug	8 "	27 "	23 "
2 Mar.	4 Feb.	10 "	31 "	25 "	3 Sept.	8 "	13 "	1 Feb.	27 "
7 "	9 "	15 "	5 Aug.	30 "	8 "	13 "	18 "	6 "	1 Jan.
12 "	14 "	20 "	10 "	6 July	13 "	18 "	23 "	11 "	6 "
17 "	19 "	25 "	15 "	11 "	18 "	23 "	28 "	16 "	11 "
22 "	24 "	30 "	20 "	16 "	23 "	28 "	3 July	21 "	16 "
27 "	1 Mar.	4 Jan.	25 "	22 "	28 "	2 Sept.	8 "	26 "	21 "
1 April	6 "	9 "	30 "	26 "	3 Oct.	7 "	13 "	3 Mar.	26 "
6 "	11 "	14 "	4 Sept.	31 "	8 "	12 "	18 "	7 "	31 "
11 "	16 "	19 "	9 "	5 Aug.	13 "	17 "	23 "	13 "	5 Feb.
16 "	21 "	24 "	14 "	10 "	18 "	22 "	28 "	18 "	11 "
21 "	26 "	29 "	19 "	15 "	23 "	27 "	2 Aug.	23 "	16 "
26 "	31 "	3 Feb.	24 "	19 "	28 "	2 Oct.	7 "	28 "	21 "
1 May	5 Apr.	8 "	29 "	25 "	2 Nov.	7 "	12 "	2 Apr.	26 "
6 "	10 "	13 "	4 Oct.	30 "	7 "	12 "	17 "	7 "	2 Mar.
11 "	15 "	18 "	9 "	4 Sept.	12 "	17 "	22 "	12 "	7 "
16 "	20 "	23 "	14 "	9 "	17 "	22 "	27 "	17 "	12 "
21 "	25 "	28 "	19 "	14 "	22 "	27 "	1 Sept.	22 "	17 "
26 "	30 "	5 Mar.	24 "	19 "	27 "	1 Nov.	6 "	27 "	22 "
31 "	5 May	10 "	29 "	24 "	2 Dec.	6 "	11 "	2 May	28 "
5 June	10 "	15 "	3 Nov.	29 "	7 "	11 "	16 "	7 "	2 Apr.
10 "	15 "	20 "	8 "	4 Oct.	12 "	16 "	21 "	12 "	7 "
15 "	20 "	25 "	13 "	8 "	17 "	21 "	26 "	17 "	12 "
20 "	25 "	30 "	18 "	14 "	22 "	26 "	1 Oct.	22 "	17 "
25 "	30 "	4 April	23 "	18 "	27 "	1 Dec.	6 "	27 "	22 "
30 "	4 June	9 "	28 "	2 "	31 "	5 "	11 "	1 June	27 "

Early Newspapers

Rome and China had from an early period issues similar to newspapers. These contained reports of great military achievements and interesting news events. They were written and posted in public places. The Pekin Gazette has appeared since about 741. It is printed in a government edition and sent to officials. It consists exclusively of official news. It is probably the earliest daily in existence. The newspaper, as known to-day, is of composite origin. In the 16th century it was represented by news sheets. The first dated examples of these appeared in 1498. These small news sheets appeared in Augsburg, Vienna, Antwerp and many other cities in Europe. The first newspaper in the United States was the Publick Occurrences, in Boston. It was published in 1690. It was suppressed by the government of Massachusetts for containing "reflections of a very high nature." The Boston News-Letter appeared in 1709. The first penny paper in the United States was in the New York Daily Sun, which was established in the year 1833.

The expression, lead pencils, is misleading. Lead pencils contain no lead, properly so called, but are composed of graphite or plumbago, an allotropic form of carbon. They received their name from the leaden plummets which were used for ruling faint lines on paper, before the discovery of the mines of graphite in Cumberland.

Spraying Calendar for Fruit and Vegetables

PLANT	FIRST APPLICATION	SECOND APPLICATION	THIRD APPLICATION	FOURTH APPLICATION	FIFTH APPLICATION
APPLE (For scab, codling moth, bud moth, tent caterpillar, canker worm, plum curculio.)	Spray before buds swell with copper sulphate.	Just before blossoms open, bordeaux and paris green.	When blossoms have fallen, bordeaux and paris green.	Eight to 10 days later, bordeaux and paris green.	Use ammoniacal copper carbonate in Sept. for scab if season is wet.
BEAN (Anthracnose, leaf blight.)	When third leaf expands, bordeaux.	10 days later, bordeaux.	14 days later, bordeaux.	14 days later, bordeaux.	Spraying with bordeaux after pods are half grown will injure them for market.
CABBAGE AND CAULIFLOWER (Worms, aphis.)	When worms first appear, kerosene emulsion or paris green.	Repeat the first application when necessary.	If plants are heading, use hellebore.	After heads form, use saltpeter for worms, teaspoonful to 1 gallon water; emulsion for aphis.	
CHERRY (Rot, aphis, slug, plum curculio, black knot.)	As buds break, hordeaux; when aphis appear, kerosene emulsion.	When fruit has set, bordeaux and arsenate of lead. If slugs appear, dust leaves with air-slaked lime or hellebore.	10-14 days if rot appears, bordeaux. Arsenate of lead for plum curculio.	10-14 days later, weak solution of copper sulphate, 3 oz to 50 gals water.	Repeat after every rain when fruit begins to color.
CURRANT (Worms, leaf blight.)	Bordeaux before leaves start. At first appearance of worms, paris green.	Repeat with paris green when necessary. Ammoniacal copper carbonate for blight.	Bordeaux for blight after fruit is picked.	Use whale-oil soap for the San Jose scale if necessary.	Cut canes close if pests are bad.
GRAPE (Fungous diseases, rose-bug, etc.)	In spring when buds swell, bordeaux.	Just before flowers unfold, bordeaux and paris green.	When fruit has set, bordeaux and paris green.	2-4 weeks later, bordeaux.	Weak solution of copper sulphate.
NURSERY STOCK (Fungous diseases, San Jose scale.)	When first leaves appear, bordeaux and paris green or arsenate of lead.	Repeat at intervals of 10-14 days through the summer.	For scale, burn or fumigate with hydrocyanic acid gas.	Cut out leaf blight as fast as it appears.	Dig all trees that have crown galls.
PEACH, NECTARINE, APRICOT (Rot, mildew, scab.)	Before the buds swell, hordeaux.	Just before blossoms open, weak bordeaux (2-4-50) and arsenate of lead for curculio.	When fruit is set, weak bordeaux.	As fruit shows color, potassium sulphide, 1 lb to 50 gals water.	Repeat once or twice until fruit is ripe.
PEAR (Leaf blight, scab, psylla, codling moth, blister mite.)	As buds are swelling, bordeaux.	Just before blossoms open, bordeaux and paris green. If necessary, kerosene emulsion or whale-oil soap, when leaves open for psylla.	After blossoms have fallen, bordeaux and paris green. If necessary, kerosene emulsion or soap.	8-12 days later, repeat third.	10-20 days later ammoniacal copper carbonate.
RASPBERRY, BLACKBERRY, DEWBERRY (Rust, anthracnose, leaf blight, saw fly.)	Before buds break, bordeaux.	Bordeaux and paris green just before the blossoms open.	(Orange or red rust is treated best by destroying the plants attacked in its early stages.)	Spray after fruit is gathered with bordeaux.	10-20 days later, repeat.
STRAWBERRY (Rust, leaf blight, mildew.)	As soon as growth begins, bordeaux. Dip plant in bordeaux before setting.	When fruits are setting, bordeaux.	Spray new plantation bordeaux.	Repeat if weather is moist.	Dig the worst diseased plants.
TOMATO (Rot, blight, flea beetle.)	Soon after planting use bordeaux.	Repeat as soon as fruit is formed. Fruit can be wiped if disfigured by bordeaux.	Repeat first when necessary.	Keep the rotting fruit picked closely.	Clean up infected vines if remedies fail.
POTATO (Beetles, blight and rot.)	Spray with paris green and bordeaux when vines are small.	Repeat before insects become too numerous.	Repeat for blight and rot at intervals of 2 or 3 weeks during summer.	Spray with paris green for late bugs.	Dig early if rot is prevalent.

Formulas for Spraying

Fungicides

Bordeaux Mixture—Dissolve 6 lbs copper sulphate (blue vitriol) in 25 gals of water. Slake 4 lbs fresh stone lime and dilute to 25 gals. Strain carefully. Mix just before spraying. For peaches, plums and other tender foliage add 25 to 30 gals more water. Dissolve sulphate by hanging it in a cheesecloth bag in water.

Copper Sulphate Solution—Dissolve 2 to 4 lbs of copper sulphate in 50 gals of water as recommended for making bordeaux mixture. Use as a spray before the foliage appears. When used on foliage dilute to about 1 lb to 200 to 300 gals of water.

Ammoniacal Copper Carbonate—Mix copper carbonate 6 ozs, ammonia 3 pts and water 50 gals together as follows: Make a paste of the copper carbonate with a little water, dilute the ammonia 7 or 8 times with water and add to the paste mixture, stirring until dissolved, add the rest of the water and then use only the clear blue liquid. It loses strength if allowed to stand and should not be mixed with insecticides.

Insecticides

Paris Green—Paris green 1 lb to 100 to 200 gals water, to which add 1 lb slaked quicklime to prevent burning foliage. For tender foliage, such as that of peach trees, use the solution 1 lb to 300 gals of water. For use as a dust, mix 1 part paris green to 10 to 20 parts flour, ashes or road dust. Use london purple the same as paris green.

Kerosene Emulsion—Dissolve 1½ lbs hard soap in 1 gal boiling water and add 2 gals of kerosene or coal oil. Mix thoroughly with a pump for 5 to 10 minutes and dilute from 8 to 10 times before using. For spraying young leaves use a mixture containing 15% kerosene.

Lime Sulphur-Wash for winter application to destroy insects is made by placing 20 lbs lime and 15 lbs sulphur in a barrel containing 30 gals water and boiling them together with steam for 3 or 4 hrs. Before using, this mixture should be diluted to make 45 gals. It is most effective when sprayed warm.

Tobacco is effective against plant lice and other small insects, especially in greenhouses. Indoors they can be killed by burning tobacco stems and fumigating with the vapors. Tobacco dust and broken stems may be buried in the soil around trees infested with aphis. Make a strong decoction by soaking the stems in water and diluting the resulting solution until it is the color of ordinary tea and spray on plants affected with lice.

Arsenites of Lime and Soda—Boil 1 lb white arsenic in 4 qts water until it is dissolved, slake in this solution 2 lbs quicklime, adding water if necessary, and when slaked dilute to 2 gals. Use 1 qt to 40 gals water. Arsenite of soda is made by boiling 1 lb arsenic with 4 lbs of salsoda crystals in 2 gals of water until dissolved. Use 1 qt to 40 gals of water.

Formalin, also called formaldehyde, is used chiefly for grain smuts and potato scab. It is not poisonous, although somewhat irritating to the skin. The commercial form contains a 40% solution of the gas in water. For potatoes a solution of ½ lb in 15 gals water is best, and for grains 1 lb in 50 gals water.

Treatment for Smut and Scab

Wheat Smut—For ordinary loose smut soak the seed 4 hrs in cold water and let stand 4 hrs more in wet sacks. Then immerse 5 min in water at 133 degrees F and spread out to dry. For stinking smut use above method or immerse 10 minutes in a solution of 2 lbs blue stone to 10 gals water. Dry the grain by shoveling it over with air-slaked lime several times and then running through a fanmill.

Oat Smut—Soak seed in ¾% solution of potassium sulphide for 2 hrs, stirring slightly and then dry. Another method is to sprinkle the pile of seed with a solution of copper sulphate 1 gal to 1 bu. After 3 or 4 hrs spread it out to dry.

Potato Scab—Soak the seed for 1 hr in corrosive sublimate or for 2 hrs in formalin. Then dry and plant on a soil which is free from scab.

The Telephone

Feb 12, 1877, Prof Alexander Graham Bell's articulating telephone was tested by experiments at Boston and Salem, Mass, and was found to convey sounds distinctly from one place to the other, a distance of 18 miles. This telephone was exhibited widely in this country and Europe during that year, and telephone companies were established to bring it into general use. Edison's carbon "loud-speaking" telephone was brought out in 1878.

Planting Table for the Most Important Field Crops

A digest of the experience of the best farmers of the United States as to methods of handling the various crops. The seeding season is given for localities within 100 miles north or south of the latitude of the Ohio river.

P=Potash, Ph=Phosphoric acid and N=Nitrogen

Crop	Seed per acre	Seeding Season	Seeding Methods	Soil Requirements	Tillage Requirements	Fertilizer Requirement	Common range of yields per acre
Alfalfa	15-25 lbs	May or Aug	Broadcast	Fertile loam	Clip weeds	P and Ph	3-8 tons hay
Artichokes	2-3 bu	May-June	Hills 3 ft wide	Well-drained loam	Surface plowing	Manure	275-1000 bu
Barley	6-10 pks	Apr-May	Drill, broadcast	Fertile clay	After clover	Strong P	30-60 bu
Broom Corn	4-6 pks	May-June	6 in apart in row	Rich loam	Frequent, shallow	Manure	400-700 lbs brush
Buckwheat	2-3 pks	June-July	Drill, broadcast	Medium loam		P and lime	15-40 bu
Beans (field)	6-8 pks	June-July	4 in apart in row	Dry clay loam	Frequent, shallow	P and Pu	20-35 bu
Clover, red	10-16 lbs	Feb-May	Broadcast	Clay loam	After grain	Lime and P	1-3 tons hay, 1-4 f u seed
" alsike	12-291bs	Apr-May	Broadcast	Moist loam	With grasses	Lime and P	1-4 tons hay
" crimson	12-15 lbs	July-Sept	Broadcast	Clay loam	Fine seedbed	Lime and P	1-3 tons hay, 8-12 bu seed
Corn	1-4 pks	Apr-June	Rows or hills, 3½ ft wide	Fertile loam	Frequent, shallow	N, P and Ph	25-75 bu
Cotton	10-14 lbs	Apr-June	Rows 4 ft wide	Deep loam	Surface plowing	Ph and P and N	200-500 lbs
Cowpeas	½-2 bu	May-Aug	Rows 30 in wide or broadcast	Loose loam	Light plowing	Ph and P	8-40-bu seed,
Flax	2-8 pks	May-June	Rows 30 in wide or broadcast	Rich deep loam		N, no manure	8-16 bu seed
Grasses, orchard	2-3 bu	Apr or Oct	Broadcast	Rich clay loam		Manure	Pasture
" blue	10-15 lbs	Sept or Oct	Broadcast	Limestone clays		Manure	Pasture
Hemp	4-6 pks	Apr-July	Broadcast	Loose loam		Manure	500-1500 lbs fibre
Hops	2000 roots	Apr-May	Hills 7x7 ft	Rich loam	Light plowing	Manure	600-1200 lbs
Kaffir corn	6-7 lbs	May-July	3 ft in rows	Rich loam	Frequent, shallow	Manure	30-50 bu
Millet	2-6 pks	May-July	Broadcast	Sandy loam		Manure	20-25 bu
Oats	4-8 pks	Apr-June	Drill, broadcast	Rich loam		N, Ph or manure	30-60 bu
Peanuts	4-8 pks	May-June	Rows 3 ft wide	Sandy loam	Shallow	P and Ph	60-100 bu
Potatoes	8-10 pks	May-Aug	Rows 3 ft wide	Rich sandy loam	Frequent, shallow	Rotted manure	75-300 bu
Rape	2-3 lbs	July-Aug	Broadcast or rows 30 in wide	Rich loam		Manure	30 tons forage
Rice	1-3 bu	Mar-May	Drill, broadcast	Clay loam	Shallow	N and Ph	25-40 bu
Rye	4-8 pks	Apr-Sept	Drill, broadcast	Dry loam	Shallow	Manure	25-30 bu
Sorghum	1-2 bu	May-June	Seed broadcast	Rich loam		Manure or	8-15 tons
Soybeans	2-3 pks	May-June	Rows 25-30 in wide	Light loam		N, Ph and P	12-30 bu
Sugar Beets	15-18 lbs	Apr-May	Rows 18 in apart	Sandy loam	Frequent, shallow	P and Ph or Manure	10-14 tons
Sugar Cane	4 tons canes	Sept-Mar	Rows 5-7 ft wide	Rich loam	Shallow plowing	N and Ph	20-30 tons
Tobacco	5000 plants	May-June	Rows 3 ft wide	Rich loam	Shallow plowing	N and P and Ph	700-1200 lbs
Vetch	4-6 pks	Aug-Sept	Broadcast	Sandy loam		P and Ph	2-3 tons hay
Wheat	5-8 pks	Oct-Nov	Broadcast	Rich loam		Ph and N and P	12-30 bu

Table of Times and Seasons for Garden Planting

This table gives the best methods of planting each garden vegetable and the common ways in which each is prepared for eating, according to the experiences of leading farmers. The time of planting is designed for sections within 100 miles north or south of the latitude of the Ohio River. For localities further north or south make an allowance in the time of planting according to seasons.

Vegetable	When to Plant	How To Plant	Days to come up	Weeks before ready to eat	How prepared for eating
Asparagus	April	2 in. rows 24 in. wide	20-30	3 (years)	Boiled
Beans, Bush, String	May-July	3 in. rows 24 in. wide	8	6 to 9 weeks	Boiled, baked
Beans, Bush, Lima	May	4-6 in. rows 30 in. wide	14	8 to 10	Boiled and seasoned
Beets	May-July	3 in. rows 12 in. wide	6	7 to 8	Boiled or baked, pickled
Brussels Sprouts	May-June	18 in. rows 24 in. wide	5	21	Boiled, pickled
Cabbage	March-June	24 in. rows 24 in. wide	6	14 to 18	Boiled, salad
Carrot	May	3 in. rows 12 in. wide	8-9	12 to 15	Soup, boiled
Cauliflower	March-June	24 in. rows 24 in. wide	8	20 to 25	Pickles, boiled, with cream
Celery	April-May	8 in. rows 48-60 in. wide	20-30	20 to 30	Raw, boiled, with cream
Corn, Sweet	May-July	6 in. rows 30 or 36 in. wide	8	10 to 20	Boil on cobs, stewed
Corn, Pop	May	6 in. rows 30 or 36 in. wide	8	30 to 40	Popped over fire
Cress	April-May	6 in. rows 12 in. wide	4	5 to 6	Relish, salad
Cucumber	May-June	4 in hill 60 in each way	4-11	8 to 10	Raw, pickled
Dandelion	April-August	8 in. rows 12 to 24 in. wide	8	Next spring	Boiled, salad, greens
Egg Plant	February-May	24 in. rows 36 in. wide	11	15 to 20 weeks	Fried, baked
Endive	March-July	12 in. rows 18 to 24 in. wide	5	8 to 10	Salad
Kale	April-June	12 in. rows 24 in. wide	6	21 to 30	Greens
Kohl Rabi	March-July	6 in. rows 12 in. wide	5	12 to 14	Boiled, mashed
Lettuce	April-May	6 in. rows 12 in. wide	4-8	8 to 10	Raw, relish, salad, boiled
Leek	May-June	3 in. rows 12 in. wide	10	16 to 20	Used in soup
Melon, Musk	May-June	4 in hill 60 in. apart	14	14 to 18	Raw
Melon, Water	May-July	4 in hill 60 in. apart	14	15 to 20	Raw
Mustard	April-May	6 in. rows 12 in. wide	4	3 to 5	Relish, salad, greens
Onions	April-May	3 in. rows 12 in. wide	10	15 to 25	Boiled, fried, baked
Okra	May	24 in. rows 36 in. wide	10-20	12 to 14	Used in soup
Parsley	March-April	12 in. rows 12 in. wide	10-20	3 to 12	Garnishing, soups and salads
Parsnip	May	3 in. rows 12 in. wide	14	20	Boiled, fried
Pepper	February-May	12 in. rows 12 in. wide	20	6 to 8	Stuffed, baked, pickled
Peas	April-July	1-2 in. rows 24 in. wide	20-40	10 to 20	Boiled
Potatoes	April-June	12 in. hill 30 in. wide	14-30	20	Boiled, baked, fried
Pumpkin	May-June	in hill 60 in. wide	20	3 to 6	Pies
Radish	April-August	1 in. rows 12 in. wide	11	25	Relish, raw
Salsify	May-June	2 in. rows 12 in. wide	4	6 to 8	Soup
Spinach	April-May	6 in. rows 12 in. wide	8	9 to 12	Boiled
Squash, Summer	May-June	4 in. hill 60 in. wide	6	16 to 18	Boiled
Tomatoes	February-May	2 in. rows 36 in. wide	11	16 to 18	Raw, sliced, stewed
Turnip	April-July	4 in. rows 12 in. wide	7	8 to 12	Boiled, mashed

Best Annuals for Cut Flowers in Summer

Common Name	Scientific Name	Height (inches)	Duration of bloom (weeks)
BLUE			
Ageratum	Ageratum conyzoides	10	All summer
Giant comet aster	Callistephus hortensis	15	Four
Victoria aster	Callistephus hortensis	18	Four
Jubilee aster	Callistephus hortensis	24	Four
Cornflower	Centaurea Cyanus	24	Till frost
Navy Blue sweet pea	Lathyrus odoratus	60	Eight
Dwarf lobelia	Lobelia Erinus	6	All summer
YELLOW			
Giant-flowering snapdragon	Antirrhinum majus	24	Eight
Klondike cosmos	Cosmos sulphureus	48	Six
Stella sunflower	Helianthus debilis, var. Stella	36	Ten
African marigold	Tagetes erecta	24	Twelve
Tom Thumb nasturtium	Tropæolum minus	12	Twelve
Double Mammoth zinnia	Zinnia elegans	24	Fifteen
PINK			
Victoria aster	Callistephus hortensis	18	Four
Branching aster	Callistephus hortensis	24	Four
Jubilee aster	Callistephus hortensis	24	Four
Clarkia	Clarkia elegans	18	Six
La Malmaison balsam	Impatiens Balsamina	20	Eight
Blanche Ferry sweet pea	Lathyrus odoratus	48	Eight
Sander's tobacco	Nicotiana Sanderæ	30	Eight
Drummond's phlox	Phlox Drummondii	18	Twelve
Mammoth verbena	Verbena hybrida	12	Ten
WHITE			
Giant-flowering snapdragon	Antirrhinum majus	24	Eight
White Bentley aster	Callistephus hortensis	24	Four
Annual chrysanthemum	Chrysanthemum coronarium	18	Twelve
Giant-flowering cosmos	Cosmos hipinnatus	48	Six
Baby's breath	Gypsophila paniculata	12	Twelve
Emily Henderson sweet pea	Lathyrus odoratus	60	Six
Ten weeks stock	Matthiola incana, var. annua	15	Ten
White swan poppy	Papaver somniferum	24	Five
RED AND SCARLET			
Giant aster	Callistephus hortensis	24	Four
Jubilee aster	Callistephus hortensis	24	Four
Victoria aster	Callistephus hortensis	18	Four
Giant-flowering cosmos	Cosmos hipinnatus	48	Six
Salopian sweet pea	Lathyrus odoratus	60	Eight
Ten weeks stock	Matthiola incana, var. annua	15	Ten
Bonfire salvia	Salvia splendens	24	Twelve
LILAC AND PURPLE			
Peony-flowered aster	Callistephus hortensis	24	Four
Late-branching aster	Callistephus hortensis	24	Four
Victoria aster	Callistephus hortensis	18	Four
Double purple balsam	Impatiens Balsamina	20	Eight
Ten weeks stock	Matthiola invana, var. annua	15	Ten
Giant Bluebird petunia	Petunia hybrida	18	Ten
Carnation-flowered poppy	Papaver somniferum, var. fimbriatum	24	Five
Peony-flowered poppy	Papaver somniferum, var. pæoniæflorum	24	Five
Mammoth verbena	Verbena hybrida	12	Ten

Profitable Life of Fruit Plants

Apple	25 to 40 years
Blackberry	6 to 12 "
Currant	20 "
Gooseberry	20 "
Orange and Lemon	50 or more "
Peach	8 to 12 "
Pear	50 to 75 "
Persimmon, or Kaki	25 to 40 "
Plum	20 to 25 "
Raspberry	6 to 12 "
Strawberry	1 to 3 "

The Feet of the World

The French foot is narrow and long.

The Spanish foot is small and elegantly curved.

The Arab's foot is proverbial for its high arch. The Koran says that a stream of water can run under the true Arab's foot without touching it.

The foot of the Scotch is high and thick.

The Irish foot is flat and square.

The English foot is short and fleshy.

When Athens was in her zenith, the Grecian foot was the most perfectly formed and exactly proportioned of that of any of the human race.

Swedes, Norwegians and Germans have the largest feet; Americans the smallest. Russian toes are "webbed" to the first joint, it is said. Tartarian toes are all the same length.

Speed Records of Horses

The world records of speed by horses were made by the following, for 1 mile:

Running—Salvator, at Monmouth Park, N J, Aug 28, 1890; 1.35½.

Pacing—Dan Patch, at St Paul, Minn, Sept 8, 1906; 1.55.

Trotting—Lou Dillon, at Memphis, Tenn, Oct 24, 1905; 1.58½.

Why should a farmer be satisfied with a small crop, when a little effort and a little common-sense use of what the experiment stations are teaching us will produce a big one? This picture talks.

SOME INTERESTING FERTILIZATION COMPARISONS

Produce of one ton of untreated yard manure. *Produce of one ton of untreated stall manure.* *Produce of one ton of stall manure treated with floats.*

Religion of Presidents

The religious connection of the various presidents of the United States follows: George Washington, Episcopalian; John Adams, Congregationalist; Thomas Jefferson, Liberal (Randall, the biographer of Jefferson, declares that he was a believer in Christianity, although not a sectarian); James Madison, Episcopalian; James Monroe, Episcopalian; John Quincy Adams, Congregationalist; Andrew Jackson, Presbyterian; Martin Van Buren, Reformed Dutch; William Henry Harrison, Episcopalian; John Tyler, Episcopalian; James Knox Polk, Presbyterian; Zachary Taylor, Episcopalian; Millard Fillmore, Unitarian; Franklin Pierce, Episcopalian; James Buchanan, Presbyterian; Abraham Lincoln, Presbyterian; Andrew Johnson, Methodist (not a church member, but was a Christian believer; his wife was a Methodist); Ulysses S. Grant, Methodist; Rutherford B. Hayes, Methodist; James Abram Garfield, Disciples; Chester Alan Arthur, Episcopalian; Grover Cleveland, Presbyterian; Benjamin Harrison, Presbyterian; William McKinley, Methodist; Theodore Roosevelt, Reformed Dutch; William H. Taft, Unitarian.

Carnegie Hero Fund

Andrew Carnegie created a fund of $5,000,000 in April, 1904, for the benefit of "the dependents of those losing their lives in heroic effort to save their fellow men, or for the heroes themselves if injured only." Provision was also made for medals to be given in commemoration of heroic acts. The endowment known as "The Hero Fund" was placed in the hands of a commission composed of 20 persons, residents of Pittsburg, Pa. of which F. M. Wilmot is secretary and manager of the fund.

Abraham Lincoln was the first president of the United States to name the last Thursday in November as a day for general Thanksgiving. This was done in 1863. Since that time, the presidents have issued proclamations for the same purpose, although Thanksgiving day is not, in fact, a legal national holiday. But all states observe it.

The geographical center of the United States is Kanopolis, Kan. The center of population in the United States, according to the last census, that of 1900, was 6 miles southeast of Columbus, Ind.

Wages and Cost of Living Comparisons

Tables Showing How the Income and Expenses of Workingmen Have Advanced in Recent Years — Wages

Per cent of increase or decrease in wages per hour and in hours of labor per week in 1907 as compared with the average for 1890-99, by industries:

INDUSTRY	WAGES PER HR. % of increase	HOURS PER WK. % of decrease
Agricultural implements.........	30.9	3.7
Bakery, bread....................	28.9	8.4
Blacksmithing and horseshoeing..	26.4	5.9
Boots and shoes.................	24.3	4.0
Brick...........................	22.7	1.5
Building trades.................	44.6	9.4
Candy...........................	24.4	none
Carpets.........................	17.1	1.4
Carriages and wagons............	18.3	4.0
Cars, steam railroad............	24.4	4.1
Clothing, factory product.......	15.8	3.3
Cotton goods....................	57.5	3.2
Dyeing, finishing, and printing textiles........................	11.3	.7
Electrical apparatus and supplies.	22.6	6.7
Flour...........................	16.0	3.3
Foundry and machine shop........	21.4	5.4
Furniture.......................	27.1	4.3
Gas.............................	7.7	3.9
Glass...........................	29.4	1.4
Harness.........................	23.5	4.1
Hats, fur.......................	26.4	8.4
Hosiery and knit goods..........	33.4	2.3
Iron and steel, bar.............	40.4	2.1
Iron and steel, Bessemer converting.	32.6	9.5
Iron and steel, blast furnace....	19.8	inc. 0.6
Leather.........................	11.8	.1
Liquors, malt...................	32.9	13.0
Lumber..........................	27.6	3.1
Marble and stone work...........	25.7	6.4
Paper and wood pulp.............	33.3	10.2
Planing mill....................	24.6	3.6
Pottery.........................	13.8	.2
Printing and binding, book and job	31.0	9.9
Printing, newspaper.............	22.6	5.2
Shipbuilding....................	20.9	4.3
Silk goods	16.9	2.4
Slaughtering and meat packing...	16.0	—
Streets and sewers, contract work	54.7	7.3
Streets and sewers, municipal work	21.6	9.5
Tobacco, cigars.................	32.4	.5
Woolen and worsted goods.......	31.9	2.0
All industries.................	28.8	5.0

Cost of Living

Average expenditures of 2,567 workingmen's families for each of the principal terms entering into cost of living, and per cent of average total expenditure, 1901.

ITEMS OF EXPENDITURE	Expenditure based on all families Average total	% of exp.
Food...........................	$326.90	42.54
Rent...........................	99.49	12.95
Mortgage:		
Principal..................	8.15	1.06
Interest...................	3.96	.52
Fuel...........................	32.23	4.19
Lighting......................	8.15	1.06
Clothing:		
Husband...................	33.73	4.39
Wife......................	26.03	3.39
Children..................	48.08	6.26
Taxes.........................	5.79	.75
Insurance:		
Property..................	1.53	.20
Life......................	19.44	2.53
Organizations:		
Labor.....................	3.87	.50
Other.....................	5.18	.67
Religious Purposes............	7.62	.99
Charity.......................	2.39	.31
Furniture and utensils........	26.31	3.42
Books and newspapers..........	8.35	1.08
Amusements and vacation.......	12.28	1.60
Intoxicating liquors..........	12.44	1.62
Tobacco.......................	10.93	1.42
Sickness and death............	20.54	2.67
Other purposes................	45.13	5.87
Total........................	$768.54	$100.00

Food Cost

Average cost per workingman's family in the United States of principal articles of food during one year is shown in the following table:

Fresh beef.......$50.05		Coffee...........	10.74
Salt beef	5.26	Sugar...........	15.76
Fresh hog products........	14.02	Molasses.......	1.69
		Flour and meal..	16.76
Salt hog products	13.89	Bread..........	12.44
Other meat......	9.78	Rice...........	2.05
Poultry.........	9.49	Potatoes.......	12.93
Fish...........	8.01	Other vegetables.	18.85
Eggs...........	16.79	Fruit..........	16.52
Milk...........	21.32	Vinegar, pickles and condiments	4.12
Butter.........	28.76	Other food......	20.40
Cheese.........	2.62		
Lard...........	9.35		
Tea............	5.30	Total.....$326.90	

Food Cost in Different Parts of the Country

Average food cost per workingman's family, by geographical divisions, for each year, 1890 to 1907. (Based on the average cost per family in 1901 and the course of retail prices of food as indicated by the relative prices weighted according to family consumption.)

	North Atlantic Division	South Atlantic Division	North Central Division	South Central Division	Western Division	United States
1890....	$330.35	$282.72	$310.08	$279.54	$332.61	$318.20
1891....	333.26	285.72	316.75	283.64	335.72	322.55
1892....	329.70	282.44	308.57	275.71	324.90	316.65
1893....	337.13	288.30	319.48	283.37	317.60	324.41
1894....	320.34	279.36	304.93	273.79	306.68	309.81
1895....	315.50	275.73	297.05	268.59	296.65	303.91
1896....	313.23	270.42	286.74	263.11	287.84	296.76
1897....	312.91	271.26	289.77	266.40	286.29	299.24
1898....	319.05	277.41	298.26	270.50	294.01	306.70
1899....	321.31	280.76	299.78	273.51	304.21	09.19
1900....	326.80	286.07	305.54	276.80	302.97	314.16
1901....	338.10	298.64	321.60	298.60	306.53	326.90
1902....	356.83	312.33	338.57	310.75	322.43	344.61
1903....	355.54	310.65	336.45	310.75	320.27	342.75
1904....	360.70	312.61	339.79	314.86	323.97	347.10
1905....	362.00	315.68	342.82	317.32	326.44	349.27
1906....	370.72	324.62	353.12	325.81	340.03	359.53
1907....	385.57	341.66	367.37	341.14	358.87	374.75

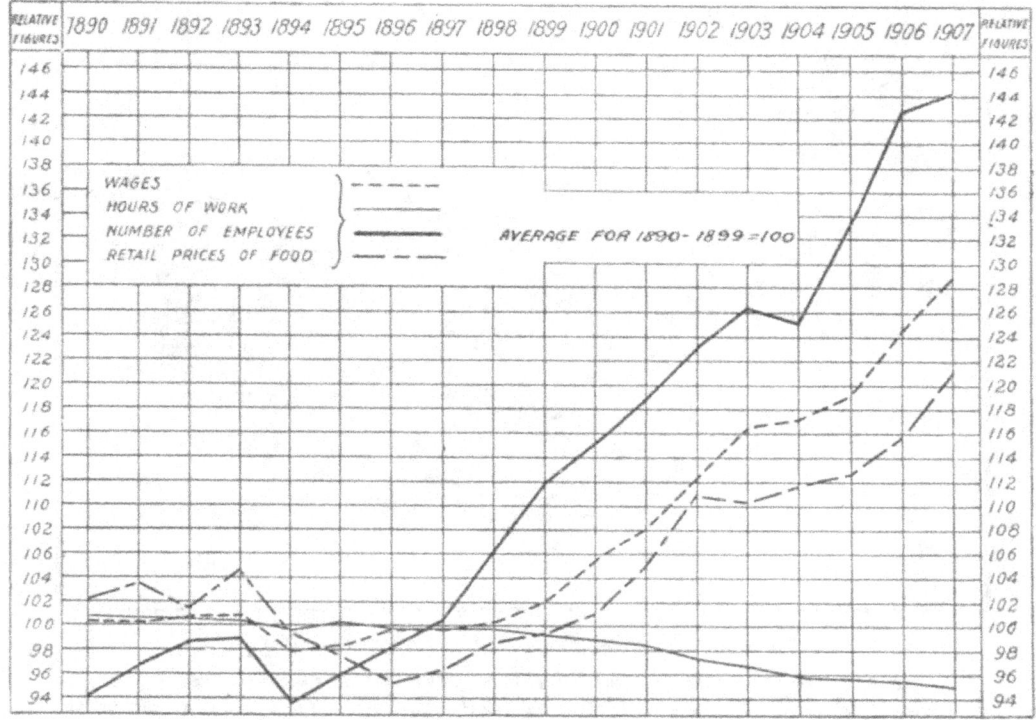

RELATIVE RANGE OF WAGES AND FOOD PRICES 1890-1907

Chart scale is based on the average figures for the years from 1890 to 1900, which is represented by 100. Note how workers increased and hours grew less

Great Automobile Race

The greatest automobile race ever undertaken was from New York to Paris in 1908. An American automobile won. The first lap of the race, across the United States, was won by the American car. It distanced all competitors. There was a misunderstanding about the route to be taken and the American car ran away up into Alaska, under the theory that the course was to be by way of Behring Strait to Siberia. The course finally taken was by steamer from San Francisco to Vladivostok, Siberia, thence to European Russia and through Germany to Paris.

The German car had bad luck all the way across the United States. It was overloaded with supplies. A blizzard was encountered in Indiana that bothered all the automobiles in the race and especially the overloaded German car. In Iowa its tires gave out and extra ones were lost. In Utah the frame of the car was broken and two cylinders were blown out.

In order to catch the same steamer upon which the American car sailed, the German car was sent by freight from Idaho to the coast. The judges of the race penalized the German contestants for taking to the railroad by imposing a 15-day handicap. The American car was given a 15-day allowance for its journey from Seattle to Alaska and back. In order to win over the American car, the German car would thus have had to reach the destination 30 days ahead. This it could not do.

Across Asia, some of the time, the American car led, but good luck attended the German car most of the time after reaching Siberia, and it finally gained the lead, which it held to the finish, arriving in Paris on July 30. About 11,000 miles were traversed by the German car in 130 days; the American car covered 12,000 miles in 108 days. In Russia the Germans had the advantage of knowledge of the roads, while the Americans lost their way and were delayed by broken gear. Otherwise, the American car would have reached Paris ahead of the German car.

We are placed on earth to be happy a little while to prepare for the eternal happiness above.

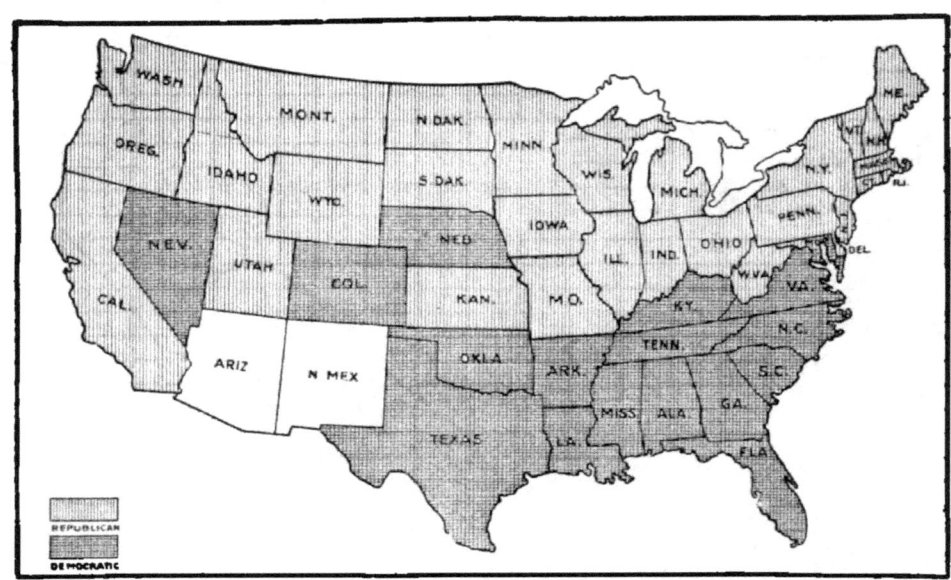

POLITICAL MAP OF THE UNITED STATES IN 1908

The dark states were carried by Bryan, the light shaded ones by Taft. The electoral vote of Maryland was divided, although Taft had a plurality of the popular vote of the state.

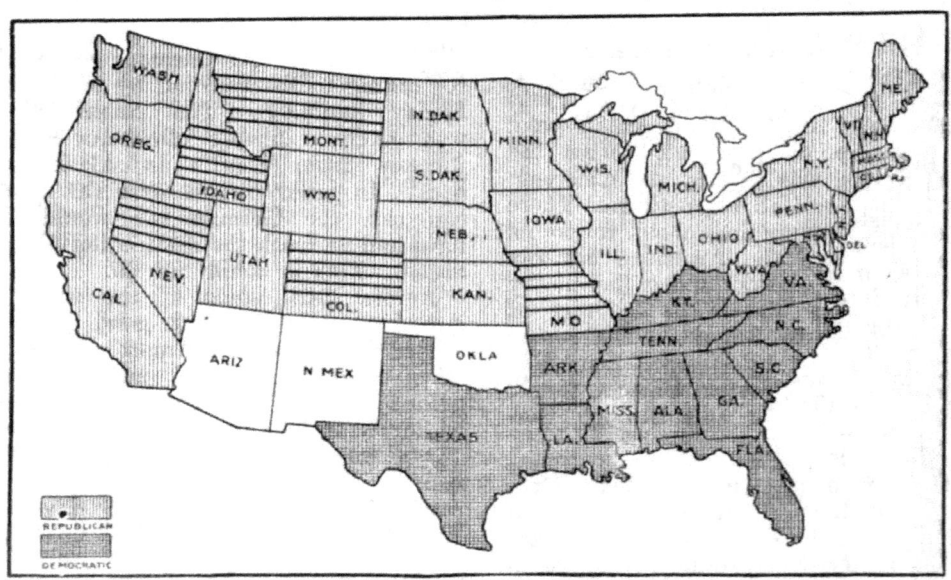

POLITICAL MAP OF THE UNITED STATES IN 1904

The dark states were classed as democratic. Parker carried all of them in 1904, while the light shaded states were all carried by Roosevelt. The states marked with heavy parallel lines—Colorado, Idaho, Missouri, Montana and Nevada—were carried by Bryan in 1900 and lost by Parker in 1904. Oklahoma had not been admitted to the union in 1904, and so had no electoral vote.

The National Elections of 1908 in Brief Comprehensive Review

Electoral and Popular Votes Tabulated and Compared—Ups and Downs of the Minor Parties—Interesting and Mixed Results in Various States

The national election, November 3, 1908, resulted in the election of electors favorable to William H. Taft, republican, for president by a popular plurality over William J. Bryan, democrat, of 1,244,494. Mr Taft's electoral vote plurality was 167. The following table shows the electoral vote of the various states in 1908, compared with the vote of the three last preceding elections:

Electoral Vote

	—1908—		—1904—		—1900—	
	Taft	Bryan	Roosevelt	Parker	McKinley	Bryan
Alabama..........	—	11	—	11	—	11
Arkansas..........	—	9	—	9	—	8
California........	10	—	10	—	9	—
Colorado.........	—	5	5	—	—	4
Connecticut......	7	—	7	—	6	—
Delaware.........	3	—	3	—	3	—
Florida..........	—	5	—	5	—	4
Georgia..........	—	13	—	13	—	13
Idaho............	3	—	3	—	—	3
Illinois..........	27	—	27	—	24	—
Indiana..........	15	—	15	—	15	—
Iowa.............	13	—	13	—	13	—
Kansas...........	10	—	10	—	10	—
Kentucky........	—	13	—	13	—	13
Louisiana........	—	9	—	9	—	8
Maine...........	6	—	6	—	6	—
Maryland........	6	2	1	7	8	—
Massachusetts....	16	—	16	—	15	—
Michigan........	14	—	14	—	14	—
Minnesota.......	11	—	11	—	9	—
Mississippi......	—	10	—	10	—	9
Missouri.........	18	—	18	—	—	17
Montana.........	3	—	3	—	—	3
Nebraska........	—	8	8	—	—	8
Nevada..........	—	3	3	—	—	3
New Hampshire..	4	—	4	—	4	—
New Jersey.......	12	—	12	—	10	—
New York........	39	—	39	—	36	—
North Carolina...	—	12	—	12	—	11
North Dakota....	4	—	4	—	3	—
Ohio.............	23	—	23	—	23	—
Oklahoma........	—	7	—	—	—	—
Oregon..........	4	—	4	—	4	—
Pennsylvania.....	34	—	34	—	32	—
Rhode Island.....	4	—	4	—	4	—
South Carolina...	—	9	—	9	—	9
South Dakota....	4	—	4	—	4	—
Tennessee........	—	12	—	12	—	12
Texas...........	—	18	—	18	—	15
Utah............	3	—	3	—	3	—
Vermont.........	4	—	4	—	4	—
Virginia.........	—	12	—	12	—	12
Washington......	5	—	5	—	4	—
West Virginia....	7	—	7	—	6	—
Wisconsin.......	13	—	13	—	12	—
Wyoming........	3	—	3	—	3	—
Totals.........	325	158	336	140	292	155

Total electoral vote, 1908.................... 483
Necessary for choice......................... 242
Taft's plurality over Bryan.................. 167

The total popular vote of the various presidential candidates at the national election gave Taft a plurality over Bryan of 1,244,494. The totals show the following votes cast:

Taft, Republican	7,637,676
Bryan, Democrat	6,393,182
Debs, Socialist	448,453
Chafin, Prohibition	241,252
Hisgen, Independence	83,186
Watson, Populist	33,871
Gilhaus, Socialist-Labor	15,421
Total for all candidates	14,852,239

This grand total exceeds by 1,341,531 the total number of votes cast in the presidential election of 1904, when the grand total was 13,510,708. Compared with that election, the candidates of the republican, democratic and socialist parties increased their vote in 1908. The reverse is true of the candidates of the prohibition, populist and socialist-labor parties. The independence party did not figure in the presidential election of 1904. T..e biggest difference in a party vote is shown in an increase for Bryan of 1,315,211 over the total vote cast in 1904 for Alton B. Parker, the democratic candidate. Taft received 14,190 votes more than were polled for Roosevelt in 1904, and Debs ran 45,368 ahead of his predecessor on the socialist ticket.

Small Parties Gain Little

The heaviest loss is shown by the populists, who, with the same candidate, registered 83,312 votes less this year than in 1904, when their total was 117,183. The prohibitionists' candidate, Chafin, ran 17,284 votes behind the 1904 mark of his party, in spite of a vigorous campaign.

New York 870,070, Pennsylvania 745,779 and Illinois 629,932, gave Taft the greatest number of votes among the states, while as to Bryan the order was, New York 667,468, Ohio 502,721 and Illinois 450,810. Debs received his largest vote in Pennsylvania, 39,913; Illinois 39,711 and New York 38,451. Hisgen's vote in New York, 35,817, and in his own state of Massachusetts, 19,237, together formed nearly two-thirds of all the votes received in the whole country.

Results in Close States

Most of the so-called doubtful states, including New York, New Jersey, Connecticut, West Virginia, Ohio, Indiana, South Dakota, Missouri and Montana, were carried by Taft. Bryan had the satisfaction of carrying his own state of Nebraska, the new state of Oklahoma, Colorado, Nevada and all the southern states that Parker carried in 1904. The electoral vote in Maryland was divided, Taft securing 2 electors and Bryan 6. At the same time Taft received a plurality of the popular vote of 136 over Bryan.

The election in Indiana was interesting; Taft carried the state by a safe plurality, but the state ticket of the democratic party was victorious, and 11 democratic congressmen were elected to only 2 republican, a democratic gain of 7 members. The governor elected was Thomas R. Marshall. As the legislature is apparently democratic, United States Senator Hemenway will probably be succeeded by a democrat.

The democratic victory in Nebraska was complete. Shallenberger was elected governor and the legislature is now democratic.

In addition to Indiana, states that were carried by Taft and at the same time elected democratic governors, were Ohio, which elected Harmon; Minnesota, which elected Johnson; and North Dakota which elected Burke.

Changes in United States Senate

The election of members of the state legislatures has its important bearing on the United States senate. Of the 92 members of that body 31 will have their terms expire March 3, 1909. Of these 19 are republicans and 12 democrats. Alabama, Arkansas, Louisiana and Maryland have already chosen democrats, and Kentucky and Vermont republicans. Republican successors are practically certain to be

The Popular Vote for President by States

STATES	Taft (Rep.)	Bryan (Dem.)	Debs (Soc.)	Hisgen (Ind.)	Chafin (Pro.)	Watson (Pop.)	Gilhaus (Soc. L.)	Pres. plurality in 1904 Rep.	Dem.
Alabama	25,308	74,374	——	495	665	1,565	——	——	57,385
Arkansas	56,947	87,043	5,000	500	1,000	500	——	——	17,574
California	182,064	107,770	18,736	4,377	6,443	——	——	115,822	——
Colorado	123,700	126,649	7,974		5,559	——	——	34,582	
Connecticut	112,815	68,255	5,113	728	2,380	——	608	38,180	——
Delaware	25,200	22,134	75	50	650	——	——	4,354	——
Florida	10,654	31,104	3,747	553	1,356	1,946	——	——	18,733
Georgia	41,692	72,350	584	77	1,059	16,965	——	——	59,460
Idaho	50,091	34,609	6,305	207	1,704	——	——	29,303	——
Illinois	629,932	450,810	39,711	7,724	29,364	633	1,680	305,039	——
Indiana	348,993	338,262	13,476	514	18,045	1,193	——	93,944	——
Iowa	275,210	200,771	8,287	404	9,837	261	——	158,766	——
Kansas	197,166	161,209	12,240		5,032	——	——	126,093	——
Kentucky	235,711	249,092	4,000	200	5,887	333	404	——	11,893
Louisiana	9,958	63,568	2,538	73	——	——	——	——	42,542
Maine	66,987	35,463	1,758	790	1,487	——	——	36,791	——
Maryland	111,253	111,117	2,500	450	3,000	——	643	51	——
Massachusetts	265,966	155,543	10,659	19,237	4,374	——	1,011	92,076	——
Michigan	333,313	174,313	11,527	734	16,705	——	1,086	227,715	——
Minnesota	195,786	109,433	14,469	523	10,114	——	——	161,454	——
Mississippi	4,463	64,250	1,408			1,309	——	——	50,189
Missouri	346,915	345,884	15,398	397	4,222	1,165	867	13,159	——
Montana	32,333	29,326	5,855	443	827	——	——	13,159	——
Nebraska	126,608	130,781	3,524			5,179	——	86,682	——
Nevada	10,214	10,655	2,029	415		——	——	2,385	——
New Hampshire	53,144	33,655	1,299	584	905	——	——	20,185	——
New Jersey	265,298	182,522	10,249	2,916	4,930	——	2,916	80,598	——
New York	870,070	667,468	38,451	35,817	22,667	——	3,877	175,552	——
North Carolina	114,887	136,928	345			——	——	——	41,679
South Dakota	57,771	32,909	2,405	38	1,453	——	——	38,322	——
Ohio	572,312	502,721	33,795	439	11,402	160	721	255,421	——
Oklahoma	110,550	123,007	21,752	274		434	——	——	——
Oregon	62,350	38,049	7,430	289	2,682	——	——	42,934	——
Pennsylvania	745,779	448,785	39,913	1,057	36,394	——	1,222	505,519	——
Rhode Island	43,942	24,706	1,365	1,005	1,016	——	——	16,766	——
South Carolina	3,847	62,289	101	43	——	——	——	——	50,009
South Dakota	67,466	40,266	2,846	88	4,309	——	——	50,114	——
Tennessee	118,287	135,630	1,878	332	360	1,081	——	——	26,284
Texas	69,229	227,264	8,524	164	1,792	1,042	3,361	——	115,958
Utah	61,028	42,601	4,895	87		——	——	29,033	——
Vermont	39,552	11,496	820	804	799	——	——	30,682	——
Virginia	52,573	82,946	255	51	1,111	105	25	——	32,768
Washington	106,062	58,383	14,777	248	4,700	——	——	73,442	——
West Virginia	137,869	111,418	3,676	46	5,107	——	——	31,765	——
Wisconsin	248,673	166,707	28,146		11,579	——	——	156,057	——
Wyoming	17,708	12,772	1,396	63		——	——	11,559	——
Totals	7,637,676	6,393,182	448,471	83,186	241,452	33,871	15,421	3,069,992	524,432

All candidates, 14,852,239.

elected to succeed the republicans, the late Senator Allison of Iowa, Ankeny of Washington, Brandegee of Connecticut, Gallinger of New Hampshire, Hansbrough of North Dakota, Heyburn of Idaho, Hopkins of Illinois, Kittridge of South Dakota, Long of Kansas, Penrose of Pennsylvania, Perkins of California, Platt of New York, Smoot of Utah, Stephenson of Wisconsin.

Democrats will continue to hold the places of the democrats Clay of Georgia, Gary of South Carolina, Gore of Oklahoma, Milton of Florida and Overman of North Carolina. The senatorships in doubt are those of Ohio, Indiana and Oregon, now held by Foraker, Hemenway and Fulton, republicans, and Colorado, Missouri and Nevada, held by Teller, Stone and Newlands, democrats.

Republicans Control Congress

The 61st congress will contain about the same republican majority as the 60th congress. In spite of a bitter fight against Speaker Cannon, he was re-elected from his district in Illinois. Overstreet of Indiana and Hepburn of Iowa were, perhaps, the most conspicuous among the republicans who failed of re-election.

Failed to Deliver Labor Vote

One of the interesting features of the election was the absolute failure of Samuel Gompers, president of the American federation of labor, to deliver the labor vote to Bryan. The trades union people turned down their leader, and there are many expressions of resentment that he should have attempted to commit the national organization to any candidate or to any political party.

The Landslide in New York

The election in New York was the subject of widespread interest. Taft carried the state by 203,000, more than 28,-000 over Pres Roosevelt's plurality four years before. He even carried the strongly democratic city of New York. At the same time Gov Hughes was re-elected by over 71,000 plurality.

Politics of Governors in 1909

Republican

California	New Jersey
Connecticut	New York
Delaware	Oregon
Idaho	Pennsylvania
Illinois	Rhode Island
Iowa	South Dakota
Kansas	Utah

Kentucky	Vermont
Maine	Washington
Massachusetts	West Virginia
Michigan	Wisconsin
Missouri	Wyoming
New Hampshire	Total, 25.

Democratic

Alabama	Nebraska
Arkansas	Nevada
Colorado	North Carolina
Florida	North Dakota
Georgia	Ohio
Indiana	Oklahoma
Louisiana	South Carolina
Maryland	Tennessee
Minnesota	Texas
Mississippi	Virginia
Montana	Total, 21.

Legislatures and Elections

The following table shows when the next sessions of the various state legislatures begin and the time of the next state election. Sessions of legislatures are held biennially except where *, indicating annually, follows date of next session, and Alabama's, whose sessions are quadrennial.

State	Next Session Begins	Next State Election
Alabama	Jan. 9, 1911	Nov. 8, 1910
Arizona	Jan. 18, 1909	Nov. 8, 1910
Arkansas	Jan. 11, 1909	Sept. 12, 1910
California	Jan. 5, 1909	Nov. 8, 1910
Colorado	Jan. 12, 1909	Nov. 8, 1910
Connecticut	Jan. 6, 1909	Nov. 8, 1910
Delaware	Jan. 5, 1909	Nov. 8, 1910
Florida	April 2, 1909	Nov. 8, 1910
Georgia	June 23, 1909*	Oct. 1, 1909
Idaho	Jan. 4, 1909	Nov. 8, 1910
Illinois	Jan. 6, 1909	Nov. 8, 1910
Indiana	Jan. 7, 1909	Nov. 8, 1910
Iowa	Jan. 11, 1909	Nov. 8, 1910
Kansas	Jan. 12, 1909	Nov. 8, 1910
Kentucky	Jan. 4, 1910	Nov. 2, 1909
Louisiana	May 9, 1910	April 19, 1910
Maine	Jan. 6, 1909	Sept. 12, 1910
Maryland	Jan. 5, 1910	Nov. 8, 1910
Massachusetts	Jan. 6, 1909*	Nov. 8, 1909
Michigan	Jan. 6, 1909	Nov. 8, 1910
Minnesota	Jan. 5, 1909	Nov. 2, 1909
Mississippi	Jan. 4, 1910	Nov. 2, 1909
Missouri	Jan. 6, 1909	Nov. 8, 1910
Montana	Jan. 4, 1909	Nov. 8, 1910
Nebraska	Jan. 5, 1909	Nov. 8, 1910
Nevada	Jan. 16, 1909	Nov. 8, 1910
New Hampshire	Jan. 6, 1909	Nov. 8, 1910
New Jersey	Jan. 5, 1909*	Nov. 2, 1909
New Mexico	Jan. 18, 1909	Nov. 8, 1910
New York	Jan. 6, 1909*	Nov. 2, 1909
North Carolina	Jan. 6, 1909	Nov. 8, 1910
North Dakota	Jan. 5, 1909	Nov. 8, 1910
Ohio	Jan. 6, 1909	Nov. 8, 1910
Oklahoma	Jan. 5, 1909	Nov. 8, 1910
Oregon	Jan. 5, 1909	June 6, 1910
Pennsylvania	Jan. 5, 1909	Nov. 8, 1910
Rhode Island	Jan. 12, 1909*	Nov. 2, 1909
South Carolina	Jan. 5, 1909*	Nov. 2, 1909
South Dakota	Jan. 4, 1909	Nov. 8, 1910
Tennessee	Jan. 4, 1909	Nov. 8, 1910
Texas	Jan. 12, 1909	Nov. 8, 1910
Utah	Jan. 11, 1909	Nov. 8, 1910
Vermont	Oct. 5, 1910	Sept. 6, 1910
Virginia	Jan. 12, 1910	Nov. 2, 1909
Washington	Jan. 11, 1909	Nov. 8, 1910
West Virginia	Jan. 13, 1909	Nov. 8, 1910
Wisconsin	Jan. 14, 1909	Nov. 2, 1910
Wyoming	Jan. 12, 1909	Nov. 1, 1910

PROF B. W. KILGORE.

A southern man, long identified with southern agriculture, is Director Kilgore of North Carolina. He is one of the first men to establish test farms that every part of the state might receive the advantages resulting from experiment stations under home environments.

State Fair Dates

The following 1909 fair dates have been set: August 30, Iowa, Ohio and Topeka, Kan. September 6, Minnesota, Nebraska, Indiana and Michigan. September 13, Wisconsin, Kentucky, South Dakota, New York, Oregon, Hutchinson, Kan, and Colorado. September 20, Spokane, Portland, Sioux City, Ia, St Joseph, Mo, and Nashville, Tenn. September 27, Memphis, Tenn, Montana, Seattle, Washington and Illinois. October 4, Missouri, Oklahoma and Utah.

Women Voting

In the United States women possess suffrage upon equal terms with men at all elections in four states—in Wyoming, established in 1869; in Colorado, in 1893; in Utah, in 1896, and in Idaho, in 1896. In 1905 the Kansas and Montana legislatures, and in 1906 the Rhode Island legislature rejected bills giving women full suffrage. In June, 1906, Oregon refused to adopt a woman suffrage amendment to its constitution by a vote of 47,075 to 36,902. In Kansas

PROF FRANK WILLIAM RANE

Massachusetts never had an abler State Forester than she now has in Prof Rane, who is one of the foremost authorities in his profession in the country. For many years Prof Rane has been identified with horticulture and forestry, and his work has borne fruit of national prominence.

women possess school suffrage, established in 1861, and municipal suffrage, established in 1887. In 18 additional states women possess school suffrage —in Michigan and Minnesota, established in 1875; in New Hampshire and Oregon, in 1878; in Massachusetts, in 1879; in New York and Vermont, in 1880; in Nebraska, in 1883; in Wisconsin, in 1900; in Washington, in 1886; in Arizona, Montana, New Jersey, North Dakota and South Dakota, in 1887; in Illinois, in 1891; in Connecticut, in 1893, and in Ohio, in 1894.

Two states permit women to vote upon the issuance of municipal bonds—Montana, established in 1887, and Iowa, in 1894. Louisiana gave all women taxpayers the suffrage upon all questions submitted to the taxpayers in 1898. In 1901 the New York legislature passed a law providing that "a woman who possesses the qualifications to vote for village or for town officers, except the qualification of sex, who is the owner of property in the village assessed upon the last preceding assessment roll thereof, is entitled to vote upon a proposition to raise money by tax or assessment."

DR W. H. JORDAN

Dr Jordan is director of the New York Agricultural Experiment Station and for a long time has been a leader in agricultural progress. His fight for reliable fertilizers, wholesome feeding stuffs and good seed has been of untold worth to the farmers of New York and other states.

PROF WM R. LAZENBY

The first professor of horticulture in this country was Prof Lazenby, who for the past 25 years has been closely identified with agricultural progress in Ohio. He has made a broad study of forestry problems, and now holds the Chair of Forestry in the Ohio State University.

Silver Prices Run Low

Silver is being produced in excess of the demand. The price was lower the latter part of 1908 than it had been at any time during the past five years. The United States government bought silver at $1 an ounce under the Bland and Sherman act. When it re-entered the market to buy more silver in 1907, the price was 70 cents, and it tumbled along down to 57 cents at the end of November, 1907. The latter price followed the break caused by the panic in October. May, 1908, the price of silver was about 54 cents an ounce, and now it is from 50¼ to 51¾ cents an ounce. The trouble with silver is that the supply is always in excess of the demand.

The whole market is dependent upon one buyer, India. Until the outside supply is used up and a shortage is felt, silver is liable to slump whenever anything goes wrong in India or the Straits Settlements, where there are at present large accumulations. Something in the nature of a corner in silver has been attempted at Bombay. The Indian government is holding an enormous stock of silver for coinage purposes, whenever the government feels that the country requires more money for circulation. There was said to be about $55,000,000 worth of available silver in India more than a year ago.

There is no trust powerful enough to arbitrarily fix silver prices. Silver, like gold, is hurried to market as soon as it comes from the mines. Most of the silver produced in the United States is handled by the Guggenheim smelters, and shipments are made to Europe regularly, irrespective of fluctuation in quotations. Prices are fixed arbitrarily in the London markets every morning and are accepted by every other country as a basis for the day's business.

After a long period of lawlessness, during which bandits have had more actual power than the government, the sultan of Morocco, Abd-el-Aziz, was deposed in 1908 by his brother Mulai-Hafid, and the usurper drove Abd-el-Aziz from the country.

PROF EUGENE DAVENPORT

The Dean of the College of Agriculture and Director of the Agricultural Experiment Station of Illinois is Prof Davenport, who has organized one of the most wonderful agricultural colleges and experiment stations in the entire country. Prof Davenport is a clear thinker, a brilliant writer and an agricultural scientist without a superior.

PROF C. F. CURTISS

One of the most famous agricultural colleges in the land is in Iowa. For many years the Division of Agriculture and the Agricultural Experiment Station of Iowa have been directed by Prof Curtiss, a man of broad sympathy for and high attainments in agricultural science. Improvement of the live stock industry is one of his special purposes.

Overcoming Tuberculosis

Science has made great progress in dealing with tuberculosis. The national congress held at Washington in October, 1908, demonstrated the great and growing interest in the prevention, care and cure of the disease. The most heated discussion occurred over the question whether or not tuberculosis in cattle is communicable to human beings. The great German scientist, Robert Koch, declared, as he had previously done, that the germ of tuberculosis in animals is different from the germ causing tuberculosis of the lungs in man. He suggested the possibility of disease being taken from tuberculous animals by persons, but insisted that such cases were so rare, if they existed at all, that none had been authenticated. He stood alone in this opinion. The veterinarians present, and all other medical scientists who expressed their opinions, insisted that there is great danger of tuberculosis spreading from animals to human beings, and insisted that it had not been

proved that there are different kinds of tuberculosis germs.

Scientists in the United States and several foreign countries reported encouraging experiments with serum used for the prevention and cure of tuberculosis in a manner similar to vaccination for smallpox. Much attention was given to consideration of methods of sanitarium treatment and of general hygienic, social, industrial, and economic aspects of tuberculosis. The importance of instructing children in the public schools with reference to taking proper care of themselves was emphasized.

The veterinarians reported the value, in the work of controlling bovine tuberculosis, of the tuberculin test, which is the most accurate method of diagnosing the disease. The urgent need of stamping out bovine tuberculosis was reiterated, and veterinarians were a unit in declaring the danger from the consumption of flesh or milk of tuberculous cattle. The American delegates urged establishment of a federal bureau of health.

A Directory of Officers— National, State, Agricultural

Showing Who Represents the People in Running the Government—Many State and Territorial Officers—Colleges, Stations and Associations Promoting Agricultural and Kindred Interests

NATIONAL

President, Theodore Roosevelt of New York.
Vice-President, Charles W. Fairbanks of Indiana.

THE CABINET

Secretary of State, Robert Bacon of New York.
Secretary of the Treasury, George B. Cortelyou of New York.
Secretary of War, Luke E. Wright of Tennessee.
Attorney-General, Charles J. Bonaparte of Maryland.
Postmaster-General, George L. Meyer of Massachusetts.
Secretary of the Navy, Truman H. Newberry of Michigan.
Secretary of the Interior, James R. Garfield of Ohio.
Secretary of Agriculture, James Wilson of Iowa.
Secretary of Commerce and Labor, Oscar S. Straus of New York.
The salary of the president is $50,000 per year, and the vice-president and members of the cabinet $12,000.

Executive Departments

STATE

Assistant Secretary, Robert Bacon of New York; Second Assistant Secretary, Alvey A. Adee of District of Columbia; Chief of the Diplomatic Bureau, Sydney C. Hengstler of Ohio; Chief of the Bureau of Citizenship, Gaillard Hunt of Virginia; Chief of the Bureau of Trade Relations, John Ball Osborne of Pennsylvania.

TREASURY

Assistant Secretaries, James B. Reynolds, Beekman Winthrop, Louis A. Coolidge; Treasurer, Charles H. Treat; Commissioner of Internal Revenue, John G. Capers; Comptroller of the Currency, Lawrence O. Murray; Director of the Mint, Frank A. Leach; Register of the Treasury, William T. Vernon; Chief of Secret Service, John E. Wilkie.

WAR

Assistant Secretary, Robert Shaw Oliver; Chief of General Staff, Maj. Gen. J. Franklin Bell; Adjutant-General, Maj. Gen. F. C. Ainsworth; Inspector-General, Brig. Gen. E. A. Garlington; Judge-Advocate-General, Brig. Gen. George B. Davis; Quartermaster-General, Brig. Gen. James B. Aleshire; Commissary-General, Brig. Gen. Henry G. Sharpe; Surgeon-General, Brig. Gen. Robert M. O'Reilly; Chief of Bureau of Engineers, Brig. Gen. Alexander Mackenzie; Chief of Bureau of Ordinance, Brig. Gen. William Crozier; Chief Signal Officer, Brig. Gen. James Allen; Chief of Artillery, Brig. Gen. Arthur Murray; Chief Bureau of Insular Affairs, Brig. Gen. Clarence R. Edwards; President Board of Engineers for Rivers and Harbors, Col. Daniel W. Lockwood; President Board of Ordnance and Fortification, Maj. Gen. J. Franklin Bell; President Army War College, Brig. Gen. Wm. W. Wotherspoon.

JUSTICE

Solicitor-General, Henry M. Hoyt; Assistant to the Attorney-General, Milton D. Purdy; Assistant Attorneys-General, John G. Thompson, Charles W. Russell, Alford M. Cooley, Edward T. Sanford, William Wallace Brown, John Q. Thompson.

POSTOFFICE

First Assistant Postmaster-General, Charles P. Grandfield; Second Assistant Postmaster-General, James T. McCleary; Third Assistant Postmaster-General, Abraham L. Lawshe; Fourth Assistant Postmaster-General, P. V. DeGraw; Superintendent Division of Rural Delivery, William R. Spilman.

NAVY

Assistant Secretary, Herbert L. Satterlee; President General Board, Admiral George Dewey; Chief Bureau of Yards and Docks, Civil Engineer R. C. Hollyday; Chief Bureau of Equipment, Rear-Admiral William S. Cowles; Chief Bureau of Navigation, Rear-Admiral John E. Pillsbury; Chief Bureau of Ordnance, Rear-Admiral N. E. Mason; Chief Bureau of Construction and Repair, Chief Constructor Washington Lee Capps; Chief Bureau of Steam Engineering, Engineer in Chief and Rear-Admiral Charles W. Rae; Chief Bureau of Supplies and Accounts, Paymaster-General Eustace B. Rogers; Chief Bureau of Medicine and Surgery, Surg. Gen. P. M. Rixey; Judge Advocate-General, Capt. Edward H. Campbell; President Board of Inspection and Survey, Rear-Admiral Richardson Clover; President Naval Examining Board, Rear-Admiral Albert R. Couden; President Naval Retiring Board, Rear Admiral Albert R. Couden.

INTERIOR

First Assistant Secretary, Frank Pierce; Assistant Secretary, Jesse E. Wilson; Commissioner of General Land Office, Fred Dennett; Commissioner of Indian Affairs, Francis E. Loupp; Commissioner of Pensions, Vespasian Warner; Commissioner of Patents, Edward B. Monroe; Commissioner of Education, Elmer E. Brown; Director of the Geological Survey, George Otis Smith; Director of Reclamation Service, Frederick H. Newell.

AGRICULTURE

Assistant Secretary, Willett M. Hays; Chief of Weather Bureau, Willis L. Moore.
Bureau of Animal Industry: Chief, Alonzo D. Melvin; Chiefs of Divisions: Inspection, Rice P. Steddom; Dairy, Ed. H. Webster; Quarantine, Richard W. Hickman; Biochemic, M. Dorset; Pathological, John R. Mohler; Zoology, B. H. Ransom; Experiment Station, E. C. Schroeder; Animal Husbandry, George M. Rommell; Editor, James M. Pickens.
Bureau of Plant Industry: Chief, Beverly T. Galloway; In Charge of Laboratory of

Plant Pathology, Plant Pathologist, Erwin F. Smith; Diseases of Fruits, Merton B. Waite; Forest Pathologist, Haven Metcalf; Plant Life History, Walter T. Swingle; Cotton-Breeding, A. D. Shamel and D. N. Shoemaker; Tobacco Investigations, A. D. Shamel, W. W. Garner, and E. H. Mathewson; Corn Investigations, C. P. Hartley; Alkali and Drouth Resistant Plant-Breeding, T. H. Kearney; Soil Bacteriology and Water Purification, Karl F. Kellerman; Economic Investigations of Tropical and Subtropical Plants, O. F. Cook; Drug and Poisonous Plant Investigations and Tea Culture, Rodney H. True; Physical, Lyman J. Briggs; Crop Technology and Fiber Plant, N. A. Cobb; Taxonomic, Frederick V. Coville; Farm Management, William J. Spillman; Grain, Mark A. Carleton; Arlington Experimental Farm and Vegetable Testing Gardens, Lee C. Corbett; Sugar-Beet, C. O. Townsend; Western Agricultural Extension, Carl S. Schofield, Dry Land Agriculture, E. Channing Chilcott; Pomological Collections, Gustavus B. Brackett; Field Pomology, William A. Taylor and G. Harold Powell; Experimental Gardens and Grounds, Edward M. Byrnes; Seed and Plant Introduction, David Fairchild; Forage Crop, C. V. Piper; Congressional Seed Distribution, Lisle Morrisson, Assistant; Seed Laboratory, Edgar Brown; Grain Standardization, John D. Shanahan; Subtropical Laboratory and Garden, Ernst A. Bessey; Plant Introduction Garden, August Mayer; South Texas Garden, Edward C. Green; Cotton Culture Farms and Farmers' Co-operative Demonstration Work. Seaman A. Knapp.

Forest Service: Forester, Gifford Pinchot; Assistant Foresters, in charge of Operation: James B. Adams; Silviculture, William T. Cox; Grazing, Albert F. Potter; Products, William L. Hall.

Bureau of Chemistry: Chemist and Chief of Bureau, Harvey W. Wiley; Chief Food Division, W. D. Bigelow; Chief Sugar Laboratory, H. W. Wiley; Chief Dairy Laboratory, G. E. Patrick; Chief Miscellaneous Laboratory, J. K. Haywood; Chief Division of Drugs, L. F. Kebler; Chief Leather and Paper Laboratory, E. P. Veitch; Chief Microchemical Laboratory, B. J. Howard.

Chief Bureau of Soils, Milton Whitney.

Chief Bureau of Entomology, L. O. Howard.

Chief Bureau of Biological Survey, Hart Merriam.

Chief Division of Accounts and Disbursements, A. Zappone.

Chief Division of Publications, George William Hill, Editor.

Chief Bureau of Statistics, Victor H. Olmsted.

Director of Experiment Stations, A. C. True.

Director Office of Public Roads, Logan W. Page.

COMMERCE AND LABOR

Assistant Secretary, William R. Wheeler; Commissioner of Corporations, Herbert Knox Smith; Commissioner of Labor, Charles P. Neill; Chief of Bureau of Manufactures, John M. Carson; Director of the Census, S. N. D. North; Superintendent of Coast and Geodetic Survey, O. H. Tittmann; Chief Bureau of Statistics, Oscar P. Austin; Supervising Inspector-General of Steamboat-Inspection Service, George Uhler; Commissioner of Fisheries, George M. Bowers; Commissioner of Navigation, E. T. Chamberlain; Commissioner-General of Immigration and Naturalization, Daniel G. Keefe; Director of Standards, S. W. Stratton; Chairman Light-House Board, Rear-Admiral Adolph Marix, U. S. N.

Supreme Court of the United States

Chief Justice, Melville W. Fuller of Illinois; Associate Justices, John Marshall Harlan of Kentucky, David Josiah Brewer of Kansas; Edward Douglass White of Louisiana; Rufus W. Packham of New York, Joseph McKenna of California, Oliver Wendell Holmes of Massachusetts, William R. Day of Ohio, and William Henry Moody of Massachusetts.

Circuit Courts of the United States

DISTRICT AND CIRCUIT JUDGES

1st Judicial Circuit, Mr Justice Holmes—Maine, New Hampshire, Massachusetts and Rhode Island; judges, LeBaron B. Colt, Providence, R I; William L. Putnam, Portland, Me; Francis C. Lowell, Boston, Mass.

2d, Mr Justice Peckham—Vermont, Connecticut, Northern New York, Southern New York, and Western New York; judges, E. Henry Lacombe, New York, N Y; Alfred C. Coxe, Utica, N Y; Henry G. Ward, New York, N Y; Walter C. Noyes, New London, Conn.

3d, Mr Justice Moody—New Jersey, Eastern Pennsylvania, Middle Pennsylvania, Western Pennsylvania, and Delaware; judges, George M. Dallas, Philadelphia, Pa; George Gray, Wilmington, Del; Joseph Buffington, Pittsburg, Pa.

4th, Mr Chief Justice Fuller—Maryland, Northern West Virginia, Southern West Virginia, Eastern Virginia, Western Virginia, Eastern North Carolina, Western North Carolina, and South Carolina; judges, Nathan Goff, Clarksburg, W Va; Peter C. Pritchard, Asheville, N C.

5th, Mr Justice White—Northern Georgia, Southern Georgia, Northern Florida, Southern Florida, Northern Alabama, Middle Alabama, Southern Alabama, Northern Mississippi, Southern Mississippi, Eastern Texas, and Western Texas; judges, Don A. Pardee, Atlanta, Ga; Andres P. McCormich, Dallas, Tex; David D. Shelby, Huntsville, Ala.

6th, Mr Justice Harlan—Northern Ohio, Southern Ohio, Eastern Michigan, Western Michigan, Eastern Kentucky, Western Kentucky, Eastern Tennessee, Middle Tennessee, and Western Tennessee; judges, Horace H. Lurton, Nashville, Tenn; Henry F. Severens, Kalamazoo, Mich; John K. Richards, Cincinnati, O.

7th, Mr Justice Day—Indiana, Northern Illinois, Eastern Illinois, Southern Illinois, Eastern Wisconsin, and Western Wisconsin; judges, Peter S. Grosscup, Chicago, Ill; Francis E. Baker, Indianapolis, Ind. William H. Seaman, Sheboygan, Wis; Christian C. Kohlsaat, Chicago, Ill.

8th, Mr Justice Brewer—Minnesota, Northern Iowa, Southern Iowa, Eastern Missouri, Western Missouri, Eastern Arkansas, Western Arkansas, Nebraska, Colorado, Kansas, North Dakota, South Dakota, Eastern Oklahoma, Western Oklahoma, Wyoming, Utah, and Territory of New Mexico; judges, Walter H. Sanborn, St Paul, Minn; Willis Van Devanter, Cheyenne, Wyo; William C. Hook, Leavenworth, Kan; Elmer B. Adams, St Louis, Mo.

9th, Mr Justice McKenna—Northern California, Southern California, Oregon, Nevada, Montana, Eastern Washington, Western Washington, Idaho, and Territories of Alaska, Arizona and Hawaii; judges, William B. Gilbert, Portland, Ore; Erskine M. Ross, Los Angeles, Cal; William W. Morrow, San Francisco, Cal.

Interstate Commerce Commission

Chairman, Martin A. Knapp of New York. Secretary, Edward A. Moseley of Massachusetts. Judson C. Clements of Georgia; Charles A. Prouty of Vermont; Francis M. Cockrell of Missouri; Franklin K. Lane of California; Edgar E. Clark of Iowa; James S. Harlan of Illinois.

Isthmian Canal Commission

Lieut. Col. George W. Goethals, Corps of Engineers, U. S. A., Chairman and Chief Engineer, Culebra; Maj. D. D. Gaillard, Corps of Engineers, U. S. A., head of the department of excavation and dredging, Culebra; Maj. William L. Sibert, Corps of Engineers, U. S. A., head of the department of lock and dam construction, Culebra; H. H. Rousseau, U. S. N., head of department of municipal engineering, motive power and machinery, and building construction, Culebra; J. C. S. Blackburn, head of the department of civil administration, Ancon; Col. William C. Gorgas, Medical department, U. S. A., head of the department of sanitation, Ancon; Jackson Smith, head of the department of labor, quarters, and sub- sistence, Culebra; Secretary, Joseph Buck- lin Bishop, Ancon.

Public Printer

Samuel B. Donnelly of New York.

Bureau of American Republics

Director—John Barrett; secretary, Fran- cisco J. Yanes.

American National Red Cross

President, William H. Taft; treasurer, Beekman Winthrop, assistant secretary of the treasury; counselor, Alford W. Cooley, assistant attorney-general; secretary, Charles L. Magee, 116 Tennessee avenue, N. E., Washington, D. C.

Congress

THE SENATE

Members of the 60th Con- gress which ends March 3, 1909. The senators whose terms expire in 1909 will retire March 3 unless re- elected by the state legis- latures. A majority will be re-elected. Republicans are marked R (61) and Democrats D (31). Total, 92.

ALABAMA
John H. Bankhead, D..1913
Joseph F. Johnston, D.1915

ARKANSAS
James P. Clark, D....1909
Jeff Davis, D.........1913

CALIFORNIA
George C. Perkins, R..1909
Frank P. Flint, R.....1911

COLORADO
Henry M. Teller, D....1909
Simon Guggenheim, R.1913

CONNECTICUT
Morgan G. Bulkeley, R 1911
Frank B. Brandegee, R 1909

DELAWARE
Henry A. duPont, R...1911
H. A. Richardson, R...1913

FLORIDA
James P. Taliaferro, D 1909
William H. Milton, D..1911

GEORGIA
Augustus O. Bacon, D..1913
Alexander S. Clay, D..1909

IDAHO
Weldon B. Heyburn, R.1909
William E. Borah, R...1913

ILLINOIS
Shelby M. Cullom, R..1913
Albert J. Hopkins, R..1909

INDIANA
Albert J. Beveridge, R.1911
James A. Hemenway, R 1909

IOWA
Albert B. Cummins, R..1909
Jonathan P. Dolliver, R 1913

KANSAS
Chester I. Long, R....1909
Charles Curtis, R......1913

KENTUCKY
James B. McCreary, D.1909
Thomas H. Paynter, D.1913

LOUISIANA
Samuel D. McEnery, D.1909
Murphy J. Foster, D...1913

MAINE
Eugene Hale, R.......1913
William P. Frye, R....1911

MARYLAND
Isidor Rayner, D......1911
John W. Smith, D.....1913

MASSACHUSETTS
Henry Cabot Lodge, R.1911
W. Murray Crane, R..1913

MICHIGAN
Julius C. Burrows, R..1911
William A. Smith, R..1913

MINNESOTA
Knute Nelson, R......1913
Moses E. Clapp, R....1911

MISSISSIPPI
Hernando D. Money, D 1911
Anselm J. McLaurin, D 1913

MISSOURI
William J. Stone, D...1909
William Warner, R....1913

MONTANA
Thomas H. Carter, R..1911
Joseph M. Dixon, R....1913

NEBRASKA
Elmer J. Burkett, R...1911
Norris Brown, R......1913

NEVADA
Francis G. Newlands, D 1909
George S. Nixon, R...1911

NEW HAMPSHIRE
Jacob H. Gallinger, R.1909
Henry E. Burnham, R..1913

NEW JERSEY
John Kean, R.........1911
Frank O. Briggs, R....1913

NEW YORK
Thomas C. Platt, R....1909
Chauncey M. Depew, R 1911

NORTH CAROLINA
F. M. Simmons, D.....1913
Lee S. Overman, D....1909

NORTH DAKOTA
H. C. Hansbrough, R...1909
Porter J. McCumber, R 1911

OHIO
Joseph B. Foraker, R..1909
Charles Dick, R.......1911

OKLAHOMA
Thomas P. Gore, D....1909
Robert L. Owen, D....1913

OREGON
Charles W. Fulton, R..1909
Jonathan Bourne, Jr, R 1913

PENNSYLVANIA
Boies Penrose, R......1909
Philander C. Knox, R..1911

RHODE ISLAND
Nelson W. Aldrich, R..1911
George P. Wetmore, R.1913

SOUTH CAROLINA
Benj. R. Tillman, D...1913
Frank B. Gary, D.....1915

SOUTH DAKOTA
Robert J. Gamble, R...1913
Alfred B. Kittredge, R.1909

TENNESSEE
James B. Frazier, D...1911
Robert L. Taylor, D...1913

TEXAS
Chas. A. Culberson, D.1911
Joseph W. Bailey, D...1913

UTAH
Reed Smoot, R........1909
George Sutherland, R..1911

VERMONT
Wm. P. Dillingham, R.1915
Carroll S. Page, R.....1911

VIRGINIA
John W. Daniel, D.....1911
Thomas S. Martin, D..1913

WASHINGTON
Levi Ankeny, R.......1909
Samuel H. Piles, R....1911

WEST VIRGINIA
Stephen B. Elkins, R..1913
Nathan B. Scott, R....1911

WISCONSIN
Robert M. LaFollette, R1911
Isaac Stephenson, R...1909

WYOMING
Francis E. Warren, R..1913
Clarence D. Clark, R...1911

THE HOUSE

Members-elect of the 61st Congress, the 2-years' term of whom will begin March 4, 1909. Republicans are marked R (219); Democrats D (172). Total 391. Those marked * did not serve in the 60th Congress.

ALABAMA
1 George W. Taylor, D
2 S. H. Dent, Jr.,* D
3 Henry D. Clayton, D
4 W. B. Craig, D
5 J. Thomas Heflin, D
6 Richmond P. Hobson, D
7 John L. Burnett, D
8 William Richardson, D
9 Oscar W. Underwood, D

ARKANSAS
1 Robert Bruce Macon, D
2 W. A. Oldfield,* D
3 John C. Floyd, D
4 Ben Cravens, D
5 Charles C. Reid, D
6 Joseph T. Robinson, D
7 Robert M. Wallace, D

CALIFORNIA
1 W. F. Englebright, R
2 Duncan E. McKinlay, R
3 Joseph R. Knowland, R
4 Julius Kahn, R
5 Everis A. Hayes, R
6 James C. Needham, R
7 James McLachlan, R
8 Sylvester C. Smith, R

COLORADO

At Large

Edward T. Taylor,* D
1 Atterson W. Rucker,* D
2 John A. Martin,* D

CONNECTICUT

At Large

John Q. Tilson,* R
1 E. Stevens Henry, R
2 Nehemiah D. Sperry, R
3 Edwin W. Higgins, R
4 Ebenezer J. Hill, R

DELAWARE

At Large

William H. Heald,* R

FLORIDA

1 S. M. Sparkman, D
2 Frank Clark, D
3 J. Walter Kehoe,* D

GEORGIA

1 Charles G. Edwards, D
2 James M. Griggs, D
3 Dudley M. Hughes,* D
4 William C. Adamson, D
5 L. F. Livingston, D
6 Charles L. Bartlett, D
7 Gordon Lee, D
8 William M. Howard, D
9 Thomas M. Bell, D
10 Thomas W. Hardwick, D
11 William G. Brantley, D

IDAHO

Thomas R. Hamer,* R

ILLINOIS

1 Martin B. Madden, R
2 James R. Mann, R
3 William W. Wilson, R
4 James T. McDermott, D
5 Adolph J. Sabath, D
6 William Lorimer, R
7 Fred Lundin,* R
8 Thomas Gallagher,* D
9 Henry S. Boutell, R
10 George Edmund Foss, R
11 Howard M. Snapp, R
12 Charles E. Fuller, R
13 Frank O. Lowden, R
14 James McKinney, R
15 George W. Prince, R
16 Joseph V. Graff, R
17 John A. Sterling, R
18 Joseph G. Cannon, R
19 William B. McKinley, R
20 Henry T. Rainey, D
21 James M. Graham,* D
22 Wm. A. Rodenberg, R
23 Martin D. Foster, D
24 Pleasant T. Chapman, R
25 N. B. Thistlewood, R

INDIANA

1 John W. Boehne,* D
2 William Cullop,*D
3 William E. Cox, D
4 Lincoln Dixon, D
5 Ralph W. Moss,* R
6 W. O. Barnard,* R
7 Charles A. Korbly,* D
8 John A. M. Adair,* D
9 Martin A. Morrison,* D
10 E. D. Crumpacker, R
11 George W. Rauch, D
12 Cyrus Kline,* D
13 Henry A. Barnhart,* D

IOWA

1 Charles A. Kennedy, R
2 Albert F. Dawson, R
3 Charles Pickett,* R
4 Gilbert N. Haugen, R
5 James W. Good,* R
6 N. E. Kendall,* R
7 John A. T. Hull, R
8 W. D. Jamieson,* R
9 Walter I. Smith, R
10 Frank P. Woods,* R
11 Elbert H. Hubbard, R

KANSAS

1 D. R. Anthony, Jr, R

2 Charles F. Scott, R
3 Philip P. Campbell, R
4 James M. Miller, R
5 Wm. A. Calderhead, R
6 William A. Reeder,
7 E. H. Madison, R
8 Victor Murdock, R

KENTUCKY

1 Ollie M. James, D
2 Augustus O. Stanley, D
3 R. Y. Thomas,* D
4 Ben Johnson, D
5 Swagar Sherley, D
6 Joseph L. Rhinock, D
7 J. Campbell Cantrill,* D
8 Harvey Helm, D
9 Joseph B. Bennett, R
10 John W. Langley, R
11 Don C. Edwards, R

LOUISIANA

1 Albert Estopinal,* D
2 Robert C. Davey, D
3 Robert F. Broussard, D
4 John T. Watkins, D
5 Joseph E. Ransdell, D
6 Robert C. Wickliffe,* D
7 Arsene P. Pujo, D

MAINE

1 Amos L. Allen, R
2 John P. Swasey,* R
3 Edwin C. Burleigh, R
4 Frank E. Guernsey,* R

MARYLAND

1 J. H. Covington,* D
2 Joshua F. C. Talbott, D
3 John Kronmiller,* R
4 John Gill, Jr, D
5 Sydney E. Mudd, R
6 George A. Pearre, R

MASSACHUSETTS

1 George P. Lawrence, R
2 Frederick H. Gillett, R
3 Charles G. Washburn, R
4 Charles Q. Tirrell, R
5 Butler Ames, R
6 Augustus P. Gardner, R
7 Ernest W. Roberts, R
8 Samuel W. McCall, R
9 John A. Keliher, D
10 Joseph F. O'Connell, D
11 Andrew J. Peters, D
12 John W. Weeks, R
13 William S. Greene, R
14 William C. Lovering, R

MICHIGAN

1 Edwin Denby, R
2 Chas. E. Townsend, R
3 Washington Gardner, R
4 Edw. L. Hamilton, R
5 Gerrit J. Diekema, R
6 Samuel W. Smith, R
7 Henry McMorran, R
8 Joseph W. Fordney, R
9 Jas. C. McLaughlin,* R
10 George A. Loud, R
11 Francis H. Dodds,* R
12 H. Olin Young, R

MINNESOTA

1 James A. Tawney, R
2 W. S. Hammond, D
3 Charles R. Davis, R
4 Frederick C. Stevens, R
5 Frank M. Nye, R
6 Chas. A. Lindbergh, R
7 Andrew J. Volstead, R
8 Clarence B. Miller,* R
9 Halvor Steenerson, R

MISSISSIPPI

1 E. S. Candler, Jr, D
2 Thomas Spight, D
3 Benj. G. Humphreys, D
4 T. U. Sisson, * D
5 Adam M. Byrd, D
6 Eaton J. Bowers, D
7 W. J. Dickson,* D
8 J. W. Collier,* D

MISSOURI

1 James T. Lloyd, D
2 William W. Rucker, D

3 Joshua W. Alexander, D
4 Charles F. Booher, D
5 William P. Borland,* D
6 David A. DeArmond, D
7 Courtney W. Hamlin, D
8 D. W. Shackleford, D
9 Champ Clark, D
10 Richard Bartholdt, R
11 Patrick F. Gill,* D
12 Harry M. Coudrey, R
13 Politte Elvins,* R
14 Charles A. Crow,* R
15 Charles H. Morgan,* R
16 Arthur P. Murphy,* R

MONTANA

At Large

Charles N. Pray, R

NEBRASKA

1 John A. Maguire,* D
2 Gilbert M. Hitchcock, D
3 James P. Latta,* D
4 Edmund H. Hinshaw, R
5 George W. Norris, R
6 Moses P. Kinkaid, R

NEVADA

At Large

George A. Bartlett, D

NEW HAMPSHIRE

1 Cyrus A. Sulloway, R
2 Frank D. Currier, R

NEW JERSEY

1 H. C. Loudenslager, R
2 John J. Gardner, R
3 Benjamin F. Howell, R
4 Ira W. Wood, R
5 Charles N. Fowler, R
6 William Hughes, D
7 Richard W. Parker, R
8 William H. Wiley,* R
9 Eugene F. Kinkead,* D
10 James A. Hamill, D

NEW YORK

1 William W. Cocks, R
2 George H. Lindsay, D
3 Otto G. Foelker,* R
4 Charles B. Law, R
5 Richard Young,* R
6 William M. Calder, R
7 John J. Fitzgerald, D
8 Daniel J. Riordan, D
9 Henry M. Goldfogle, D
10 William Sulzer, D
11 Charles V. Fornes, D
12 Michael F. Conroy,* D
13 Herbert Parsons, R
14 William Willett, Jr, D
15 J. Van Vechten Olcott, R
16 Francis B. Harrison, D
17 William S. Bennet, R
18 Joseph A. Goulden, D
19 John E. Andrus, R
20 Thomas W. Bradley, R
21 Hamilton Fish,* R
22 William H. Draper, R
23 George N. Southwick, R
24 George W. Fairchild, R
25 Cyrus Durey, R
26 George R. Malby, R
27 Chas. S. Millington,* R
28 Charles L. Knapp, R
29 Michael E. Driscoll, R
30 John W. Dwight, R
31 Sereno E. Payne, R
32 James Breck Perkins, R
33 J. Sloat Fassett, R
34 James S. Simmons,* R
35 Daniel A. Driscoll,* D
36 De Alva S. Alexander, R
37 Edward B. Vreeland, R

NORTH CAROLINA

1 John H. Small, D
2 Claude Kitchin, D
3 Charles R. Thomas, D
4 Edward W. Pou, D
5 J. M. Morehead,* R
6 H. L. Godwin, D
7 Robert N. Page, D
8 Charles H. Cowles,* R
9 Edwin Y. Webb, D
10 J. G. Grant,* R

NORTH DAKOTA
At Large
L. B. Hanna,* R
Asle J. Gronna, R

OHIO
1 Nicholas Longworth, R
2 Herman P. Goebel,* R
3 James M. Cox,* D
4 William E. Tou Velle, D
5 Timothy T. Ansberry, D
6 Matt R. Denver, D
7 J. Warren Keifer, R
8 Ralph D. Cole, R
9 Isaac R. Sherwood, D
10 A. R. Johnson,* R
11 Albert Douglas, R
12 Edw. L. Taylor, Jr, R
13 Carl Anderson,* D
14 William G. Sharpe,* D
15 James Joyce,* R
16 D. A. Hollingsworth,* R
17 William A. Ashbrook, D
18 James Kennedy, R
19 W. Aubrey Thomas, R
20 Paul Howland, R
21 Theodore E. Burton, R

OKLAHOMA
1 Bird S. McGuire, R
2 Richard T. Morgan,* R
3 C. E. Creager,* R
4 Charles D. Carter, D
5 Scott Ferris, D

OREGON
1 Willis C. Hawley, R
2 W. R. Ellis, R

PENNSYLVANIA
1 Henry H. Bingham, R
2 Joel Cook, R
3 J. Hampton Moore, R
4 Reuben O. Moon, R
5 W. W. Foulkrod, R
6 George D. McCreary, R
7 Thomas S. Butler, R
8 Irving P. Wanger, R
9 William W. Griest,* R
10 T. D. Nicholls, D
11 Henry W. Palmer,* R
12 Alfred B. Garner,* R
13 John H. Rothermel, D
14 Charles C. Pratt,* R
15 William B. Wilson, D
16 John G. McHenry, D
17 Benjamin K. Focht, R
18 Marlin E. Olmsted, R
19 John M. Reynolds, R
20 Daniel F. Lafean, R
21 Charles F. Barclay, R
22 George F. Huff, R
23 Allen F. Cooper, R
24 John K. Tener,* R
25 Arthur L. Bates, R
26 A. Mitchell Palmer,* D
27 J. N. Langham,* R
28 Nelson P. Wheeler, R
29 William H. Graham, R
30 John Dalzell, R
31 James F. Burke, R
32 Andrew J. Barchfeld, R

RHODE ISLAND
1 William P. Sheffield,* R
2 Adin B. Capron, R

SOUTH CAROLINA
1 George S. Legare, D
2 James O. Patterson, D
3 Wyatt Aiken, D
4 Joseph T. Johnson, D
5 David E. Finley, D
6 J. Edwin Ellerbe, D
7 Asbury F. Lever, D

SOUTH DAKOTA
At Large
Charles H. Burke,* R
Eben W. Martin,* R

TENNESSEE
1 Walter P. Brownlow, R
2 R. W. Austin,* R
3 John A. Moon, D
4 Cordell Hull, D

5 William C. Houston, D
6 J. W. Byrnes,* D
7 Lemuel P. Padgett, D
8 Thetus W. Sims, D
9 Finis J. Garrett, D
10 George W. Gordon,* D

TEXAS
1 Morris Sheppard, D
2 Martin Dies,* D
3 Gordon Russell, D
4 Choice B. Randell, D
5 Jack Beall, D
6 Rufus Hardy, D
7 Alexander W. Gregg, D
8 John M. Moore, D
9 George F. Burgess, D
10 Albert S. Burleson, D
11 Robert L. Henry, D
12 Oscar W. Gillespie, D
13 John H. Stephens, D
14 James L. Slayden, D
15 John N. Garner, D
16 William R. Smith, D

UTAH
At Large
Joseph Howell, R

VERMONT
1 David J. Foster, R
2 Frank H. Plumley,* R

VIRGINIA
1 William A. Jones, D
2 Harry L. Maynard, D
3 John Lamb, D
4 Francis R. Lassiter, D
5 E. W. Saunders, D
6 Carter Glass, D
7 James Hay, D
8 Charles C. Carlin, D
9 C. Bascom Slemp, R
10 Henry D. Flood, D

WASHINGTON
1 Wm. E. Humphrey, R
2 Francis W. Cushman, R
3 Miles Poindexter, R

WEST VIRGINIA
1 William P. Hubbard, R
2 George C. Sturgiss, R
3 Joseph Holt Gaines, R
4 Harry C. Woodyard, R
5 James A. Hughes, R

WISCONSIN
1 Henry A. Cooper, R
2 John M. Nelson, R
3 A. W. Kopp,* R
4 William J. Cary, R
5 William H. Stafford, R
6 Charles H. Weisse, D
7 John J. Esch, R
8 James H. Davidson, R
9 Gustav Kustermann, R
10 E. A. Morse, R
11 Irvine L. Lenroot,* R

WYOMING
Frank W. Mondell, R

ALASKA
James Wickersham,* R

ARIZONA
Ralph H. Cameron,* R

NEW MEXICO
William H. Andrews, R

HAWAII
Jonah K. Kalanianaole, R

PORTO RICO
Resident Commissioner
Tulio Larrinaga, R

PHILIPPINE ISLANDS
Resident Commissioners
Benito Legarda,* R
Pablo Ocampo de Leon,* R

REPRESENTATIVES WHO RETIRE MARCH 3, 1909

ALABAMA
2 Ariosoto A. Wiley, D

ARKANSAS
2 S. Brundidge, Jr., D

COLORADO
1 Robert W. Bonynge, R
2 Warren A. Haggott, R

DELAWARE
At Large
Hiram R. Burton, R

FLORIDA
8 William B. Lamar, D

GEORGIA
8 Elijah B. Lewis, D

IDAHO
Burton L. French, R

ILLINOIS
7 Philip Knopf, R
8 Charles McGavin, R
21 Ben. F. Caldwell, D

INDIANA
1 John H. Foster, R
2 John C. Chaney, R
5 Elias S. Holliday, R
6 James E. Watson, R
7 Jesse Overstreet, R
12 Clarence C. Gilhams, R

IOWA
3 Benj P. Birdsall, R
5 Robert G. Cousins, R
6 Daniel W. Hamilton, D
8 William P. Hepburn, R
10 James P. Conner, R

KENTUCKY
3 Addison D. James, R
7 W. P. Kimball, D

LOUISIANA
6 George K. Favrot, D

MAINE
2 Chas. E. Littlefield, R
4 Llewellyn Powers, R

MARYLAND
1 William H. Jackson, R
3 Harry B. Wolf, D

MICHIGAN
11 Arch B. Darragh, R

MINNESOTA
8 J. Adam Bede, R

MISSISSIPPI
4 Wilson S. Hill, D
7 Frank M. McLain, D
8 John S. Williams, D

MISSOURI
5 Edgar C. Ellis, R
11 Henry S. Caulfield, R
13 Madison R. Smith, D
14 Joseph J. Russell, D
15 Thomas Hackney, D
16 Robert Lamar, D

MONTANA
At Large
Charles N. Pray, R

NEBRASKA
1 Ernest M. Pollard, R
3 John F. Boyd, R

NEW JERSEY
8 Le Gage Pratt, D
9 Eugene W. Leake, D

NEW YORK
3 Charles T. Dunwell, R
5 George E. Waldo, R
12 W. Bourke Cockran, D
21 Samuel McMillan, R
27 James S. Sherman, R
34 Peter A. Porter, R
35 William H. Ryan, D

NORTH CAROLINA
5 William W. Kitchin, D
8 Richard N. Hackett, D
10 William T. Crawford, D

NORTH DAKOTA
At Large
Thomas F. Marshall, R

OHIO
13 Grant E. Mouser, R

14	J. Ford Laning, R		
15	Beman F. Dawes, R		
16	Capell L. Weems, R		

RHODE ISLAND

1 Daniel L. D. Granger, D

OKLAHOMA

2 Elmer L. Fulton, D
3 James S. Davenport, D

SOUTH DAKOTA
At Large
Philo Hall, R
William H. Parker, R

PENNSYLVANIA

9 H. Burd Cassel, R
11 John T. Lenahan, D
12 Charles N. Brumm, R
24 Ernest F. Acheson, R
26 J. Davis Brodhead, D
27 Joseph G. Beale, R

TENNESSEE

2 Nathan W. Hale, R
6 John W. Gaines, D

TEXAS

2 Sam B. Cooper, D

VERMONT

2 Kittredge Haskins, R

WASHINGTON
At Large
Wesley L. Jones, R
Wm E. Humphrey, R

WISCONSIN

3 James W. Murphy, D
11 John J. Jenkins, R

DELEGATES

ALASKA
Thomas Cale, Ind.

ARIZONA
Marcus A. Smith, D

Chairmen of Committees

SENATE

Agriculture and Forestry—Hansbrough of North Dakota.
Appropriations—Eugene Hale of Maine.
Canadian Relations—W. Murray Crane of Massachusetts.
Census—Chester I. Long of Kansas.
Civil Service and Retrenchment—George C. Perkins of California.
Commerce—William P. Frye of Maine.
Cuban Relations—Henry E. Burnham of New Hampshire.
District of Columbia—Jacob H. Gallinger of New Hampshire.
Education and Labor—Jonathan P. Dolliver of Iowa.
Finance—Nelson W. Aldrich of Rhode-Island.
Fisheries—Jonathan Bourne, Jr, of Oregon.
Foreign Relations—Shelby M. Cullom of Illinois.
Forest Reservations and the Protection of Game—Frank B. Brandegee of Connecticut.
Immigration—William P. Dillingham of Vermont.
Indian Affairs—Moses E. Clapp of Minnesota.
Interoceanic Canals—Alfred B. Kittredge of South Dakota.
Interstate Commerce—Stephen B. Elkins of West Virginia.
Irrigation—Levi Ankeny of Washington.
Judiciary—Clarence D. Clark of Wyoming.
Manufactures—Weldon B. Heyburn of Idaho.
Military Affairs—Francis E. Warren of Wyoming.
Mines and Mining—Charles Dick of Ohio.
Mississippi River and its Tributaries—William Warner of Missouri.
Naval Affairs—Eugene Hale of Maine.
Pacific Islands and Porto Rico—Joseph B. Foraker of Ohio.
Patents—Reed Smoot of Utah.
Pensions—Porter J. McCumber of North Dakota.
Philippines—Henry Cabot Lodge of Massachusetts.
Postoffices and Post Roads—Boies Penrose of Pennsylvania.
Printing—Thomas C. Platt of New York.
Privileges and Elections—Julius C. Burrows of Michigan.
Public Buildings and Grounds—Nathan B. Scott of West Virginia.
Public Health and National Quarantine—John W. Daniel of Virginia.
Public Lands—Knute Nelson of Minnesota.
Railroads—Morgan G. Bulkeley of Connecticut.
Rules—Philander C. Knox of Pennsylvania.
Territories—Albert J. Beveridge of Indiana.

SELECT COMMITTEES

Expenditures in the Department of Agriculture—Isaan Stephenson of Wisconsin.

Expenditures in the Interior Department—Harry A. Richardson of Delaware.
Investigate Trespassers upon Indian Lands—George Sutherland of Utah.
National Banks—William Alden Smith of Michigan.
Standards, Weights and Measures—William E. Borah of Idaho.
Transportation and Sale of Meat Products—Samuel D. McEnery of Louisiana.
Woman Suffrage—Alexander S. Clay of Georgia.

COMMITTEES OF THE HOUSE

Agriculture—Charles F. Scott of Kansas.
Alcoholic Liquor Traffic—Nehemiah D. Sperry of Connecticut.
Appropriations—James A. Tawney of Minnesota.
Banking and Currency—Charles N. Fowler of New Jersey.
Coinage, Weights and Measures—William B. McKinley of Illinois.
District of Columbia—Samuel W. Smith of Michigan.
Education—George N. Southwick of New York.
Election of President, Vice-President and Representatives in Congress—Joseph H. Gaines of West Virginia.
Expenditures in the Department of Agriculture—Charles E. Littlefield of Maine.
Expenditures on Public Buildings—E. Stevens Henry of Connecticut.
Foreign Affairs—Robert G. Cousins of Iowa.
Immigration and Naturalization—Benjamin F. Howell of New Jersey.
Indian Affairs—James S. Sherman of New York.
Insular Affairs—Henry A. Cooper of Wisconsin.
Interstate and Foreign Commerce—William P. Hepburn of Iowa.
Invalid Pensions—Cyrus A. Sulloway of New Hampshire.
Irrigation of Arid Lands—William A. Reeder of Kansas.
Judiciary—John J. Jenkins of Wisconsin.
Labor—John J. Gardner of New Jersey.
Levees and Improvements of the Mississippi River—George W. Prince of Illinois.
Manufactures—Henry McMorran of Michigan.
Merchant Marine and Fisheries—William S. Greene of Massachusetts.
Military Affairs—John A. T. Hull of Iowa.
Mines and Mining—George F. Huff of Pennsylvania.
Naval Affairs—George Edmund Foss of Illinois.
Patents—Frank D. Currier of New Hampshire.
Postoffice and Post Roads—Jesse Overstreet of Indiana.
Public Building and Grounds—Richard Bartholdt of Missouri.
Public Lands—Frank W. Mondell of Wyoming.
Railways and Canals—James H. Davidson of Wisconsin.

Reform in the Civil Service—Frederick H. Gillett of Massachusetts.

Rivers and Harbors—Theodore E. Burton of Ohio.

Rules—The Speaker.

Territories—Edward L. Hamilton of Michigan.

War Claims—Kittredge Haskins of Vermont.

Ways and Means—Sereno E. Payne of New York.

STATES AND TERRITORIES

Capitals and Principal Officers, with Salaries and Terms of Office

ALABAMA

Montgomery

Governor, B. B. Comar............ $5,000
Lieut.-Gov., Henry B. Gray....$4 per day
Secretary of State, Frank N. Julian 1,800
Treasurer, Walter D. Seed........ 2,200
Auditor, William W. Brandon...... 2,400
Chief Justice, John R. Tyson...... 5,000
Att'y-Gen., Alex M. Garber........ 2,500
  Officers elected for four years. Present term expires, January, 1911. Chief Justice, January, 1913.

ALASKA

Juneau

Governor, Wilford B. Hoggatt..... $5,000
Secretary, William L. Distin....... 4,000
  Governor's term expires March 21, 1910; secretary's term December 14, 1908. Appointed by the president.

ARIZONA

Phoenix

Governor, Joseph H. Kibbey....... $3,000
Secretary, John H. Page........... 1,800
Chief Justice, Edward Kent........ 3,000
  Term of governor and chief justice expire February 27, 1909; secretary, January 14, 1912. Appointed by the president.

ARKANSAS

Little Rock

Governor, George W. Donaghey.... $3,000
Secretary of State, O. C. Ludwig... 2,250
Treasurer, James L. Yates.......... 2,250
Auditor, John E. Jobe............. 2,250
Chief Justice, Joseph M. Hill...... 3,000
Att'y-Gen., Hal. L. Norwood........ 2,500
  Officers elected for two years. Present term expires January, 1911.

CALIFORNIA

Sacramento

Governor, J. N. Gillett............ $6,000
Lieut.-Gov., Warren R. Porter $10 per day
Secretary of State, Charles F. Curry 3,000
Treasurer, W. R. Williams........ 3,000
Comptroller, A. B. Lye............ 3,000
Chief Justice, W. H. Beaty........ 8,000
Attorney-General, U. S. Webb..... 3,000
  Officers elected for four years. Present term expires January, 1911.

COLORADO

Denver

Governor, John F. Shafroth....... $5,000
Lieut.-Gov., Stephen R. Fitzgerald 1,000
Secretary of State, James B. Pearce 4,000
Treasurer, W. J. Galligan.......... 6,000
Auditor, Roady Kenahan.......... 4,000
Chief Justice, Robert W. Steele.... 5,000
Att'y-Gen., John T. Barnett....... 3,000
  Officers elected for two years. Present term expires January, 1911.

CONNECTICUT

Hartford

Governor, George L. Lilley........ $4,000
Lieut.-Gov., Frank B. Weeks....... 500
Sec'y of State, Matthew H. Rogers 1,500

Treasurer, Freeman F. Patten,..... 1,500
Comptroller, Thomas D. Bradstreet 1,500
Chief Justice, Simeon E. Baldwin.. 8,000
Att'y-Gen., Marcus H. Holcomb.... 4,000
  Officers elected for two years, except chief justice. Present term expires January, 1911; chief justice, January, 1915.

DELAWARE

Dover

Governor, Simeon S. Pennewill..... $4,000
Lieut.-Gov., John M. Mendenhall... Fees
Secretary of State, *Jos. L. Cahall.. 4,000
Treasurer, David O. Moore $1,950 and fees
Auditor, Theodore F. Clark........ 2,000
Chief Justice, Charles B. Lore..... 4,500
Attorney-General, Andrew C. Gray 2,500
  Officers elected for four years, except treasurer and auditor, two years, and chief justice 12 years. Present term expires, January, 1913, except treasurer and auditor, January, 1911; chief justice, June, 1909.
  *Successor to be appointed by governor January, 1909.

DISTRICT OF COLUMBIA

National Capital—Washington

Commissioner, Henry B. F. Macfarland, President $5,000
Commissioner, Henry L. West...... 5,000
Commissioner, Maj Jay J. Morrow 5,000
Secretary of Board, William Tandall 2,160
  Appointed by the president.

FLORIDA

Tallahassee

Governor, Albert W. Gilchrist...... $5,000
Secretary of State, H. Clay Crawford 2,500
Treasurer, W. V. Knott............ 2,500
Auditor, Ernest Amos............. 2,500
Chief Justice, Thos. M. Shackelford 3,000
Attorney-General, Park M. Trammell 2,500
  Officers elected for four years, except chief justice, elected for six years. Present term expires January, 1913, except state auditor, May, 1911, and chief justice, January, 1915.

GEORGIA

Atlanta

Governor, Joseph M. Brown........ $5,000
Secretary of State, Philip Cook.... 2,000
Treasurer, Robert E. Park........ 2,000
Chief Justice, William H. Fish.... 4,000
Attorney-General, John C. Hart.... 2,000
  Officers elected for two years. Present term expires November, 1910.

GUAM

Governor, Captain E. J. Dorn. $3,000
  Appointed by the President.

HAWAII

Honolulu

Governor, Walter F. Frear........ $5,000
Secretary, Ernest A. Mott-Smith... 3,000
  Terms of governor and secretary expire December 18, 1911. Appointed by the president.

IDAHO

Boise

Governor, James H. Brady........ $5,000
Lieut.-Gov, Lewis H. Sweatser..... 300
Secretary of State, Robert Lansdon 2,400
Treasurer, Charles A. Hastings.... 4,000
Auditor, Stephen D. Taylor........ 2,400
Chief Justice, James F. Ailshie.... 4,000
Att'y-Gen, Daniel C. McDougal..... 3,000
  Officers elected for two years except chief justice, elected for six years. Present term expires January, 1911.

ILLINOIS

Springfield

Governor, Charles S. Deneen......$12,000
Lieut.-Gov., John G. Oglesby....... 2,500

PROF W. F. MASSEY

For many years Prof Massey was Horticulturist of the North Carolina Station, where he did invaluable work. His reputation as a horticultural authority is world-wide, and he has been for years the great southern agricultural leader.

PROF CHAS E. BESSEY

One of the men long identified with agricultural progress is Dr Bessey, botanist of the Nebraska Agricultural Experiment Station. Farmers of that new state owe a big debt to Dr Bessey for the long service in the interest of their work.

A PRIZE-WINNING AYRSHIRE COW

Annie Bedford, No 19556, is the fine Ayrshire cow in this picture. She is owned by H. M. Kimball of New Hampshire, and won first honors at the New Hampshire state fair at Concord last fall.

Dr W. J. Beal

Dr Beal has almost for a lifetime been identified with Michigan Agricultural College. He has always been in the front, fighting for the interest of the farm and agricultural education.

Prof H. J. Waters

Prof Waters is dean of the College of Agriculture and director of the Agricultural Experiment Station of Missouri. His experiments in feeding cattle have given him a wide reputation.

Typical White-Crested Black Polish Fowls

Few breeds of fowls are so radiantly beautiful as the Polish, and none are better adapted for pets in village quarters. But their merits do not end with beauty and docility; they are truly wonderful layers, and thus earn their keep while giving pleasure to the eye. The White-Crested Black is probably the most popular variety and the most generally admired. The specimens shown above were raised by Charles L. Seely of Chenango County, N Y.

Secretary of State, James A. Rose.. $7,500
Treasurer, Andrew Russel........ 10,000
Auditor, James S. McCullough..... 7,500
Chief Justice, Guy C. Scott....... 10,000
Att'y-Gen., William H. Stead...... 10,000

Officers elected for four years, to January, 1913, except state treasurer, two years, to January, 1911, and chief justice, 1 year, to January, 1909.

INDIANA
Indianapolis

Governor, Thomas R. Marshall..... $8,000
Lieut.-Gov., Frank J. Hall......... 1,000
Secretary of State, Fred A. Sims.. 6,500
Treasurer, Oscar Hadley.......... 7,500
Auditor, John C. Billhelmer....... 7,500
Chief Justice, J. H. Jordan........ 6,000
Att'y-Gen., James Bingham........ 7,500

Governor, lieutenant-governor and chief justice elected for four years, to January, 1913; secretary of state and state auditor elected for 1 year, to November, 1910; state treasurer and attorney-general elected for two years, to January, 1911.

IOWA
Des Moines

Governor, B. F. Carroll........... $5,000
Lieut.-Gov., George W. Clarke..... 1,100
Secretary of State, H. C. Hayward 2,200
Treasurer, W. W. Morrow......... 2,200
Auditor, J. L. Bleakly............. 2,200
Chief Justice, W. D. Evans........ 6,000
Attorney-General, H. W. Byers.... 4,000

Officers elected for two years. Present term expires January, 1911, except chief justice, December 31, 1909.

KANSAS
Topeka

Governor, W. R. Stubbs............ $5,000
Lieut.-Gov., W. J. Fitzgerald....... 700
Sec'y of State, Charles E. Denton.. 2,500
Treasurer, Mark Tulley............ 2,500
Auditor, James M. Nation......... 2,500
Chief Justice, W. A. Johnson...... 3,000
Attorney-General, Fred S. Jackson 2,500

Officers elected for two years. Present term expires January, 1911.

KENTUCKY
Frankfort

Governor, Augustus E. Wilson..... $6,500
Lieut.-Gov., W. H. Cox........$10 per day
Secretary of State, Ben L. Bruner 3,000
Treasurer, Edward Farley......... 4,200
Auditor, F. P. James.............. 3,600
Chief Justice, E. C. O'Rear....... 5,000
Att'y-Gen., James Breathitt....... 4,000

Officers elected for four years. Present term expires January, 1912, except governor, January, 1911, and chief justice, elected for 12 years, to January, 1919.

LOUISIANA
Baton Rouge

Governor, Jared Y. Sanders........ $5,000
Lieut.-Gov., Paul M. Lambremont.. 1,800
Secretary of State, John T. Michel 5,000
Treasurer, C. B. Steele........... 4,000
Auditor, Paul Capdevielle......... 5,000
Chief Justice, Joseph A. Breaux.... 6,000
Attorney-General, Walter Guion.... 5,000

Officers elected for four years. Present term expires May, 1912, except chief justice, 1914.

MAINE
Augusta

Governor, Bert M. Fernald........ $3,000
Sec'y of State, *Arthur H. Brown.. 1,500
Treasurer, *P. S. Gilmore.......... 2,000
Auditor, Charles P. Hatch......... 2,500
Chief Justice, Lucilius A. Emery.. 5,000
Att'y-Gen., *Hannibal E. Hamlin.. 1,000

Officers elected for two years. Present term expires January, 1911, except chief justice, December, 1913.　*Successors to

secretary of state, state treasurer and attorney-general to be appointed by the legislature.

MARYLAND
Annapolis

Governor, Austin L. Crother...... $4,500
Sec'y of State, N. Winslow Williams 2,000
Treasurer, Murray Vandiver....... 2,500
Auditor, George R. Ash........... 1,800
Chief Justice, A. Hunter Boyd.... 4,500
Att'y-Gen., Isaac Lobe Straus.... 2,000

Officers elected for four years. Present term expires January, 1912, except state treasurer, 1910.

MASSACHUSETTS
Boston

Governor, Eben F. Draper $8,000
Lieut.-Gov., Louis A. Frothingham 2,000
Secretary of State, William M. Olin 3,500
Treasurer, Arthur B. Chapin....... 5,000
Auditor, Henry E. Turner......... 3,500
Chief Justice, Marcus P. Knowlton 9,000
Attorney-General, Dana Malone.... 5,000

Officers elected for one year. Present term expires January, 1910.

MICHIGAN
Lansing

Governor, Fred M. Warner........ $5,000
Lieut.-Gov., Patrick H. Kelley.... 800
Sec'y of State, Frederick C. Martindale 2,500
Treasurer, Albert E. Sleeper...... 2,500
Auditor-General, Oramel B. Fuller.. 2,500
Chief Justice, Charles A. Blair..... 7,000
Att'y-Gen., John E. Bird......... 5,000

Officers elected for two years. Present term expires December, 1910.

MINNESOTA
St Paul

Governor, John A. Johnson........ $7,000
Lieut.-Gov., A. O. Eberhart....$10 per day
Sec'y of State, Julius A. Schmahl.. 3,500
Treasurer, C. C. Dinehart.......... 3,500
Auditor, S. G. Iverson............. 4,000
Chief Justice, C. M. Start......... 6,000
Att'y-General, George T. Simpson.. 4,800

Officers elected for two years, except chief justice, four years. Present term expires, January, 1911; chief justice, January, 1913.

MISSISSIPPI
Jackson

Governor, E. F. Noel.............. $4,500
Lieut.-Gov., Luther Manship...... 500
Sec'y of State, Joseph W. Power.. 2,000
Treasurer, George R. Edwards.... 3,000
Auditor, E. J. Smith.............. 2,500
Chief Justice, A. H. Whitfield...... 4,500
Att'y-Gen., R. V. Fletcher......... 2,500

Officers elected for four years. Present term expires January, 1912, except chief justice, May 10, 1912.

MISSOURI
Jefferson City

Governor, Herbert S. Hadley....... $5,000
Lieutenant-Governor* 1,000
Secretary of State, Cornelius Roach 3,000
Treasurer, James Cawgill.......... 3,000
Auditor, John P. Gordon.......... 3,000
Chief Justice, **James B. Gantt.... 4,500
Att'y-Gen., Elliot W. Major........ 3,000

Officers elected for four years. Present term expires January, 1913. *Lieutenant-governorship contested—to be determined by legislature in January, 1909. **Successor to chief justice to be elected in January.

MONTANA
Helena

Governor, Edwin L. Norris........ $5,000
Lieut.-Gov., W. R. Allen......$6 per day
Sec'y of State, Abraham N. Yoder.. 3,000

Treasurer, E. Esslestyn $3,000
Auditor, Harry R. Cunningham.... 3,000
Presiding Justice, William L. Hollo-
way 5,000
Attorney-General, A. J. Galen...... 3,000
Officers elected for four years. Present
term expires January, 1918.

NEBRASKA
Lincoln

Governor, A. C. Shallenberger $2,500
Lieut.-Gov., M. R. Hopewell.. $10 per day
Secretary of State, George C. Junkin 2,000
Treasurer, L. G. Brian 2,500
Auditor, C. A. Barton............. 2,000
Chief Justice, M. B. Reese......... 4,500
Att'y-Gen., N. T. Thompson........ 2,000
Officers elected for two years, except
chief justice, for four years. Present term
expires January, 1911; chief justice, Janu-
ary, 1911.

NEVADA
Carson City

Governor, D. S. Dickerson (acting
governor) $3,333.40
Lieut.-Gov., D. S. Dickerson..... 1,800.00
Secretary of State, W. G. Douglas 2,400.00
Treasurer, D. M. Ryan.......... 2,400.00
Auditor 2,400.00
Ttt'y-Gen., C. R. Stoddard....... 2,000.00
Officers elected for four years. Present
term expires January, 1911.

NEW JERSEY
Trenton

Governor, John Franklin Post.....$10,000
Secretary of State, S. D. Dickinson 6,000
Treasurer, Daniel S. Voorhees...... 6,000
Auditor, William E. Drake........ 3,000
Chief Justice, William S. Gummere 10,000
Att'y-Gen., Robert H. McCarter.... 7,000
Officers elected for three years, except
chief justice, elected for seven years.
Present term expires, January, 1911; chief
justice, November, 1917; secretary of state,
April, 1912; state treasurer, March, 1910;
and attorney-general, May, 1913.

NEW HAMPSHIRE
Concord

Governor, Henry B. Quimby....... $2,000
Sec'y of State, Edward N. Pearson 3,000
Treasurer, Solon A. Carter... 2,500
Chief Justice, Frank N. Parsons... 4,200
Att'y-Gen., Edwin G. Eastman..... 2,500
Officers elected for two years, except
chief justice. Regular term expires Janu-
ary, 1911; chief justice, January, 1924.

NEW MEXICO
Santa Fe

Governor, George Curry........... $3,000
Secretary, Nathan Jaffa............. 1,800
Treasurer, J. H. Vaughn........... 2,400
Auditor, William H. Sargent...... 3,000
Chief Justice, William J. Mills..... 3,000
Att'y-Gen., James M. Hervey...... 3,000
Appointed by the president. Terms of
officers expire January 12, 1912, except at-
torney-general, March 19, 1909.

NEW YORK
Albany

Governor, Charles E. Hughes......$10,000
Lieut.-Gov., Horace White........ 5,000
Sec'y of State, Samuel S. Koenig.. 5,000
Treasurer, George B. Dunn........ 5,000
Comptroller, Charles H. Gans...... 6,000
Chief Justice, Edgar M. Cullen..... 10,500
Att'y-Gen., Edward R. O'Malley.... 5,000
Officers elected for two years, excepting
chief justice. Present term expires De-
cember, 1910. Chief justice, December,
1913.

NORTH CAROLINA
Raleigh

Governor, William Walton Kitchin.. $4,000

Lieut.-Gov., W. C. Newland.... $6 per day
Secretary of State, J. Bryan Grimes $3,500
Treasurer, B. R. Lacy............. 3,500
Auditor, B. F. Dixon............. 3,000
Chief Justice, Walter Clark........ 3,500
Att'y-Gen., T. W. Bickett.......... 3,000
Officers elected for four years. Present
term expires January, 1918.

NORTH DAKOTA
Bismarck

Governor, John Burke............. $3,000
Lieut.-Gov., R. S. Lewis........... 1,000
Secretary of State, Alfred Blaisdell 2,000
Treasurer, C. L. Bickford......... 2,000
Auditor, D. K. Brightbill............ 2,000
Chief Justice, D. E. Morgan....... 5,000
Att'y-Gen., Andrew Miller......... 2,000
Officers elected for two years. Terms of
officers expire January, 1911, except chief
justice, December, 1910.

OHIO
Columbus

Governor, Judson Harmon........$10,000
Lieut.-Gov., Francis W. Treadway.. 1,500
Sec'y of State, Carmi A. Thompson 6,500
Treasurer, David S. Creamer...... 6,500
Auditor, Edward M. Fullington.... 6,500
Chief Justice, John A. Shanck..... 6,000
Att'y-Gen., Ulysses G. Denman..... 6,500
Officers elected for two years. Present
term expires January, 1911.

OKLAHOMA
Guthrie

Governor, C. N. Haskell........... $4,500
Lieut.-Gov., George W. Bellamy... 1,000
Secretary of State, Bill Cross...... 2,500
Treasurer, J. A. Menepee.......... 3,000
Auditor, M. E. Trapp.............. 2,500
Chief Justice, R. L. Williams...... 4,000
Att'y-Gen., Charles West.......... 4,000
Officers elected for three years. Present
term expires January, 1911.

OREGON
Salem

Governor, George E. Chamberlain.. $5,000
Secretary of State, Frank W. Benson 4,500
Treasurer, George A. Steel......... 4,500
Chief Justice, Robert S. Bean...... 4,500
Att'y-Gen., A. M. Crawford........ 3,600
Officers elected for four years. Present
term expires January, 1911, except chief
justice, January 11, 1915.

PENNSYLVANIA
Harrisburg

Governor, Edwin S. Stuart........$10,000
Lieut.-Gov., Robert S. Murphy.... 5,000
Secretary of State, Robert McAfee 8,000
Treasurer, John O. Sheatz........ 8,000
Auditor, Robert K. Young... 4,000
Chief Justice, James T. Mitchell... 10,500
Att'y-Gen., M. Hampton Todd...... 12,000
Term of governor and lieutenant-gover-
nor expires January, 1911; term of treas-
urer, auditor and chief justice, 1910; sec-
retary of state and attorney-general, term
pleasure of governor. Terms of office four
years.

PHILIPPINE ISLANDS
Manila

Gov.-Gen., James F. Smith........$20,000
Sec'y of the In'r, Dean C. Worcester 15,500
Sec'y of Commerce and Police, W.
Cameron Forbes 15,500
Sec'y of Public Instruction, W.
Morgan Shuster 15,500
Sec'y of Finance and Justice, Greg-
orio Araneta 15,500
Other Commissioners: T. H. Pardo
de Tavera 7,500
Rafael Palma 7,500
Newton W. Gilbert 7,500
Jose R. de Luzuriaga............ 7,500
Appointed by the president.

PORTO RICO

San Juan

Governor, Regis H. Post............. $8,000
Secretary, William F. Willoughby.. 4,000
Commissioner of Interior, Laurance
　H. Grahame 4,000
Commissioner of Education, Ed-
　ward Grant Dexter............ 4,000
　Terms of governor and secretary expire
January 16, 1912; term of commissioner of
interior, December 19, 1909; term of com-
missioner of education expires January
16, 1912. Appointed by the president.

RHODE ISLAND

Providence

Governor, Aaron J. Pothier........ $3,000
Lieut.-Gov., Arthur W. Dennis..... 500
Sec'y of State, Charles P. Bennett.. 4,500
Treasurer, Walter A. Read........ 4,000
Auditor, Charles C. Gray.......... 4,000
Chief Justice, Edward C. Dubois... 6,500
Att'y-Gen., William B. Greenough.. 4,500
　Officers elected for one year. Present
term expires 1910, except state auditor,
January 21, 1909, and chief justice, for life.

SOUTH CAROLINA

Columbia

Governor, Martin F. Ausel........ $3,000
Lieut.-Gov., T. F. McLeod...... $4 per day
Sec'y of State, R. M. McCowen..... 1,900
Treasurer, R. H. Jennings......... 1,900
Comptroller General, A. D. Jones... 1,900
Chief Justice, Y. J. Pope.......... 3,000
Attorney-General, J. Fraser Lyon.. 1,900
　Officers elected for two years. Present
term expires January, 1911.

SOUTH DAKOTA

Pierre

Governor, R. S. Vessey........... $3,000
Lieut.-Gov., H. C. Shober...... $5 per day
Secretary of State, S. C. Polley.... 1,800
Treasurer, George W. Johnson..... 1,800
Auditor, John Hirning............ 1,800
Chief Justice, Dick Haney......... 3,000
Attorney-General, S. W. Clark..... 1,000
　Officers elected for two years. Present
term expires January, 1911.

TENNESSEE

Nashville

Governor, Malcolm R. Patterson... $4,000
Secretary of State, J. W. Morton.. 3,000
Treasurer, Reau E. Folk........... 3,500
Comptroller, Frank Debrell........ 4,000
Chief Justice, W. D. Beard........ 3,500
Att'y-Gen., Charles F. Cates, Jr.... 3,000
　Officers elected for two years, except sec-
retary of state, four years, chief justice
and attorney-general, eight years. Present
term expires January, 1911, except secre-
tary of state, February, 1909, and chief
justice and attorney-general, January, 1910.

TEXAS

Austin

Governor, T. M. Campbell......... $4,000
Lieut.-Gov., A. B. Davidson
　　　　$5 during legislature
Secretary of State. W. R. Davie.... 2,500
Treasurer, Sam. Sparks........... 2,500
Comptroller, J. W. Stephens...... 2,500
Att'y-Gen., R. V. Davidson........ 4,000
Chief Justice, R. R. Gaines........ 4,000
　Officers elected for two years, except
chief justice, six years. Present term ex-
pires January, 1911, chief justice, January,
1913.

TUTUILA

(Samoa)

Governor, Captain John F. Parker. $3,500
　Appointed by the president.

UTAH

Salt Lake City

Governor, William Spry.......... $4,000
Sec'y of State, Charles S. Tingey.. 2,000
Treasurer, David Mattson........ 1,500
Auditor, Jesse D. Jewkes.......... 2,000
Chief Justice, Daniel N. Straup.... 5,000
Att'y-Gen., Albert R. Barnes....... 2,000
　Officers elected for four years. Present
term expires January, 1913.

VERMONT

Montpelier

Governor, George H. Prouty....... $1,500
Lieut.-Gov., John A. Mead.... $6 per day
Secretary of State, Guy W. Bailey 1,700
Treasurer, Edward H. Deavitt..... 1,700
Auditor, Horace F. Graham....... 2,000
Chief Justice, John W. Rowell..... 3,500
Att'y-Gen., John G. Sargeant..... 2,500
　Officers elected for two years. Present
term expires October, 1910.

VIRGINIA

Richmond

Governor, Claude A. Swanson...... $5,000
Lieut.-Gov., J. Taylor Ellyson. Reg-
　ular sessions........$720; extra $360
Secretary of Commonwealth, D. Q.
　Eggleston 2,800
Treasurer, A. W. Harman, Jr...... 2,000
Auditor of Public Accounts, Morton
　Marye 4,000
Chief Justice, James Keith........ 4,700
Att'y-Gen., William A. Anderson.. 4,000
　Officers elected for four years, except
chief justice, elected for 12 years, to Feb-
ruary, 1917. Present term expires Feb-
ruary, 1910, except auditor of public ac-
counts, March 1, 1912.

WASHINGTON

Olympia

Governor, Samuel G. Cosgrove..... $6,000
Lieut.-Gov., M. E. Hay........... 1,200
Secretary of State, Sam H. Nichols 3,000
Treasurer, John G. Lewis.......... 3,000
Auditor, C. W. Clausen........... 3,000
Attorney-General, W. P. Bell...... 3,000
Chief Justice, Wallace Mount...... 5,000
　Officers elected for four years, except
chief justice, six years. Present term ex-
pires January, 1913; chief justice, Janu-
ary, 1909.

WEST VIRGINIA

Charleston

Governor, W. E. Glasscock........ $5,000
Secretary of State, Stuart F. Reed 4,000
Treasurer, E. L. Long............. 2,500
Auditor, J. S. Darst.............. 4,500
Chief Justice, Henry Brannon..... 4,500
Att'y-Gen., W. G. Conley......... 2,500
　Officers elected for four years. Present
term expires March 4, 1913.

WISCONSIN

Madison

Governor, James O. Davidson...... $5,000
Lieut.-Gov., John Strange.......... 1,000
Secretary of State, James A. Frear 5,000
Treasurer, Andrew H. Dahl....... 5,000
Supreme Court Justice, John B.
　Winslow 5,000
Att'y-Gen., Frank L. Gilbert...... 3,000
　Officers elected for two years. Present
term expires January, 1911, except chief
justice, elected for eight years; present
term expires January, 1916.

WYOMING

Cheyenne

Governor, Bryant B. Brooks...... $2,500
Sec'y of State, Wm. R. Schnitger 2,000
Treasurer, Edward Gillette 2,000

Auditor, Leroy Grant $2,000
Chief Justice, Charles N. Potter... 3,000
Att'y-Gen., William E. Mullen..... 3,000
Officers elected for four years, except justice of supreme court, elected for eight years; present term expires January, 1911, except chief justice, January, 1913.

Dominion of Canada

Governor-General, Earl Gray.

THE CABINET

Premier, Sir Wilfred Laurier.
Minister of Trade and Commerce, Sir Richard John Cartwright.
Secretary of State, Charles Murphy.
Minister of Militia and Defense, Sir Frederick William Borden.
Minister of Agriculture, Sydney Arthur Fisher.
Minister of Finance, William Stevens Fielding.
Minister of Customs, William Paterson.
Minister of Inland Revenue, William Templeman.
Minister of Marine and Fisheries, Louis Philippe Brodeur.
Minister of the Interior, Frank Oliver.
Minister of Justice, Allen Bristol Aylesworth.
Postmaster-General and Minister of Labor, Rodolphe Lemieux.
Minister of Public Works, William Pugsley.
Minister of Railways and Canals, George Perry Graham.
Salary of governor-general, $48,667; of premier, $12,000; of cabinet ministers, $7,000.

Latin-American Republics

Argentine Republic	Sr. Don José Figueroa Alcorta
Bolivia	Col. Ismael Montes
Brasil	Dr. Affonso A. Moreira Penna
Chile	Pedro Montt
Colombia	Gen. Rafael Reyes
Costa Rica	Cleto Gonzales Viquez
Cuba	Gen. Jose Miguel Gomes
Dominican Republic	Ramón Cáceres
Ecuador	Gen. Eloy Alfaro
Guatemala	Manuel Estrada Cabrera
Haiti	Gen. Antoine Simon
Honduras	Gen. Miguel R. Dávila
Mexico	Gen. Porfirio Dias
Nicaragua	Gen. José Santos Zelaya
Panama	Dr. T. Domingo de Obaldía
Paraguay	Dr. Benigno Ferreira
Peru	Dr. Augusto B. Leguia
Salvador	Gen. Fernando Figueroa
Uruguay	Dr. Claudio Williman
Venezuela	Juan Vicente Gomes

Rulers of the Old World

Country	Official Head	Title	Acceded
Abyssinia	Menelik II	Emperor	1889
Belgium	Leopold II	King	1865
Bulgaria	Ferdinand	Emperor	1908
China	Hsuan Tung	Emperor	1908
	Pu Chun	Regent	1908
Denmark	Frederick VIII	King	1906
Egypt	Abbas Pacha	Khedive	1892
France	Armand Fallieres	President	1906
Germany	William II	Emperor	1888
Great Britain	Edward VII	King	1901
Greece	George	King	1863
Italy	Victor Emmanuel III	King	1900
Japan	Mutsuhito	Mikado	1867
Korea	Yi Hiung	Emperor	1864
Liberia	Arthur Barclay	President	1904
Morocco	Mulai Hafid	Sultan	1908
Netherlands	Wilhelmina	Queen	1898
Norway	Haakon VII	King	1905
Persia	Mohammed Ali Mirza	Shah	1907
Portugal	Manuel	King	1908
Roumania	Charles	King	1881
Servia	Peter (Karageorgevitch)	King	1903
Russia	Nicholas II	Emperor	1894
Siam	Khoulalonkorn	King	1886
Spain	Alfonso XIII	King	1886
Sweden	Gustav V	King	1827
Switzerland	Edouard Muller	President	1906
Turkey	Abdul Hamid II	Sultan	1876
Zanzibar	Seyyid Ali	Sultan	1902

United States Diplomatic Representatives

(1) Ambassador extraordinary and plenipotentiary.
(2) Envoy extraordinary and minister plenipotentiary.
(3) Minister resident and consul general.

Country	Name		From	Salary
Abyssinia	Hoffman Philip	(3)	N. Y.	
Argentine Republic	Spencer F. Eddy	(2)	Ill.	$12,000
Austria-Hungary	Charles S. Francis	(1)	N. Y.	17,500
Belgium	Henry Lane Wilson	(2)	Wash.	12,000
Bolivia	James F. Stutesman	(2)	Ind.	10,000
Brazil	Irving B. Dudley	(1)	Cal.	17,500
Chile	John Hicks	(2)	Wis.	12,000
China	William W. Rockhill	(2)	D. C.	12,000
Colombia	Thomas C. Dawson	(2)	Iowa	10,000
Costa Rica	William L. Merry	(2)	Cal.	10,000
Cuba	Edwin V. Morgan	(2)	N. Y.	12,000
Denmark	Maurice F. Egan	(2)	D. C.	10,000
Dominican Republic	Fenton R. McCreery	(3)	Mich.	10,000
Ecuador	William C. Fox	(2)	N. J.	10,000
France	Henry White	(1)	R. I.	17,500
German Empire	David J. Hill	(1)	N. Y.	17,500
Great Britain	Whitelaw Reid	(1)	N. Y.	17,500
Greece	Richmond Pearson	(2)	N. C.	10,000
Guatemala	Richard Heimke	(2)	Kan.	10,000
Haiti	Henry W. Furniss	(2)	Ind.	10,000
Honduras	William B. Sorsby	(2)	Miss.	10,000
Italy	Lloyd C. Griscom	(1)	Pa.	17,500
Japan	Thomas J. O'Brien	(1)	Mich.	17,500
Liberia	Ernest Lyon	(3)	Md.	5,000
Luxembourg	Arthur M. Beaupre	(2)	Ill.	12,000
Mexico	David E. Thompson	(1)	Neb.	17,500
Montenegro	Richmond Pearson	(2)	N. C.	10,000
Morocco	Samuel R. Gummere	(2)	N. J.	10,000
Netherlands	Arthur M. Beaupre	(2)	Ill.	12,000
Nicaragua	John Gardner Coolidge	(2)	Mass.	10,000
Norway	Herbert H. D. Pierce	(2)	Mass.	10,000
Panama	Herbert G. Squiers	(2)	N. Y.	10,000
Paraguay	Edward C. O'Brien	(2)	N. Y.	10,000
Persia	John B. Jackson	(2)	N. J.	10,000
Portugal	Charles Page Bryan	(2)	Ill.	10,000
Roumania	Horace G. Knowles	(2)	Del.	10,000
Russia	John W. Riddle	(1)	Minn.	17,500

Dr E. B. Voorhees

For a score of years Dr Voorhees has directed the experimental work in the state of New Jersey. One of his most important contributions to agricultural science has been with dealing with the rational use of chemical manures in land maintenance. Dr Voorhees is one of the most energetic leaders of the eastern states.

Prof L. R. Taft

For many years Professor of Horticulture in Michigan Agricultural College, Prof Taft has now become director of Farmers' Institutes in that state. Many years ago Prof Taft made a careful study of the growing of market garden crops under glass, and the results of his investigations have now become guides for growers in all parts of the country.

Some Notable Ohio Apples—Rome Beauties

GENUINE WILD HORSES FROM ASIA

The only genuine wild horses in the world live in central Asia. They are found in the Gobi desert and in a section of Mongolia on the northwest border in the vicinity of the Altai mountains. A few of these animals have been captured. The picture above shows two of the mares and a colt taken at Woburn park in England, and furnished us by Prof Robert Wallace, author of the interesting book, "Farm Live Stock of Great Britain," which is sold by the Orange Judd Company. In a wild state the animals are timid and difficult to approach. Those in captivity have been tamed so that they seemed to be as gentle as domestic horses. They are very intelligent. Although small, they are well built. It will doubtless be found profitable to cross them with some modern type of horse to produce a new and valuable breed. The picture gives a good idea of their appearance. There are several types of different colorings. Those shown have bodies of a russety brown; their legs are black, their hoofs are hard with no tendency to flatness as is often the case with some domestic horses. They have no forelock.

MIXED RATION BRINGS PROFIT

The two small pigs at the left were supplied corn meal only. The two at the right, corn meal 8 parts and tankage 1 part. The feeding was done in a dry lot, and the experiment lasted 62 days. The results show that the mixed ration is best.

Salvador	H. Percival Dodge	(2)	Mass.	10,000
Servia	Horace G. Knowles	(2)	Del.	10,000
Siam	Hamilton King	(2)	Mich.	10,000
Spain	William H. Collier	(2)	N. Y.	12,000
Sweden	Charles H. Graves	(2)	Minn.	10,000
Switzerland	Brutus J. Clay	(2)	Ky.	10,000
Turkey	John G. A. Leishman	(1)	Pa.	17,000
Bulgaria	Horace G. Knowles	(Dip. Agt.)	Del.	10,000
Egypt	Lewis M. Iddings	(3)	N. Y	6,500
Uruguay	Edward C. O'Brien	(2)	N. Y	10,000
Venezuela	Vacant	(2)		10,000

Diplomatic Representatives to United States

Country	Name	Rank
Argentine Republic	Senor Don Epifania Portela	(2)
Austria-Hungary	Baron Hengelmuller Von Hengervar	(1)
Belgium	Baron Moncheur	(2)
Bolivia	Senor Don Ignacio Calderon	(2)
Brasil	Mr. Joaquim Nabuco	(1)
Chile	Senor Don Anibal Crus	(2)
China	Dr. Wu Ting-fang	(2)
Colombia	Senor Don Enrique Cortes	(2)
Costa Rica	Senor Don Joaquin Bernardo Calvo	(2)
Cuba	Senor Don Gonsalo de Wuesada	(2)
Denmark	Mr. Constantin Brun	(2)
Dominican Republic	Senor Don Emilio C. Joubert	(3)
Ecuador	Senor Don Luis Felipe Carbo	(2)
France	Mr. J. J. Jusserand	(1)
German Empire	Count Herman von Hatzfeldt-Wildenburg	(1)
Great Britain	Right Honorable James Bryce	(1)
Greece	Mr. L. A. Coromilas	(3)
Guatemala	Senor Dr. Don Luis Toledo Herrarte	(2)
Haiti	Mr. N. J. Leger	(2)
Honduras	Dr. Angel Ugarte	(2)
Italy	Baron Edmondo Mayor des Planches	(1)
Japan	Baron Kogoro Takahira	(1)
Mexico	Senor Don Enrique C. Creel	(1)
Netherlands	Jonkheer R de Marees van Swinderen	(2)
Nicaragua	Senor Don Luis F. Corea	(2)
Norway	Mr. O. Gude	(2)
Panama	Senor Don Jose Augustin Arango	(2)
Persia	General Morteza, Khan	(2)
Peru	Mr. Felipe Pardo	(2)
Portugal	Viscount de Alte	(2)
Russia	Baron Rosen	(1)
Salvador	Senor Don Federico Mejia	(2)
Siam	Phya Akharaj Varadhara	(2)
Spain	Senor Don Ramon Pina	(2)
Sweden	Mr. Herman de Lagercrants	(2)
Switzerland	Mr. Leo Vogel	(2)
Turkey	Munji Bey	(3)
Uruguay	L.. Luis Melian Lafinur	(2)

Agricultural Colleges

LOCATIONS AND PRESIDENTS

Alabama—Alabama Polytechnic institute, Charles C. Thach, Agricultural and Mechanical college for negroes, Normal, W. H. Council. Tuskegee Normal and Industrial institute for negroes, Tuskegee, B. T. Washington.

Arizona—University of Arizona, Tucson, Kendric C. Babcock.

Arkansas—University of Arkansas, Fayetteville, John N. Tillman.

California—University of California, Berkeley, Benjamin I. Wheeler.

Colorado—State agricultural college of Colorado, Fort Collins, B. O. Aylesworth.

Connecticut—C. L. Beach, Connecticut agricultural college, Storrs, Ct.

Delaware—Delaware college, Newark, G. A. Harter. State College for Colored students, W. C. Jason, Dover.

Florida—University of Florida, Andrew Sledd, Gainesville. Florida State normal and industrial college, Tallahassee, Nathan B. Young.

Georgia—Georgia state college of agriculture and the mechanic arts, Athens, Andrew M. Soule. Georgia state industrial college, R. R. Wright, Savannah.

Idaho—University of Idaho, Moscow, James A. MacLean.

Illinois—University of Illinois, Urbana, Edmund J. James.

Indiana—Purdue university, Lafayette, Winthrop E. Stone.

Iowa—Iowa State college of agriculture and mechanic arts, Ames, Albert B. Storms.

Kansas—Kansas state agricultural college, Manhattan, Ernest R. Nichols.

Kentucky—State University of Kentucky, Lexington, James K. Patterson. The Kentucky normal and industrial institute for colored persons, Frankfort, John H. Jackson.

Louisiana—Louisiana State university and agricultural and mechanical college, Baton Rouge, Thomas D. Boyd. Southern university and agricultural and mechanical college, New Orleans, H. A. Hill.

Maine—The university of Maine, Orono, G. E. Fellows.

Maryland—Maryland agricultural college, College Park, R. W. Silvester. Princess Anne Academy, eastern branch, of Maryland agricultural college, Princess Anne, Frank Trigg.

Massachusetts—Massachusetts agricultural college, Amherst, K. L. Butterfield.

Mississippi—Mississippi agricultural and mechanical college, Agricultural College, J. C. Hardy. Alcorn agricultural and mechanical college, Alcorn, L. J. Rowan.

Missouri—University of Missouri, Columbia, Albert Ross Hill. Lincoln institute, Jefferson City, B. F. Allen.

Montana—Montana college of agriculture and mechanic arts, Bozeman, J. M. Hamilton.

Nebraska—University of Nebraska, Lincoln, E. B. Andrews.

Nevada—University of Nevada, Reno, Joseph E. Stubbs.

New Hampshire—The New Hampshire college of agriculture and the mechanic arts, Agricultural college, Luther Foster.

New Jersey—Rutgers Scientific school, New Jersey state college for the benefit of agriculture and mechanic arts, New Brunswick, W. H. S. Demarest.

New Mexico—New Mexico college of agriculture and mechanic arts, Agricultural College, W. E. Garrison.

New York—Cornell University, Ithaca, Jacob Gould Schurman.

North Carolina—North Carolina college of agricultural and mechanic arts, West Raleigh, Daniel H. Hill. The agricultural and mechanical college for the colored race, Greensboro, James B. Dudley.

North Dakota—North Dakota agricultural college, Agricultural College, J. H. Worst.

Ohio—Ohio State university, Columbus, William O. Thompson.

Oklahoma — Oklahoma agricultural and mechanical college, Stillwater; J. H. Connell, Agricultural and normal university, I. E. Page, Langston.

Oregon—Oregon state agricultural college, Corvallis, William J. Kerr.

Pennsylvania—The Pennsylvania state college, State college, Edwin E. Sparks.

Rhode Island—Rhode Island college of agriculture and mechanic arts, Kingston, Howard Edwards.

South Carolina—Clemson agricultural college of South Carolina, Clemson college, P. H. Mell. The colored normal, industrial, agricultural and mechanical college of South Carolina, Orangeburg, Thos. E. Miller.

South Dakota—South Dakota agricultural college, Brookings, Robert L. Slagle.

Tennessee — University of Tennessee, Knoxville, Brown Ayres.

Texas—Agricultural and mechanical college of Texas, College Station, R. T. Milner. Prairie View state normal and industrial college, E. L. Blackshear.

Utah—Agricultural college of Utah, Logan, John A. Widtsoe.

Vermont—University of Vermont and State agricultural college, Burlington, M. H. Buckham.

Virginia—Virginia Polytechnic institute, Blacksburg, Paul B. Barringer.

Wisconsin — University of Wisconsin, Madison, Charles R. VanHise.

Agricultural Experiment Stations

LOCATIONS AND DIRECTORS

Alabama (College), Auburn, J. F. Duggar.

Alabama (Canebrake), Uniontown, F. D. Stevens.

Alabama (Tuskegee), Tuskegee institute, G. W. Carver.

Alaska, Sitka, Charles C. Georgeson.

Arizona, Tucson, R. H. Forbes.

Arkansas, Fayetteville, W. G. Vincenheller.

California, Berkeley, E. J. Wickson.

Colorado—Fort Collins, L. G. Carpenter.

Connecticut (State), New Haven, E. H. Jenkins.

Connecticut (Storrs), Storrs, L. A. Clinton.

Delaware, Newark, Harry Hayward.

Florida, Gainesville, P. H. Rolfs.

Georgia, M. V. Calvin, Experiment.

Hawaii, Honolulu, E. V. Wilcox.

Idaho, Moscow, Hiram T. French.

Illinois, Urbana, E. Davenport.

Indiana, Lafayette, Arthur Goss.

Iowa, Ames, Charles F. Curtis.

Kansas, Manhattan, E. H. Webster.

Kentucky, Lexington, M. S. Scovell.

Louisiana (Sugar), Baton Rouge, W. R. Dodson.

Louisiana (State), Baton Rouge, W. R. Dodson.

Louisiana (North), Calhoun, W. R. Dodson.

Maine, Orono, C. D. Woods.

Maryland, College Park, H. J. Patterson.

Massachusetts, Amherst, W. P. Brooks.

Michigan, Agricultural college, R. S. Shaw.

Minnesota, St. Anthony Park, St. Paul, E. W. Randall.

Mississippi, Agricultural college, W. L. Hutchinson.

Missouri (College), Columbia, H. J. Waters.

Missouri (Fruit), Paul Evans.

Montana, Bozeman, F. B. Linfield.

Nebraska, Lincoln, E. A. Burnett.

Nevada, J. E. Stubbs, Reno.

New Hampshire, Durham, E. D. Sanderson.

New Jersey (State college), New Brunswick, E. B. Voorhees.

New York (State), Geneva, W. H. Jordan.

New York (Cornell), Ithaca, L. H. Bailey.

North Carolina, Raleigh, C. B. Williams.

North Dakota, Agricultural college, J. H. Worst.

Ohio, Wooster, Charles E. Thorne.

Oklahoma, Stillwater, B. C. Pittuck, assistant director.

Oregon, Corvallis, J. Withycombe, director.

Pennsylvania (State), State college, Thomas F. Hunt.

Porto Rico, Mayaguez, David W. May.

Rhode Island, Kingston, H. J. Wheeler.

South Carolina (State), Clemson college, J. N. Harper.

South Dakota, Brookings, James W. Wilson.

Tennessee, Knoxville, H. A. Morgan.

Texas, College Station, H. H. Harrington.

Utah, Logan, E. D. Ball.

Vermont, Burlington, J. L. Hills.

Virginia, Blacksburg, Dr. S. W. Fletcher.

Washington, Pullman, R. W. Thatcher.

West Virginia, Morgantown, J. H. Stewart.

Wisconsin, Madison, Harry L. Russell.

Wyoming, Laramie, J. D. Towar.

In Charge of Farmers' Institutes

Farmers' Institute Specialist, Department of Agriculture, John Hamilton, Washington, D. C.

Alabama—C. A. Cary, Alabama Polytechnic institute, Auburn, Director agricultural experiment station. G. W. Carver, director agricultural experiment station, Tuskegee institute.

Arizona— R. W. Clothier, superintendent farmers' institute, Tucson.

Arkansas—W. J. Vincenheller, Fayetteville agricultural experiment station.

California—Warren T. Clarke, director farmers' institute, Berkeley.

Colorado—H. M. Cottrell, Fort Collins, superintendent farmers' institute.

Connecticut—J. F. Brown, secretary state board of agriculture, No. Stonington; J. G. Schwink, secretary Connecticut dairymen's association, Meriden; H. C. C. Miles, secretary Connecticut Pomological society, Milford.

Delaware—Wesley Webb, director farmers' institute, Dover.

Georgia—Andrew M. Soule, director farmers' institute, Atlanta.

Hawaii—William Weinrich, director farmers' institute, Honolulu.

Idaho—H. T. French, director agricultural experiment station, Moscow.

Illinois—H. A. McKeene, secretary of farmers' institute, Springfield.

Indiana—W. C. Lata, superintendent farmers' institute, Lafayette.

Iowa—J. C. Simpson, secretary state board of agriculture, Des Moines.

Kansas—J. H. Miller, director farmers' institute, Manhattan.

Kentucky—M. C. Rankin, in charge farmers' institute, Frankfort.

Louisiana — Charles Schuler, commissioner of agriculture, Baton Rouge.

Maine—A. W. Gilman, commissioner of agriculture, Augusta.

Maryland—William L. Amoss, director farmers' institutes, College Park.

Massachusetts—J. L. Ellsworth, secretary state board of agriculture, Boston.

Michigan—L. R. Taft, director farmers' institute, Detroit, Mich.

Minnesota—A. D. Wilson, superintendent farmers' institute, St. Anthony Park.

Mississippi—E. R. Lloyd, Agricultural college, director farmers' institute.

Missouri—George B. Ellis, secretary state board of agriculture, Columbia.

Montana—F. B. Linfield, director farmers' institute, Bozeman.

Nebraska—E. A. Burnett, superintendent farmers' institute, Lincoln.

Nevada—Joseph E. Stubbs, president Nevada state university, Reno.

New Hampshire—N. J. Bachelder, secretary state board of agriculture, Concord.

New Jersey—Franklin Dye, secretary state board of agriculture, Trenton.

New Mexico—John D. Tinsely, director farmers' institute.

New York—R. A. Pearson, commissioner of agriculture, Albany.

North Carolina—Tait Butler, director farmers' institutes, Raleigh.

North Dakota—T. A. Hoverstad, superintendent farmers' institute, Agricultural college.

Ohio—T. L. Calvert, secretary state board of agriculture, Columbus.

Oklahoma—J. P. Connors, in charge of farmers' institute, Guthrie.

Oregon—James Withycombe, agricultural experiment station, Corvallis.

Pennsylvania—A. L. Martin, director farmers' institute, Harrisburg.

Porto Rico—D. W. May, director farmers' institute, Mayaguez.

Rhode Island—John J. Dunn, secretary state board of agriculture, Providence.

South Carolina—J. N. Harper, director farmers' institute, Clemson college.

South Dakota—A. E. Chamberlain, superintendent farmers' institute, Brookings.

Tennessee—John Thompson, commissioner of agriculture, Nashville.

Utah—Lewis A. Merrill, in charge farmers' institute, Logan.

Vermont—F. L. Davis, No. Pomfret.

Virginia—George W. Koiner, in charge of farmers' institute, Blacksburg.

Washington—R. W. Thatcher, in charge farmers' institute, Pullman.

West Virginia—J. B Garvin, superintendent farmers' institute, Charleston.

Wisconsin—George McKerrow, superintendent farmers' institute, Madison.

Wyoming—J. D. Towar, director agricultural experiment station, Laramie.

Secretaries of State Boards of Agriculture

California—J. A. Filcher, Sacramento.

Colorado—A. M. Hawley, Fort Collins.

Connecticut—J. F. Brown, North Stonington.

Delaware—Wesley Webb, Camden.

Illinois—J. K. Dickinson, Springfield.

Indiana—Charles Downing, Indianapolis.

Iowa—J. C. Simpson, Des Moines.

Kansas—F. D. Coburn, Topeka.

Maryland—W. Frank Hines, superintendent of immigration, Baltimore.

Massachusetts—J. L. Ellsworth, Boston.

Michigan—Addison M. Brown, Agricultural college.

Minnesota—C. N. Cosgrove, St. Paul, secretary state agricultural society.

Missouri—George B. Ellis, Columbia.

Nebraska—W. R. Mellor, Lincoln.

Nevada—W. G. Douglas, Carson City.

New Hampshire—N. J. Bachelder, Concord.

New Jersey—Franklin Dye, Trenton.

New Mexico—Nathan Jaffa, Santa Fe.

North Carolina—Elias Parr, Raleigh.

Ohio—T. L. Calvert, Columbus.

Oklahoma—Charles F. Barrett, Guthrie.

Oregon—Frank Welch, Salem.

Pennsylvania—N. B. Critchfield, Harrisburg.

Rhode Island—J. Dunne, Providence.

South Dakota—C. N. McIlvaine, Huron.

Vermont—F. L. Davis, No. Pomfret.

Washington—Sam H. Nichols, Olympia.

West Virginia—J. B. Garvin, Charleston.

Wisconsin—John M. True, Madison.

Wyoming—Clarence T. Johnson, Cheyenne. (State engineer.)

State Commissioners of Agriculture

Alabama—J. A. Wilkinson, Montgomery.

Arkansas—Guy B. Tucker, Little Rock.

Florida—B. E. McLin, Tallahassee.

Georgia—T. G. Hudson, Atlanta.

Hawaii—C. S. Holloway, Honolulu.

Idaho—Allen Miller, Boise.

Kentucky—M. C. Rankin, Frankfort.

Louisiana — Charles Schuler, Baton Rouge.

Maine—A. W. Gilman, Augusta.

Montana—J. A. Ferguson, Helena.

New York—Raymond A. Pearson, Albany.

North Carolina—S. L. Patterson, Raleigh.

North Dakota—W. C. Gilbreath, Bismarck.

Philippine Islands—G. E. Nesom, director of agriculture, Manila.

Porto Rico—Lawrence K. Grahame, commissioner of the interior, San Juan.

Tennessee—John Thompson, Nashville.

Texas—Ed. R. Kone, Austin.

Virginia—G. W. Koiner, Richmond.

Washington—Sam H. Nichols, secretary of state, Olympia.

Sanitary Officers in Charge of Live Stock

Alabama—C. A. Cary, Auburn, veterinarian.

Arizona—J. C. Norton, Phoenix, veterinarian; secretary, J. C. Norton, Phoenix.

Arkansas—Wilfred Lenton, Fayetteville, veterinarian.

California—Charles Keane, Sacramento, veterinarian.

Colorado—Charles G. Lamb, Denver, veterinarian; secretary, E. McCrillis, Capitol building, Denver.

Connecticut—Heman O. Averill, commissioner for domestic animals, Hartford.

Delaware—A. E. Frantz, secretary state board of health, Wilmington.

Florida—Joseph Y. Porter, Jacksonville, state health officer.

Georgia—T. G. Hudson, commissioner of agriculture, Atlanta.

Idaho—G. E. Noble, Boise, veterinarian; W. H. Philbrick, American Falls, secretary.

Illinois — W. E. Savage, secretary; Springfield; Dr. J. M. Wright, 1827 Wabash Ave., Chicago, Ill., veterinarian.

Indiana—A. W. Bitting, LaFayette, veterinarian.

Iowa—Paul O. Koto, Des Moines, veterinarian.

Kansas—J. B. Baker, live stock sanitary commission, Topeka.

Kentucky—F. T. Eisenman, Louisville, veterinarian.

Maine—F. O. Beal, Bangor, veterinarian.

Maryland—F. H. Mackie, Baltimore, veterinarian.

Massachusetts—Austin Peters, Boston, veterinarian.

Michigan—William Morris, Cass City, Mich., veterinarian; Comfort A. Tyler, 310 E. Chicago St., Coldwater, Mich., secretary.

Minnesota—S. H. Ward, St. Paul, veterinarian.

Mississippi—J. C. Roberts, professor of veterinary science, Agricultural college.

Missouri—D. F. Luckey, Columbia, veterinarian; George B. Ellis, Columbia, secretary.

Montana—William Treacy, Helena, president; M. E. Knowles, Helena, secretary and veterinarian.

Nebraska—Charles A. McKim, Lincoln, veterinarian.
Nevada—Isaac O'Rourke, Carson City, veterinarian.
New Hampshire—Irving A. Watson, Concord, veterinarian; N. J. Bachelder, Concord, secretary.
New Jersey—T. Earle Budd, New Brunswick, veterinarian.
New Mexico—E. Godwin Austen, East Las Vegas, secretary.
New York—William H. Kelly, veterinarian, Albany.
North Carolina—Tait Butler, Raleigh, veterinarian.
North Dakota—W. F. Crewe, Devil's Lake, veterinarian.
Ohio—Paul Fischer, Columbus, veterinarian; T. L. Calvert, Columbus, secretary.
Oklahoma—G. T. Bryan, superintendent live stock inspection, Guthrie.
Oregon—William H. Lytle, Pendleton, state sheep inspector.
Pennsylvania—Leonard Pearson, Philadelphia, veterinarian.
Rhode Island—John S. Pollard, Providence, veterinarian.
South Carolina—M. Ray Powers, Clemson college, veterinarian.
South Dakota—T. H. Hicks, Pierre, veterinarian.
Tennessee—W. H. Dunn, Nashville, veterinarian.
Texas—J. H. Wilson, Quanah, chairman of board in charge of live stock.
Utah—T. B. Beatty, Salt Lake City, secretary.
Vermont—H. S. Wilson, Arlington.
Virginia—J. G. Ferneybough, veterinarian.
Washington—S. B. Nelson, Pullman, veterinarian.
West Virginia—J. B. Garvin, secretary of agriculture, Charleston.
Wisconsin—David Roberts, Waukesha, veterinarian.
Wyoming—William F. Pflaeging, Cheyenne, veterinarian; George S. Walker, Cheyenne, secretary sheep commission.

State Food Officers

California—Martin Regensburger, state board of health, Sacramento.
Colorado—A. H. Davis, state board of health, Denver; Wilber F. Cannon, Denver; Dr. Hugh L. Taylor, secretary state board of health, Denver.
Connecticut—H. F. Potter, state dairy and food commissioner, New Haven.
Delaware—Oscar C. Draper, state board of health, Wilmington.
District of Columbia—William C. Woodward, health department, Washington.
Florida—B. E. McLin, commissioner of agriculture, Tallahassee.
Georgia—Thomas G. Hudson, commissioner of agriculture, Atlanta.
Hawaii—Robert A. Duncan, 1737 Makiki St., Honolulu.
Idaho—J. R. Field, Boise.
Illinois—Alfred H. Jones, food and dairy commission, Robinson.
Indiana—Harry E. Barnard, food and drug commissioner, Indianapolis.
Iowa—H. R. Wright, food and dairy commission, Des Moines.
Kansas—S. J. Crumbine, chief food and drug inspector, secretary, Topeka.
Kentucky—M. A. Scovell, agricultural experiment station, Lexington.
Louisiana—Harvey Dillon, state board of health, New Orleans.
Maine—Charles D. Woods, agricultural experiment station, Orono.
Massachusetts—C. D. Richardson, West Brookfield.
Michigan—Arthur C. Bird, dairy and food department, Lansing.
Minnesota—E. K. Slater, dairy and food commission, St. Paul.

Missouri—M. H. Lamb, dairy and food commissioner, Columbia.
Nebraska—J. W. Johnson, food, dairy and drug commission, Lincoln.
New Hampshire—Irving A. Watson, state board of health, Concord.
New Jersey—George W. McGuire, chief division of dairies and creameries, Trenton; Bruce S. Keaton, state board of health, Trenton.
New York—Raymond A. Pearson, commissioner of agriculture, Albany; Eugene H. Porter, commissioner of health, Albany.
North Carolina—W. A. Graham, commissioner of agriculture, Raleigh; W. M. Allen, food chemist, Raleigh.
North Dakota—E. F. Ladd, agricultural experiment station, agricultural college.
Ohio—Renick W. Dunlap, dairy and food commission, Columbus.
Oklahoma—J. C. Mahr, Shawnee.
Oregon—Robert C. Yenney, state health officer, Portland; J. W. Bailey, dairy and food commissioner, Portland.
Pennsylvania—Dairy and food commissioner, James Foust, Harrisburg.
Porto Rico—R. M. Hernadaz, bureau of health, San Juan.
South Carolina—Robert Wilson, Jr., state board of health, Charleston.
South Dakota—A. H. Wheaton, food and dairy commissioner, Brookings.
Tennessee—L. P. Brown, food official and drug inspector, Nashville.
Texas—Dairy and food commissioner, Austin.
Utah—Dairy and food commissioner, John Peterson, Salt Lake City.
Vermont—Henry D. Holton, state board of health, Brattleboro.
Virginia—W. D. Saunders, dairy and food commissioner, Richmond.
Washington—L. Davies, dairy and food commission, Davenport.
Wisconsin—J. Q. Emery, dairy and food commission, Madison.
Wyoming—E. W. Burke, dairy, food and oil commission, Cheyenne.
Canada — William Templeman, department of internal revenue, Ottawa.

State Dairy Officers

California—Secretary and chemist of state dairy bureau, William H. Saylor, 16 California St., San Francisco.
Colorado—State dairy commissioner, B. G. D. Bishopp, Denver.
Connecticut—State dairy commissioner, Hubert F. Potter, 54 State capitol, Hartford; deputy dairy commissioner, Tyler Cruttenden, Hartford.
Idaho—State dairy, food and oil commissioner, J. R. Field, Boise.
Illinois—State dairy and food commissioner, Alfred H. Jones, Robinson.
Indiana—State food and drug commissioner, H. A. Barnard, Indianapolis.
Iowa—State food and dairy commissioner, H. R. Wright, Des Moines.
Kentucky—The state pure food law is enforced by the experiment station and is particularly enforced with regard to milk and dairy products. Head of food division, R. M. Allen, Kentucky agricultural experiment station.
Kansas—State dairy commissioner, D. M. Wilson, Topeka.
Massachusetts—Executive officer of the dairy bureau, the secretary of the state board of agriculture. General agent, state dairy bureau, P. M. Harwood, room 136, State House, Boston.
Michigan—Dairy and food commissioner, A. C. Bird, Lansing.
Minnesota—Dairy and food commissioner, Edward K. Slater, St. Paul.
Missouri—State dairy and food commissioner, M. H. Lamb, Columbia.
Nebraska—Food commissioner, the governor of the state. Deputy commissioner, J. W. Johnson, Lincoln.

New Jersey—Chief division of dairies and creameries, George W. McGuire, Trenton.

New York—Commissioner of agriculture (including dairy), Raymond A. Pearson, Albany.

North Dakota—Commissioner of agriculture and labor, ex officio state dairy commissioner, W. C. Gilbreath, Bismarck.

Ohio—State dairy and food commissioner, Renick W. Dunlap, Columbus.

Oregon—Dairy and food commissioner, J. W. Bailey, Portland.

Pennsylvania—Dairy and food commissioner, James Foust, Harrisburg.

South Dakota—Food and dairy commissioner, A. H. Wheaton, Brookings.

Utah—Dairy and food commissioner, John Peterson, Salt Lake City.

Virginia—State food and dairy commissioner, W. D. Saunders, Richmond.

Washington—Dairy and food commissioner, L. Davies, Davenport.

Wisconsin—State dairy and food commissioner, J. Q. Emery, Madison.

Wyoming—Dairy, food and oil commission, Ed. W. Burke, Cheyenne.

Dairy Associations

SECRETARIES

Association of Inspectors and Instructors of the National and State Dairy and Food Departments—B. D. White, Department of agriculture, Washington, D. C.

Association of State and National Food and Dairy Departments—R. M. Allen, Washington, D. C.

American Association of Medical Milk Commissions—Otto P. Geier, 124 Garfield Pl., Cincinnati, O.

Boston Co-operative Milk Producers' Association—W. A. Hunter, 9 Woodland St., Worcester, Mass.

California Creamery Operators' Association—J. H. Severin, 36 Commercial St., San Francisco, Cal.

Chicago Milk Shippers' Union—H. B. Farmer, Chicago, Ill.

Connecticut Dairymen's Association—J. G. Schwink, Jr., Meriden.

Dairy Instructors' Association—C. B. Lane, Washington, D. C.

Eastern Minnesota Dairymen's Association—J. E. Lindberg, White Bear Lake.

Eastern Iowa Buttermakers' Association—L. S. Edwards, Arlington.

Five States Creamery Association—William Junt, Great Bend, Pa.

Five States Milk Producers' Association—H. T. Coon, Homer, N. Y.

Georgia Dairy and Live Stock Association—C. L. Willoughby, Experiment.

Grand Traverse Dairymen's Association—James Harris, Traverse City, Mich.

Idaho State Dairy Association—A. E. Gipson, Caldwell, Ida.

Illinois Buttermakers' Association—George Caven, 154 Lake St., Chicago.

Illinois State Milk Producers' Institute—J. M. McVean, 184 LaSalle St., Chicago.

Indiana State Dairy Association—H. J. Fidler, Purdue university, Lafayette, Ind.

International Association of Milk Dealers—B. D. White, United States department of agriculture, Washington, D. C.

International Federation of Dairying—Edward H. Webster, chairman American committee, United States department of agriculture, Washington, D. C.

Iowa State Dairy Association—W. B. Johnson, 1337 E. 9th St., Des Moines, Ia.

Iowa Buttermakers' Association—W. B. Johnson, Des Moines.

Kansas State Dairy Association—I. D. Graham, Topeka.

Kentucky Dairy Cattle Club—J. J. Hooper, State college, Lexington.

Maine Dairymen's Association—Leon S. Merrill, Solon.

Massachusetts Creamery Association—A. M. Lyman, Montague.

Michigan Dairymen's Association—S. J. Wilson, Flint.

Minnesota—Co-operative Dairies Association—Charles A. Morse, Sauk Centre.

Minnesota State Dairymen's Association—J. R. Morley, Owatonna.

Minnesota State Butter and Cheesemakers' Association—Edwin Heed, Mankato.

Missouri State Dairy Association—Frank L. Austin, Columbia.

National Dairy Show—E. Sudendorf, Clinton.

National Creamery Buttermakers' Association—S. B. Shilling, 154 Lake St., Chicago, Ill.

National Dairy Union — Charles Y. Knight, 154 Lake St., Chicago, Ill.

Nebraska Dairymen's Association—S. C. Bassett, Gibbon.

New Hampshire Granite State Dairymen's Association—Frank Reed Sanders, Bristol.

New York State Dairymen's Association—Thomas E. Tiquin, Sherburne.

North Dakota Dairymen's Association—T. A. Hoverstad, Fargo.

Official Dairy Instructors' Association—C. B. Lane, Washington, D. C.

Ohio Dairymen's Association—Oscar Erf, Columbus.

Oklahoma Dairymen's Association—Roy C. Potts, Stillwater.

Oregon Dairymen's Association—W. L. Crissey, Portland.

Pennsylvania Creamery Association—George R. Meloney, 1809 Springfield Ave., West Philadelphia, Pa.

Pennsylvania Dairy Union—H. E. Van Norman, Niagara.

Red River Valley Dairymen's Association—O. A. Starvick, Crookston, Minn.

Southern Indiana Dairy and Co-operative Creamery Association—H. A. Reynolds, Huntingburg.

State Dairymen's Association of Montana—W. J. Elliott, Bozeman.

Virginia Dairymen's Association—A. L. French, Byrdville, R. F. D. No. 2.

Vermont Dairymen's Association—L. L. Davies, White River Junction.

Washington State Dairymen's Association—Ira P. Whitney, Pullman.

West Virginia State Dairymen's Association—W. K. Brainerd, Morgantown.

Wisconsin Buttermakers' Association—J. G. Moore, Madison.

Wisconsin Cheesemakers' Association—U. S. Baer, Madison.

Wisconsin Dairymen's Association—A. J. Glover, Ft. Atkinson.

Stock Breeders' Associations

SECRETARIES—HORSES

American Trotter; American Trotting Register Association, William H. Knight, 365 Dearborn St., Chicago, Ill.

Belgian Draft; American Association of Importers and Breeders of Belgian Draft Horses, J. D. Conner, Jr., Wabash, Ind.

Cleveland Bay; Cleveland Bay Society of America, R. P. Stericker, 80 Chestnut Ave., West Orange, N. J.

Clydesdale; American Clydesdale Association, R. B. Ogilvie, Union Stock Yards, Chicago, Ill.

French Coach; French Coach Horse Society of America, Duncan E. Willett, Maple Ave. and Harrison St., Oak Park, Ill.

French Coach; French Coach Horse Registry Company, Charles C. Glenn, 1319 Wesley Ave., Columbus, O.

French Draft; National French Draft Horse Association of America, C. E. Stubbs, Fairfield, Ia.

German Coach; German, Hanoverian and Oldenburg Coach Horse Association of America, J. Crouch, Lafayette, Ind.

Hackney; American Hackney Horse Society, Gurney C. Gue, 308 W. 97th St., New York.

Morgan; American Morgan Register Association, Thomas E. Boyce, Middlebury, Vt.

Oldenburg; Oldenburg Coach Horse Association of America, J. Crouch, Lafayette, Ind.

Percheron; Percheron Society of America, George W. Stubblefield, Union Stock Yards, Chicago, Ill.

Percheron; The Percheron Registry Company, Charles C. Glenn, 1319 Wesley Ave., Columbus, O.

Percheron; The American Breeders' and Importers' Percheron Registry Company, Jno. A. Perney, Plainfield, O.

Saddle Horse; American Saddle Horse Breeders' Association, I. B. Nall, Louisville, Ky.

Shetland Pony; American Shetland Pony Club, Mortimer Levering, Lafayette, Ind.

Shire; American Shire Horse Association, Charles Burgess, Wenona, Ill.

Suffolk; American Suffolk Horse Association, Alex. Galbraith, Janesville, Wis.

Thoroughbred; The Jockey Club, W. H. Rowe, Registrar, 571 5th Ave., New York.

Welsh Pony and Cob; The Welsh Pony and Cob Society of America, John Alexander, Aurora, Ill.

MULES

Jacks and Jennies; American Breeders' Association of Jacks and Jennets, J. W. Jones, secretary, Columbia, Tenn.

CATTLE

Aberdeen Angus; American Aberdeen Angus Breeders' Association, Charles Gray, Union Stock Yards, Chicago, Ill.

Ayrshire; Ayrshire Breeders' Association, C. M. Winslow, Brandon, Vt.

Brown Swiss (Schwytz); Brown Swiss Cattle Breeders' Association, C. D. Nixon, Owego, N. Y.

Devon; American Devon Cattle Club, L. P. Sisson, Newark, O.

Dutch Belted; Dutch Belted Cattle Association of America, H. B. Richards, Easton, Pa.

Galloway; American Galloway Breeders' Association, Robert W. Brown, Union Stock Yards, Chicago, Ill.

Guernsey; American Guernsey Cattle Club, William H. Caldwell, Peterboro, N. H.

Hereford; American Hereford Cattle Breeders' Association; C. R. Thomas, 225 W. 12th St., Kansas City, Mo.

Holstein Friesian; Holstein Friesian Association of America, Frederick L. Houghton, Brattleboro, Vt.

Jersey; American Jersey Cattle Club, J. J. Hemingway, 8 W. 17th St., New York.

Polled Durham; Polled Durham Breeders' Association, J. H. Martz, Greenville, O.

Red Polled; Red Polled Cattle Club of America (Incorporated), Harley A. Martin, Gotham, Wis.

Shorthorn; American Shorthorn Breeders' Association, John W. Groves, Union Stock Yards, Chicago, Ill.

Sussex; American Sussex Association, Overton Lea, Caldwell, N. J.

SHEEP

Cheviot; American Cheviot Sheep Society, F. E. Dawley, Fayetteville, N. Y.

Cotswold; American Cotswold Registry Association, F. W. Harding, Waukesha, Wis.

Dorset Horn; The Continental Dorset Club; Joseph E. Wing, Mechanicsburg, O.

Highland, Blackfaced; Blackfaced Highland Sheep Association, Frank Reed Sanders, Bristol, N. H.

Hampshire Down; American Hampshire Breeders' Association, Comfort A. Tyler, 310 E. Chicago St., Coldwater, Mich.

Leicester; American Leicester Breeders' Association, A. J. Temple, Cameron, Ill.

Lincoln; National Lincoln Sheep Breeders' Association, Bert Smith, Charlotte, Mich.

Merino (Delaine); Dickinson Sheep Record Company, Beulah McDowell, 49 Oak Hill Ave., Delaware, O.; Canton, O.

Merino (Delaine); National Delaine Merino Sheep Breeders' Association, J. B. Johnson, 248 W. Pike St., Canonsburg, Pa.

Merino (French); American Rambouillet Sheep Breeders' Association, Dwight Lincoln, Milford Center, O.

Merino (German); International Von Homeyer Rambouillet Club, E. N. Ball, Ann Arbor, Mich.

Merino (Spanish); Michigan Merino Sheep Breeders' Association, E. N. Ball, Ann Arbor, Mich.

Merino (Spanish); Vermont, New York and Ohio Merino Sheep Breeders' Association, Wesley Bishop, Delaware, O.

Merino (Spanish); Black Top Spanish Merino Sheep Breeders' Publishing Association, R. P. Berry, Eightyfour, Pa.

Oxford Down; American Oxford Down Record Association, W. A. Shafor, Hamilton, O.

Shropshire; American Shropshire Registry Association, Mortimer Levering, Lafayette, Ind.

Southdown; American Southdown Breeders' Association, Frank E. Springer, 510 E. Monroe St., Springfield, Ill.

Suffolk; American Suffolk Flock Registry Association, George W. Franklin, Des Moines, Ia.

GOATS

Angora Goat; American Angora Goat Breeders' Association, John W. Fulton, Kansas City, Mo.

Milch Goat; American Milch Goat Breeders' Association, W. A. Shafor, Hamilton, O.

HOGS

Berkshire; American Berkshire Congress, Frank E. Springer, 510 E. Monroe St., Springfield, Ill.

Cheshire; Cheshire Swinebreeders' Association, Ed. S. Hill, Freeville, N. Y.

Chester, Ohio Improved; O. I. C. Swine Breeders' Association, J. C. Hiles, Cleveland, O.

Chester White; Chester White Swine Breeders' Association, Frank F. Moore, Rochester, Ind.

Chester White; American Chester White Record Association, Ernest Freigau, Columbus, O.

Duroc-Jersey; American Duroc-Jersey Swine Breeders' Association, T. B. Pearson, Thorntown, Ind.

Duroc-Jersey; National Duroc-Jersey Record Association, H. C. Sheldon, Peoria, Ill.

Hampshire (Thin Rind); American Hampshire Swine Record Association, F. C. Stone, Armstrong, Ill.

Poland-China; American Poland-China Record Association, W. M. McFadden, Union Stock Yards, Chicago, Ill.

Poland-China; National Poland-China Record Company, A. M. Brown, Winchester, Ind.

Poland-China; Southwestern Poland-China Record Association, H. P. Wilson, Gadsden, Tenn.

Poland-China; Standard Poland-China Record Association, George F. Woodworth, Maryville, Mo.

Tamworth; American Tamworth Swine Record Association, E. N. Ball, Ann Arbor, Mich.

Yorkshire; American Yorkshire Club, Harry G. Krum, White Bear Lake, Minn.

DOGS

Fifty-seven Recognized Varieties; American Kennel Club, A. P. Vredenburg, 55 Liberty St., New York

CATS

Longhaired (Angora or Persian), Shorthaired (Siamese, Manx, Mexican, Abyssinian, Indian, Russian, and Japanese); United States Official Register Association (Incorporated), Mrs. S. Hazen Bond, Registrar, 310 1st St., S. E., Washington, D. C.

Longhaired (Persian or Angora), Shorthaired (Russian, Siamese, Japanese, Mexican, Manx, Abyssinian, Native); American Cat Association, Mrs. Anna L. Besse, 5534 Union Ave., Chicago, Ill.

Live Stock Associations

SECRETARIES

American Association of Live Stock Herd Book Secretaries—Charles F. Mills, Springfield, Ill.

American Berkshire Congress—C. S. Bartlett, Pontiac, Mich.

American Animal Breeders' Association Charles B. Davenport, Cold Spring Harbor, N. Y.

American Guernsey Cattle Club—W. H. Caldwell, Peterboro, N. H.

American Jersey Cattle Club—Secretary-treasurer, J. P. Hutchinson, Georgetown, N. J.

American National Live Stock Association—C. W. Tomlinson, 909 17th St., Denver, Col.

Cattle Raisers' Association of Texas—Secretary and general manager, H. E. Crowley, Fort Worth, Tex.

Central Wisconsin Holstein - Friesian Breeders' Association—J. P. Heintz, Hewitt, Wis.

Connecticut Berkshire Swine Breeders' Association—C. H. Marsh, New Milford, Ct.

Connecticut Sheep Breeders' Association —B. C. Patterson, Torrington, Ct.

Connecticut Valley Live Stock Association—Secretary-treasurer, O. C. Burt, Easthampton, Mass.

Corn Belt Meat Producers' Association— H. C. Wallace, Des Moines, Ia.

Dorset Horn Sheep Breeders' Association—M. A. Cooper, Washington, Pa.

Iowa Aberdeen-Angus Breeders' Association—H. M. Graham, Des Moines, Ia.

Iowa Dairy Cattle Improvement Association—Treasurer, H. E. Colby, Waterloo.

Iowa Sheep Breeders' Association—E. S. Leonard, Corning, Ia.

Illinois Live Stock Breeders—Fred H. Rankin, Urbana, Ill.

Interstate Breeders' Association—F. L. Witrick, Sioux City, Ia.

International Live Stock Exposition— President, J. A. Spoor, Chicago, Ill.; general superintendent, B. H. Heide, Union Stock Yards, Chicago, Ill.

Hereford Breeders' Association of Missouri and Kansas—Secretary-treasurer, John W. Rouse, Kansas City, Mo.

Kansas Improved Stock Breeders' Association—H. A. Heath, Topeka, Kan.

Kentucky Beef Cattle Breeders' Association—J. J. Hooper, Lexington, Ky.

Massachusetts Cattle Owners' Association—James L. Harrington, Lunenburg, Mass.

Montana Stock Growers' Association— H. R. Wells, Miles City, Mont.

National Live Stock Exchange—A. F. Stryker, Union Stock Yards, South Omaha, Neb.

National Wool Growers' Association— George L. Walker, Cheyenne, Wyo.

New England Live Stock Association— Ralph Blaisdell, Malden, Mass.

Nebraska Improved Live Stock Breeders' Association — Secretary-treasurer, A. T. Peters, State university, Lincoln, Neb.

North Dakota Live Stock Association— N. B. Richards, Agricultural college, N. D.

North Montana Round Up Association— Thomas A. Cummings, Fort Benton, Mont.

Ohio Live Stock Association—Charles S. Plumb, Columbus, O.

Oklahoma Live Stock Association—W. E. Bolton, Woodward, Okla.

Sheep Breeders' Association of Connecticut—B. C. Patterson, Torrington, Ct.

South Dakota Improved Live Stock Breeders' Association—James W. Wilson, Brookings, S. D.

South Dakota Sheep Breeders and Wool Growers' Association—Secretary-treasurer, W. E. Raymond, Summit, S. D.

South Dakota Swine Breeders' Association—Secretary-treasurer, P. E. Murphy, Oldham, S. D.

Southwest Michigan Pedigreed Stock Association—R. E. Jennings, Paw Paw, Mich.

West Virginia Live Stock Breeders' Association—C. E. Lewis, Maxwelton, W. Va.

Western South Dakota Stock Growers' Association—F. M. Stewart, Buffalo Gap, S. D.

Wisconsin Holstein Breeders' Association—J. B. Hintz, Hewitt, Wis.

Wisconsin Horse Breeders' Association —James G. Fuller, Madison, Wis.

Wisconsin Live Stock Breeders' Association—F. H. Scribner, Rosendale, Wis.

Wisconsin Sheep Breeders' Association— William F. Renk, Sun Prairie, Wis.

Wyoming Wool Growers' Association— George S. Walker, Cheyenne, Wyo.

Poultry Associations

SECRETARIES

American Poultry Association—Ross C. H. Hallock, St. Louis, Mo.

Boston Poultry Association—George P. Coffin, Freeport, Me.

Central Vermont Poultry and Pet Stock Association—E. J. Badger, Barre, Vt.

Connecticut Poultry Association—Harrison L. Hamilton, Ellington.

Maine Poultry and Pet Stock Association—A. L. Merrill, Merrill.

National Fanciers and Breeders' Association—Fred L. Kimmey, Morgan Park, Ill.

National Poultry and Game Association —George G. Brown, New York City.

Nebraska Poultry Association—Luther P. Ludden, Lincoln.

Northwest Branch American Poultry Association—Ralph Whitney, Stewartville, Minn.

Rhode Island Poultry Association—William L. Brown, Providence.

Beekeepers

SECRETARIES

Connecticut Beekeepers' Association—J. Arthur Smith, Hartford.

Michigan Beekeepers' Association—E. B. Tyrell, Detroit.

National Beekeepers' Association— George E. Hilton, president, Fremont, Mich.; Gus Dittmer, secretary, Augusta, Wis.

Texas Beekeepers' Association—Louis H. Nichols, New Braunfels.

State Forest Officers

Alabama—State forest commission, secretary, John Wallace, Jr., Montgomery.

California—State forester, G. B. Lull, Sacramento.

Connecticut—State forester, Austin F. Hawes, New Haven.

Hawaii—Executive officer, C. S. Holloway, Honolulu; superintendent of forestry, Ralph S. Hosmer, Honolulu.

Indiana—State board of forestry, president, F. C. Carson, Michigan City; secretary, W. H. Freeman, Indianapolis.

Kansas—Commissioner of forestry, F. H. Ridgway, Ogallah; commissioner of forestry, H. C. Cooper, Dodge City.

Kentucky—Chairman state board of agriculture, forestry and immigration, M. C. Rankin, commissioner of agriculture, Frankfort.

Louisiana—State forest commissioner, A. W. Crandell, Baton Rouge.

Maine—Land agent and forest commissioner, Edgar E. Ring, Augusta.

Massachusetts—State forester, Frank W. Rane, State House, Boston.

Maryland—State forester, F. W. Besley, Baltimore; state geologist and state highway commissioner, William Bullock Clark, Baltimore.

Michigan—Forestry commission: Secretary, Huntley Russell, Lansing; state forest warden, Filibert Roth, Ann Arbor.

Minnesota—State forestry board: President, Sidney M. Owen, Minneapolis; secretary and forestry commissioner, Gen. C. C. Andrews, St. Paul.

Mississippi—Director state geological survey, A. F. Crider, Biloxi.

New Hampshire—Secretary forest commission, R. E. Faulkner, Keene.

New Jersey—Geological survey: Forester, Alfred Gaskill, Trenton; secretary, forest park reservation commission, Alfred Gaskill, Trenton.

New York—Forest, Fish and Game Commission: Commissioner, James S. Whipple, Albany.

New York—Forester, Abraham Knechtel, Albany; superintendent state forests, William F. Fox, Albany.

North Carolina—Geological Survey: State geologist, Joseph Hyde Pratt, Chapel Hill.

Ohio—Forester, W. J. Green, Ohio Agricultural experiment station, Wooster.

Oregon—Game and forestry warden, R. O. Stevenson, Forest Grove; master fish warden, C. H. McAlister, Portland; secretary forestry commission, E. P. Sheldon, Portland.

Pennsylvania—Department of forestry: President and commissioner, Robert S. Conklin, Harrisburg; secretary, J. T. Rothrock, Harrisburg; chief forester, George H. Wirt, Harrisburg.

Rhode Island—Commissioner of forestry, Jesse B. Mowry, Chepatchet.

Washington—Chairman department of forestry, R. W. Condon, Port Gamble.

West Virginia—Geologic and economic survey, I. C. White, superintendent, Morgantown.

Wisconsin—State forester, E. M. Griffith, Madison; director state geological survey, Edward A. Birge, Madison.

Forestry Associations

SECRETARIES

American Alpine Club—President, John Muir, Martinez.

American Forestry Association—President, Hon. James Wilson, secretary of agriculture; Thomas E. Will, Washington, D. C.

American Forest Preservation Society—George Milroy Bailey, Corfu, N. Y.

American Scenic and Historic Preservation Society—President, George Frederick Kunz, Tribune Bldg., New York city; Edward Hagaman Hall, Tribune Bldg., New York City.

Appalachian Mountain Club—Rosewell B. Lawrence, Boston, Mass.

Arizona Salt River Valley Water Users' Association—Charles A. Van der Veer, Phoenix.

Association for the Protection of the Adirondacks—President, Henry E. Howland; Edward H. Hall, Tribune Bldg., New York City.

California Water and Forest Association—T. C. Friedlander, 707 Merchants Exchange, San Francisco, Cal.

Cincinnati Forest and Improvers' Association—Dr. Adolph Leue, 127 West 12th St., Cincinnati, O.

Colorado State Forestry Association—President, W. G. M. Stone, Denver.

Connecticut Forestry Association—Miss Mary Winslow, Weatogue.

Forestry, Water Storage and Manufacturing Association of New York—Chester W. Lyman, 30 Broad St., New York City.

Forestry Educational Association—J. N. F. Bischoff, San Diego, Cal.

Forest and Water Society of Southern California—William H. Knight, Los Angeles, Cal.

Franklin Forestry Society—W. G. Bowers, Chambersburg, Pa.

Georgia State Forestry Association—Secretary, Alfred Akerman, Athens.

International Society of Arboriculture—President, Gen. William J. Palmer, Colorado Springs, Col.; John P. Brown, Connersville, Ind.

Iowa Park and Forestry Association—Wesley Greene, Ames, Ia.

Lake George Forestry Association—D. V. R. Westervelt, Schenectady, N. Y.

Maine Forestry Association—E. E. Ring, Augusta; president, John Appleton, Bangor.

Massachusetts Forestry Association—President, Dr. H. P. Walcott, Cambridge; Edwin A. Start, 4 Joy St., Boston.

Michigan Forestry Association—President, John H. Bissell, Detroit; J. Fred Baker, East Lansing.

Minnesota State Forestry Association—E. G. Cheney, St. Anthony Park.

Nebraska Park and Forestry Association—President, C. S. Harrison, York; L. B. Craig, York.

New England Forest, Fish and Game Association—A. T. Harris, Boston, Mass.

New Hampshire Society for the Protection of New Hampshire Forests—Allen Hollis, Concord.

New York Fish, Game and Forest League—D. V. R. Westervelt, 1126 State St., Schenectady, N. Y.

North Dakota State Sylvaton Society—President, A. M. Powell, Devil's Lake; chief forester and secretary, W. W. Barrett, Church's Ferry.

Northern New York Forestry Association—Director O. H. Tappan, Potsdam.

Ohio State Forestry Association—C. W. Wald, Wooster.

Oregon Forestry Association—Arthur D. Monteith, Portland.

Pennsylvania Forestry Association—F. L. Bitler, 1012 Walnut St., Philadelphia, Pa.

Pacific Coast Forest, Fish and Game Association—William Greer Harrison, San Francisco, Cal.

Saginaw Fishing and Yachting Association—President, L. H. Goodwin, Saginaw, Mich.; John McPhillips, Saginaw, Mich.

Sierra Club—President, John Muir, Martinez, Cal.; William E. Colby, San Francisco, Cal.

Society for Protection of New Hampshire Forests—Allen Hollis, Concord, N. H.

Society of American Foresters—William F. Sherfesee, Washington, D. C.; president, Gifford Pinchot, Washington, D. C.

The Appalachian National Forest Association—President, D. A. Tompkins, Charlotte, N. C.; John H. Finney, Washington, D. C.

The Mazamas—Frank B. Riley, Portland, Ore.

Tri-Counties Reforestation Committee—L. A. Finch, Riverside, Cal.

Vermont Forestry Association—President, L. R. Jones, Burlington; Ernest Hitchcock, Pittsford, Vt.

Washington Conservation Association—C. H. Bailey, 325 New York Block, Seattle.

West Virginia Forestry Association—A. W. Nolan, Morgantown.

Horticultural and Kindred Societies

SECRETARIES

American Federation of Horticultural Societies—Charles E. Bassett, Fennville, Mich.

American Seed Trade Association—C. E. Kendel, 115 Ontario St., Cleveland, O.

American Association of Nurserymen—George C. Seager, Rochester, N. Y.

American Apple Growers' Congress—T. C. Wilson, 5633 Clemens Ave., St. Louis, Mo.

American Plant Breeders' Association—Willett M. Hays, department of agriculture, Washington, D. C.

American Carnation Society—Albert M. Herr, Lancaster, Pa.

American Cranberry Growers' Association—A. J. Rider, Hammonton, N. J.

American Nurserymen's Protective Association—Thomas B. Meehan, Dresher, Pa.

American Pomological Society—John Craig, Ithaca, N. Y.

American Retail Nurserymen's Protective Association—Guy A. Bryant, Princeton, Ill.

American Rose Society—Benjamin Hammond, Fishkill-on-Hudson, N. Y.

Arkansas Horticultural Society—John P. Logan, Grannis, Ark.

Cape Cod Cranberry Growers' Association—William M. Marsh, Wareham, Mass.

Chrysanthemum Society of America—David Fraser, Pittsburg, Pa.

Central Illinois Horticultural Society—J. B. Burrows, Decatur.

Cider and Cider Vinegar Makers' Association of the Northwest—George Miltenberger, St. Louis, Mo.

Delaware Corn Growers' Association—A. E. Grantham, treasurer, Newark.

Columbus Horticultural Society—William R. Lazenby, president, Columbus, O.

Connecticut Pomological Society—H. C. C. Miles, Milford.

Eastern Nurserymen's Association—Wm. Pitkin, Rochester, N. Y.

Fruit Growers and Shippers' Association—Ernest R. Ostrom, Siloam Springs, Ark.

Grand River Valley Horticultural Society—Almond Griffin, Grand Rapids, Mich.

Horticultural Science—C. P. Close, College Park, Md.

Horticultural Society of Chicago—Edwin A. Kanst, 5700 Cottage Grove Ave., Chicago, Ill.

Hudson Valley Fruit Growers' Association—J. R. Cornell, president, Newburgh, N. Y.

Idaho State Horticultural Association—Frank E. Price, Payette.

Illinois Corn Growers' Association—Leigh F. Maxcy, Curran.

Indiana Corn Growers' Association—Edward Christie, La Fayette.

International Apple Shippers' Association—C. P. Rothwell, Martinsburg, W. Va.

International Society of Arboriculture—John P. Brown, Connersville, Ind.

Iowa Corn Growers' Association—Miller S. Nelson, Goldfield, Ia.

Iowa State Corn Growers' Association—B. W. Crossley, Ames.

Iowa State Horticultural Society—Wesley Greene, Des Moines.

Maine Pomological Society—William J. Ricker, Turner.

Maryland State Horticultural Society—C. P. Close, secretary-treasurer, College Park.

Massachusetts Fruit Growers' Association—C. A. Whitney, Upton.

Massachusetts Horticultural Society—William P. Rich, Horticultural Hall, Boston, Mass.

Michigan Bean Jobbers' Association—G. F. Allmindinger, Ann Arbor, Mich.

Michigan State Horticultural Society—Charles E. Bassett, Fennville.

Mississippi Valley Apple Growers' Association—James Handly, Quincy, Ill.

Missouri Valley Horticultural Society—A. V. Wilson, Muncie, Kan.

National Association Retail Nurserymen—Frederick E. Grover, Rochester, N. Y.

National Corn Exposition—G. W. Wattles, president, Omaha; T. F. Sturgess, secretary, Omaha; J. Wilkes Jones, general manager, 606 Bee Bldg., Omaha.

National Council of Horticulture—H. C. Irish, St. Louis, Mo.

National Corn Association—J. W. Wilkes Jones, Omaha, Neb.

National Fruit Exchange—C. M. Chaney, 105 Hudson St., New York City.

National Horticultural Congress—J. P. Hess, president; G. W. Reye, secretary-treasurer, Day & Hess Bldg., Council Bluffs, Ia.

National Nut Growers' Association—J. F. Wilson, Poulan, Ga.

Nebraska Corn Improvers' Association—T. G. Montgomery, Lincoln.

Nebraska State Horticultural Society—L. M. Russell, Lincoln.

New England Cranberry Sales Company—Frank N. Churchill, Middleboro, Mass.

New Hampshire Horticultural Society—E. Dwight Sanderson, Durham.

New York State Fruit Growers' Association—E. O. Gillett, Penn Yan, N. Y.

New York Evaporated Fruit Producers' Association—L. J. Tweeney, Marion.

North Dakota Horticultural Society—O. O. Churchill, Agricultural College.

Northern Illinois Horticultural Society—Jacob Friend, Nekoma, Ill.

Nurserymen's Mutual Protective Association—George C. Seager, Rochester, N. Y.

Ohio State Corn Show Association—Loring H. Goddard, Wooster.

Ohio State Horticultural Society—W. W. Farnsworth, Waterville.

Peninsula Horticultural Society—Wesley Webb, Dover, Del.

Pennsylvania Horticultural Society—David Rust, Philadelphia.

Pennsylvania Nurserymen's Association—Earl Peters, Carlisle, Pa.

Pennsylvania State Horticultural Association—Enos B. Engle, Waynesboro.

Rhode Island Horticultural Society—Charles W. Smith, 27 Exchange St., Providence.

Savannah Valley Association Farmers' Club—W. E. Mealing, North Augusta, S. C.

Society of American Florists and Ornamental Horticulturists—P. J. Hauswirth, 282 Michigan Ave., Chicago.

Southern Illinois Horticultural Society—E. G. Mendenhall, secretary-treasurer, Kinmundy.

Southern Florists' Society—Paul Abele, New Orleans, La.

Southern Nurserymen's Association—A. I. Smith, Knoxville, Tenn.

State Horticultural Society of Maryland—C. P. Close, College Park, Md.

Tennessee State Nurserymen's Association—Gordon M. Bentley, Knoxville.

Texas Farmers' Congress—T. W. Larkin, Denison.

Texas Horticultural Society—E. J. Kyle, secretary-treasurer, College Station.

Texas Nurserymen's Association—John S. Kerr, Sherman.

Texas Nut Growers' Association—H. B. Beck, Denton.

Texas Rice Growers' Association—A. E. Groves, Houston.

Vermont Horticultural Society—Prof. William F. Stuart, Burlington.

Vermont State Agricultural Society—C. M. Winslow, Brandon.

Western Association of Nurserymen—E. J. Holman, Leavenworth, Kan.

Western Fruit Jobbers' Association—E. B. Branch, Omaha, Neb.

Western N. Y. Horticultural Society—John Hall, 204 Granite Bldg., Rochester, N. Y.

Wisconsin Cranberry Growers' Association—J. W. Fitch, Cranmoor, Wis.

Wisconsin State Horticultural Society—Frederick Cranefield, Madison, Wis.

For Bird and Game Protection

SECRETARIES

American Humane Association—Mrs. Mary F. Lovell, Wyncote, Pa.

American Ornithologists' Union, Committee on Bird Protection—A. K. Fisher, chairman, Department of Agriculture, Washington, D. C.

American Society for the Prevention of Cruelty to Animals—Richard Welling, Madison Ave. and Twenty-sixth St., New York.

Bird Protective Society of America—E. C. Pease, Buffalo, N. Y.

Boone and Crockett Club—Madison Grant, secretary, 11 Wall St., New York City.

Forest, Fish and Game society of America—William F. Kimber, secretary, 505 Fifth Ave., New York City.

League of American Sportsmen—H. M. Brach, secretary, 1061 Simpson St., New York City.

Lewis and Clark Club—J. Bissell Speer, secretary, 345 Fourth Ave., Pittsburg, Pa.

National Association of Fish and Game Wardens—Charles A. Vogelsung, secretary, San Francisco, Cal.

National Association of Audubon Societies—President, William Dutcher, 141 Broadway, New York.

New York Zoological Society—Madison Grant, secretary, 11 Wall Street, New York City.

North American Fish and Game Protective Association—J. O. Reaume, president, Toronto, Ontario, Canada; E. T. D. Chambers, Toronto, Canada, secretary.

Associations of General Interest

Association of American Agricultural Colleges and Experiment Stations—President, J. L. Snyder, president of Michigan state agricultural college; secretary-treasurer, J. L. Hills, director of Vermont Experiment Station, Burlington.

Agricultural Experimenters' League—Secretary, Charles H. Tuck, Ithaca, N. Y.

American Association of Farmers' Institute Workers—President, J. Lewis Ellsworth, State House, Boston, Mass.; secretary, John Hamilton, Washington, D. C.

American Association of State Fairs and Expositions—George Downing, Indianapolis, Ind.

American Civic Association—President, J. Horace McFarland, Harrisburg, Pa.; treasurer, William B. Howland, New York City; secretary, C. R. Rogers, Philadelphia, Pa.

American Federation of Labor—President, Samuel Gompers, Washington, D. C.; secretary, Frank Morrison, 423 G St., Washington, D. C.

American Institute Farmers' Club—Secretary, W. A. Eagleson, care American Institute, New York City.

American Social Science Association—President, John Huston Finley, New York; secretary, Isaac Franklin Russell, New York.

American Veterinarian Medical Association—President, J. G. Rutherford, Ottawa, Canada; secretary, Richard P. Lyman, Hartford, Ct.

Anti-Imperialist League—President, Moorfield Storey, Boston, Mass.; secretary, Erving Winslow, Boston, Mass.

Association of Economic Entomologists—President, Dr. S. A. Forbes, Urbana, Ill.; secretary, A. F. Burgess, Bureau of Entomology, Washington, D. C.

Association of Official Agricultural Chemists—President, Harry Snyder, St. Anthony Park, Minn.; secretary, H. W. Wiley, chemist, U. S. Department of Agriculture, Washington, D. C.

American Bankers' Association—President, W. F. Keyer, Sedalia, Mo.; secretary-treasurer, F. E. Farnsworth, New York.

Bay State Agricultural Society—Secretary, N. I. Bowditch, Framingham, Mass.

Connecticut Plant Breeders' Association—Secretary, C. J. Jarvis, Storrs, Ct.

Farmers' Congress of Texas—Secretary, T. W. Larkin, Denison.

Farmers' Educational and Co-operative Union of America—President, C. S. Barrett, Union City, Ga.; secretary, R. H. McCulloch, Beebe, Ark.

Farmers' National Congress—Treasurer, W. L. Ames, Oregon, Wis.

Iowa State Drainage Association—Secretary, W. H. Stevenson, Ames.

Lakes-to-the-Gulf Waterways Association—President, William H. Kavanaugh, Memphis, Tenn.; secretary, William F. Saunders, St. Louis, Mo.

Lincoln Centennial Association—Chairman, J. Otis Humphreys, Springfield, Ill.; secretary, P. B. Warren, Springfield, Ill.

Michigan Good Roads Association—Secretary, Edward N. Hines, Detroit.

Michigan State Association of Farmers' clubs—Mrs. Myra L. Cheney, Mason.

National Association of Concrete Users—President, Richard L. Humphrey, Philadelphia; secretary, George C. Wright, Rochester, N. Y.; treasurer, H. C. Turner, 11 Broadway, N. Y.

National Drainage Commission—President, N. B. Broward, Tallahassee, Fla.; secretary, Maj. J. A. Depray, 1916 16th St., N. W., Washington, D. C.

National Irrigation Congress—President, George E. Barstow, Denver, Col.; secretary, A. Warren Patch, Boston, Mass.

National Rivers and Harbors Congress—President, Joseph E. Ransdell, Washington, D. C.; secretary, J. F. Ellison, Cincinnati, O.

National Society of Equity—President, C. O. Drayton, Indianapolis, Ind.; secretary-treasurer, S. D. Kump, Indianapolis.

New England Grain Dealers' Association—Secretary, John W. Cox, 713 Chamber of Commerce, Boston, Mass.

Ohio Canners' Association—Secretary-treasurer, James Stoops, Waynesville.

Ohio and Indiana Grain Shippers' Traffic Association—Secretary, Harry W. Krees, Piqua, O.

Ohio Millers' Association—Secretary-treasurer, C. E. Jenkins, Marion.

Pennsylvania Lumbermen's Association—Secretary, B. F. Landig, Scranton, Pa.

Postal Progress League—Secretary-treasurer, James L. Cowles, 361 Broadway, New York.

Salt River Valley Water Users' Association—Secretary, Charles A. Van der Veer, Phoenix, Ariz.

Savannah Valley Associated Farmers' Clubs—Secretary, Dr. W. E. Mealing, North Augusta, S. C.

Trans-Mississippi Commercial Congress—President, Thomas F. Walsh, Denver, Col.; secretary, Arthur F. Francis, Cripple Creek, Col.

The American Anti-Horse Thief Association—President, J. W. Wall, Parsons, Kan.; secretary-treasurer, J. M. Pence, Morrisville, Ill.

Tennessee Fair Circuit—Secretary, Frank D. Fuller, Hermitage, Tenn.

The Jewish Agricultural and Industrial Aid Society—President, Cyrus L. Sultzberger, New York; Leonard G. Robinson, general manager, New York; secretary, Percy S. Straus, 174 Second Ave., N. Y.

Vermont State Fair Commission—Secretary, F. L. Davis, Woodstock, Vt.

Patrons of Husbandry

Officers of the National Grange:
Master, N. J. Bachelder, Concord, N. H.
Overseer, T. C. Atkeson, Morgantown, W. Va.
Lecturer, G. W. F. Gaunt, Mullica Hill, N. J.
Steward, J. A. Newcomb, Golden, Col.
Asst. Steward, C. D. Richardson, North Brookfield, Mass.
Chaplain, O. S. Wood, Ellington, Conn.
Treasurer, Mrs. E. S. McDowell, Rome, N. Y.
Secretary, C. M. Freeman, Tippecanoe City, Ohio.
Gatekeeper, A. C. Powers, Beloit, Wis.
Ceres, Elizabeth Patterson, College Park, Md.
Pomona, Mrs. S. G. Baird, Edina Mills, Minn.
Flora, Mrs. Ida Judson, Balfour, Ia.
Lady Steward, Mrs. J. M. Walker, Marshallton, Del.
Chairman of Executive Committee, F. N. Godfrey, Olean, N. Y.

STATE MASTERS

W. V. Griffin, Geyserville, Cal.
J. A. Newcomb, Golden, Col.
Orson S. Wood, Ellington, Conn.
S. H. Messick, Bridgeville, Del.
D. C. Mullen, Nampa, Idaho.
Oliver Wilson, Peoria, Ill.
Aaron Jones, South Bend, Ind.
A. B. Judson, Balfour, Ia.
George Black, Olathe, Kan.
F. P. Wolcott, Covington, Ky.
Obadiah Gardner, Rockland, Me.
H. J. Patterson, College Park, Md.
C. D. Richardson, North Brookfield, Mass.
Nathan P. Hull, Dimondale, Mich.
Mrs. S. G. Baird, Edina Mills, Minn.
C. O. Raine, Canton, Mo.
H. O. Hadley, Peterboro, N. H.
George W. F. Gaunt, Mullica Hill, N. J.
F. N. Godfrey, Olean, N. Y.
T. C. Laylin, Norwalk, Ohio.
Austin T. Buxton, Forest Grove, Ore.
W. F. Creasey, Catawissa, Pa.
F. E. Marchant, West Kingston, R. I.
W. K. Thompson, Liberty Hill, S. C.
W. L. Richardson, Brownsville, Tenn.
E. R. Reid, Dublin, Tex.
C. F. Smith, Morrisville, Vt.
C. B. Kegley, Pullman, Wash.
T. C. Atkeson, Morgantown, W. Va.
A. C. Powers, Beloit, Wis.

West Point

Cadets to the United States Military academy in West Point, N Y, are appointed annually. The provisions regulating the appointments state that "each senator, congressional district and territory, also the District of Columbia and Porto Rico, is entitled to have one cadet at West Point." Appointees to the military academy must be between 17 and 21 years old, free from any infirmity which may render them unfit for military service, and able to pass a careful examination in reading, writing, spelling, English grammar, English composition. English literature, arithmetic, algebra through quadratic equations, plane geometry, descriptive geography and the elements of physical geography, especially the geography of the United States, United States history, the outlines of general history

and the general principles of physiology and hygiene; or, in lieu thereof, to submit a certificate of graduation from a public high school or state normal school, or a certificate that the candidate is a regular student of an incorporated college or university. The academic duties begin September 1 and continue until June 1. Examinations are held in each December and June. Further information may be secured by addressing the superintendent, Col Hugh L. Scott, U S A, West Point, N Y.

The first rural free delivery route in the United States was established Oct 1, 1896, between Halltown, Uvilla and Charlestown, W Va, while William L. Wilson was postmaster-general.

PROF THOMAS CLARK ATKESON

Prof Atkeson has been for many years associated with the University of West Virginia, having served, for the past several years, as he does now, as Dean of the College of Agriculture. Prof Atkeson has not only been deeply interested in agricultural education but has been closely identified with farm affairs. At the present time he is Master of West Virginia State Grange and Overseer of the National Grange.

The wise man knows enough to change his opinions with conditions; only the fool is invariably consistent.

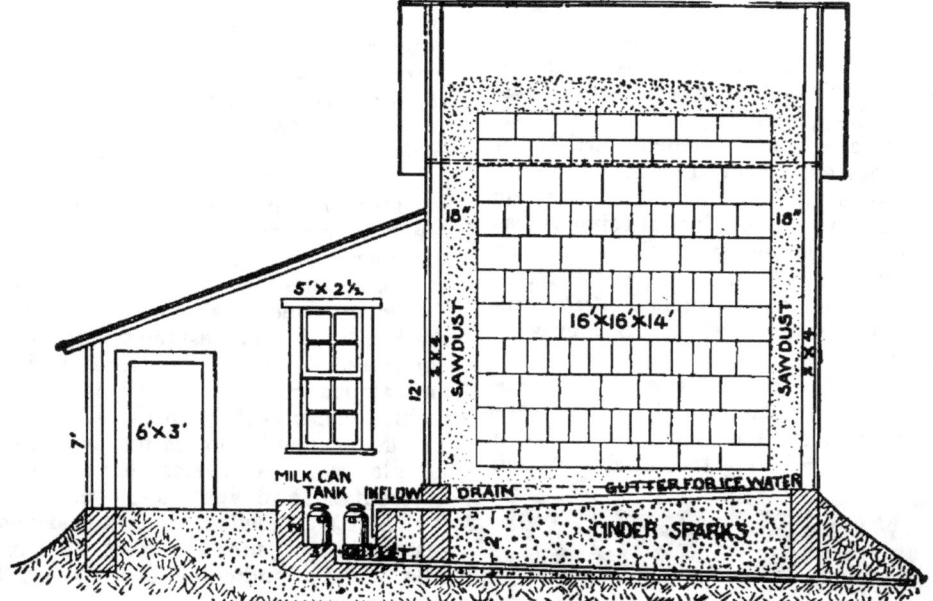

ICE AND MILK HOUSES COMBINED

Side elevation of an ice house with milk house attached, showing how the scheme has the advantage of utilizing the water from the ice house here shown for cooling the milk. No ice needs to be removed from the ice house. It operates automatically. If the weather is warm the ice melts more rapidly and keeps water in the tank at the required temperature.

A TRIO OF EMBDEN GEESE

Of all breeds of geese the Embden dresses best for market. The carcass is plump and compact. Its feathers being white, always bring a high price. The White Chinese is its closest rival in the feather market. Embdens are often considered as less prolific than the Chinese, African or Toulouse, but if this be the case in some flocks, it is the fault of the breeder.

Sunflowers Worth the Raising

Many farmers find it advantageous to grow a few sunflowers every year. Poultrymen, especially, favor sunflowers, inasmuch as the growing plants give shade and also prove an excellent article of food. Pictured is a head of Russian sunflower grown by F. J. Ward of Grafton county, N H. This fine specimen was exhibited at the last annual meeting of the Vermont state fair at White River Junction, and was awarded first premium. It measured about 18 inches in diameter.

Any land that will grow corn will produce sunflowers satisfactorily. The seeds are planted 2 or 3 inches deep and can be placed in drills 3½ feet apart. When

GOOD FOOD FOR CHICKENS

plants are about 8 inches high they should be thinned to stand 12 to 18 inches distant in the row. The plants are not injured by light frosts and may be seeded before the corn crop is put in. This gives the heads time to mature before the early frosts of fall. Shallow cultivation, the same as for corn, should prevail. Sunflowers withstand drouth well and are remarkably free from insect pests and fungous diseases. When the plants are in bloom the fields should be gone over and the excess bloom pulled off, leaving only 3 or 4 heads to develop.

District of Columbia

The District of Columbia, the national capital and its adjoining territory, is owned and administered directly by the United States government, to prevent its actions being hampered by conflicts with local jurisdiction. It consists of 60 square miles of land and 10 of water, on the eastern side of the Potomac. The commission of 1874 was a temporary government; in co-operation with congress and the citizens of the District, it framed as a permanent system the Act of June, 1878, which the Supreme court has pronounced "The Constitution of the District of Columbia." Under this act half of the expenses of the District, previously laid entirely upon the residents, are paid by the national government, as the owner of more than half the real estate. There is no popular suffrage, the entire executive government being in the hands of commissioners appointed by congress. They recommend legislation and appropriations to the latter, which in turn consults them in the same matters. The government, though not appointed by popular vote, is nevertheless one steadily deferent to and swayed by public opinion, and so excellent that agitations for restoration of suffrage meet no support. The District judiciary dates from 1801.

Naval Enlistment

The term of enlistment of all enlisted men of the navy is 4 years. Minors over the age of 18 may be enlisted without consent of parents or guardians, but minors under, but claiming to be over 18 years of age, are liable, if enlisted, to punishment for fraudulent enlistment. Only such persons shall be enlisted as can reasonably be expected to remain in the service. Every person, before being enlisted, must pass the physical examination prescribed in the medical instructions. Applicants for enlistment must be American citizens, able to read and write English, and when enlisted must take the oath of allegiance. No person under the age of 17 can be enlisted.

High Mountains

Mount McKinley in Alaska is the highest mountain in North America, its altitude being 20,464 feet. The United States geological survey has recently completed a revision of the heights of mountains in this country. The report shows that Mount Shasta in California is 17 feet higher than Mount Ranier in Washington, the former being 14,389 feet and the latter 14,363. Mount Whitney is the highest peak in California, with 14,449 feet. Mount Dickerman, in Washington, is credited with 15,776 feet. Pike's Peak is 14,147 feet high. Mount Hood, instead of having an elevation of 11,932 feet, as former measurements showed, has, according to the new report, only 11,225 feet.

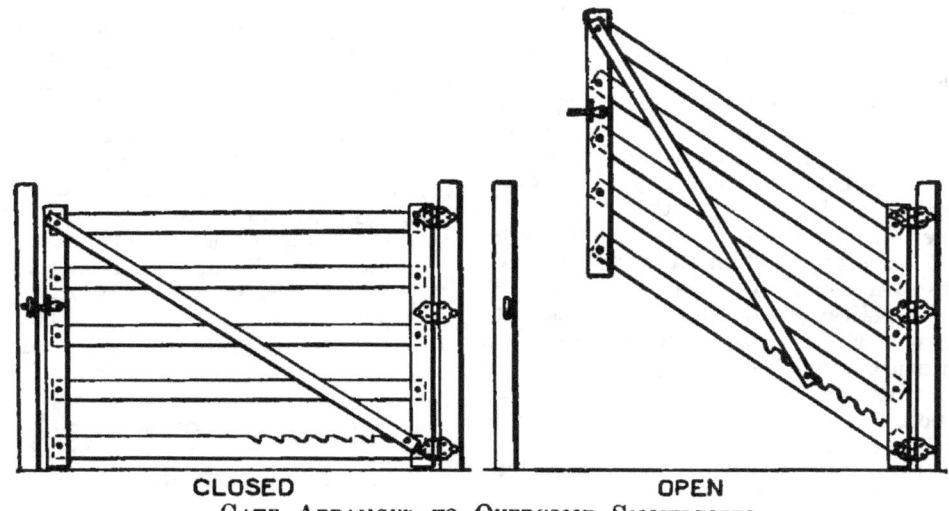

CLOSED   OPEN

GATE ARRANGED TO OVERCOME SNOWDRIFTS

In this picture is shown a gate which can be readily adjusted to overcome snowdrifts. It is easily made from ordinary lumber. A 1 by 6 upright is used for the two lower boards, 1 by 4 for the upper ones. The uprights at the hinge post are double 1 by 4, one piece outside and the other inside the bars. The upright at the latch side may be the same weight of stuff or slightly lighter, and fastened in the same way. Instead of nailing the bars to these uprights, bolts are used, one for each bar at each end. The lowest board is notched as shown, and the double brace used from the top of the latch post to the bottom of the hinge post. For the braces, 1 by 3 stuff is strong enough. These may be held in place at the top by a single bolt. They are joined near the bottom with a bolt which engages with the notches when the gate is raised.

ROSE COMB BUFF LEGHORNS

When it comes to laying, the Leghorns as a class are still in the lead, and the Buffs behave as well as the famous Whites. Except as to color, one is like the other. Leghorns do best where they can have plenty of room to run and earn their own living. They are therefore strikingly the farmer's fowl when the object is plenty of eggs.

Initiative and Referendum

Initiative and referendum are terms applied to law-making by the people. The initiative implies the recognized power of the people to initiate legislation by petition or demand addressed to the representative assembly—the congress, the legislature, the municipal council or to whatever jurisdiction it applies. The referendum signifies the right of the people to have referred to them—on the petition of a specified number of voters—for their adoption or rejection, a measure which has been pending in the representative assembly. Either device or both make the people a part of the legislative machinery, modifying a purely representative scheme of government. The initiative and referendum are in force in Switzerland.

Maine, Missouri, Montana, North Dakota, South Dakota, Oregon and Oklahoma have the initiative and referendum. Nevada has the referendum only. Illinois has a "public opinion" law which is the same as the initiative and referendum with the important exception that the popular vote is not mandatory on the legislature. The Texas primary law provides that the voters of a political party may bind the party to certain policies.

Copyrights

A printed copy of the title of the book, map, chart, dramatic or musical composition, engraving, cut, print, photograph or chromo, or a description of the painting, drawing, statue, statuary or model or design for a work of the fine arts, for which copyright is desired, must be delivered to the Librarian of Congress, or deposited in the mail, within the United States, prepaid. This must be done on or before the day of publication. The legal fee for recording each copyright claim is 50 cents, and for a copy of this record (or certificate of copyright), under seal of the office, an additional fee of 50 cents is required, making $1, or $1.50 if certificate is wanted, which will be mailed as soon as reached in the records. The payment should be made with either an express order or a money order.

Panama Canal

The idea of piercing the Isthmus of Panama is very old, and from 1828 many surveys were made with reference to it, including very complete ones by the United States government, 1872-1875. In 1877 the Columbian government granted a concession to a Frenchman named Wyse for constructing the canal. Ferdinand de Lesseps supported the scheme. The contract by which the Panama canal passed into the hands of the United States was signed, sealed and delivered April 22, 1904, in Paris. The canal will be built on the lock canal system. From lock to lock, the entire length of the canal is 38 miles. The width varies. From Gatun lock to San Pablo and Miraflores to Sosa the width at the bottom of the canal is 1,000 feet. From Boca Mindi to Bohio the canal is 150 feet wide at the bottom. The engineers in charge of the work estimate that the canal will cost $140,000,000 and that it will require from 8 to 12 years in which to complete it.

BEST EAR AT CORN SHOW

This ear won the grand champion prize in the single ear class at the 1908 National Corn Exposition at Omaha. It was grown and exhibited by J. R. Overstreet, Indiana. It is the famous Johnson County White.

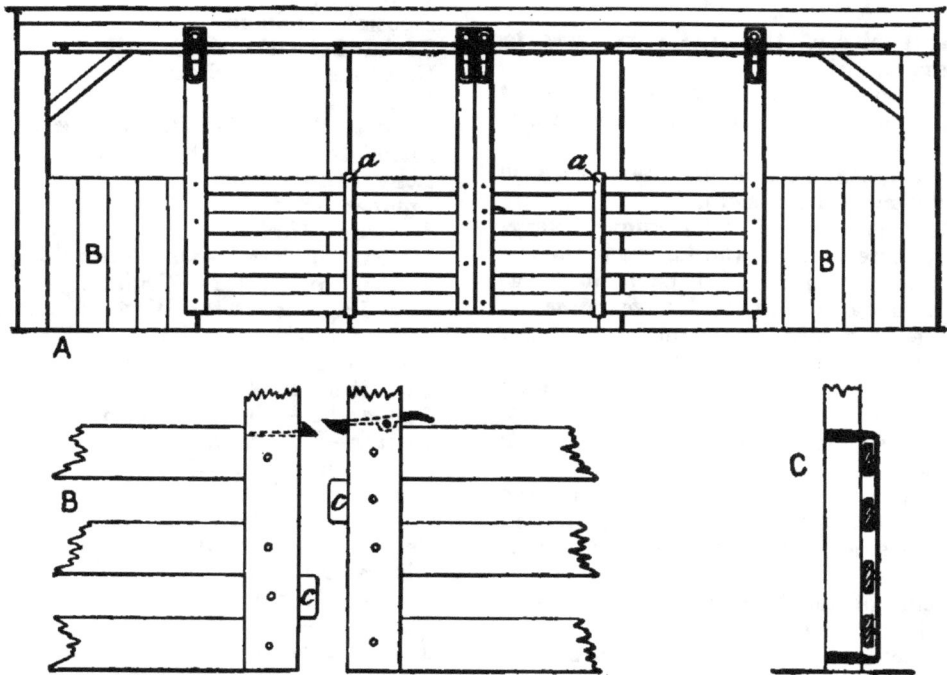

DETAILS OF GATES FOR STOCK PENS

My barn is 30 feet wide, and across one end it is divided into three pens, each 10 feet square. The gates are 10 feet long and are hung on common barn door rollers and track, as will be seen by the sketch. There is a stationary fence extending from the wall half across each outside pen (B, B, Fig 1). By shoving the gates to the right the left-hand pen is opened, and pushing them to the left opens the right-hand pen, while parting the gates in the center opens the middle pen. Fig 2 shows the form of catch I use for the center of the gates. If one has a forge it can be made very easily at home. C, C, in Fig 2 are short projections of wood beveled at the ends to pass between the bars of the opposite gates and hold them rigid when closed. Fig 3 shows the form of loop that may be made of old wagon tire. It is secured to the posts that divide the pens, shown at a, a, Fig 1, and keeps the gates from swinging.—[Geo. A. Clark, N Y.

SHED FOR CHICKEN COLONY HOUSES

Where colony houses are used during the summer, many people leave them in the fields unoccupied during the winter, and thus get only half the service from them that should be secured. In the cut shown herewith, three colony houses are pictured under a shed, which not only protects the houses themselves and thus increases their period of usefulness, but also permits the poultrymen to keep flocks of chickens in the houses themselves. All that is necessary in such a case is to have the house of uniform size and then build the shed just large enough to accommodate the number desired. In this cut three styles of openings are shown in houses, but this is a mere detail.

Hindu Proverbs

The habits of the cradle will last to the grave.

The plant that could not be bent when it was a twig will not bend when it is a branch.

To acquire science in later years is like dyeing a dirty cloth.

There is no man who knows everything; there is none who knows nothing.

Anybody will say: I have no memory; nobody will say: I have no sense.

None is ruined by telling the truth; no one prospers by telling lies.

Even when a wound is healed a scar will remain.

If a neighbor's roof catches fire one's own is in danger.

When the cattle are on one shore, the opposite looks green; when they are on the opposite shore, the other side looks green.

A slip of the tongue is worse than a slip of the foot.

Do not believe all you hear; do not say all you believe.

Nicknames of States

Badger—Wisconsin.
Bay—Massachusetts.
Bayou—Mississippi.
Bear—Arkansas.
Blackwater—Nebraska.
Blizzard—South Dakota.
Blue Grass—Kentucky.
Blue Hen—Delaware.
Blue Law—Connecticut.
Bread and Butter—Minnesota.
Buckeye—Ohio.
Bug Eater—Nebraska.
Bullion—Missouri.
Centennial—Colorado.
Central—Kansas.
Corn Cracker—Kentucky.
Cotton—Alabama.
Cracker—Georgia.
Creole—Louisiana.
Dark and Bloody Ground—Kentucky.
Diamond—Delaware.
Dominion, Ancient or Old—Virginia.
Egypt—Southern Illinois.
El Dorado—California.
Empire—New York.
Empire of the South—Georgia.
Evergreen—Washington.
Excelsior—New York.
Freestone—Connecticut.
Golden—California.
Gopher—Minnesota.
Granite—New Hampshire.
Green Mountain—Vermont.

Hawkeye—Iowa.
Hoosier—Indiana.
Jayhawker—Kansas.
Keystone—Pennsylvania.
Lake—Michigan.
Land of Steady Habits—Connecticut.
Little Rhody—Rhode Island.
Lone Star—Texas.
Lumber—Maine.
Mormon—Utah.
Mother of Presidents—Virginia.
Mother of States—Virginia.
New England of the West—Minnesota.
Old Line—Maryland.
Old North—North Carolina.
Palmetto—South Carolina.
Panhandle—West Virginia.
Pelican—Louisiana.
Peninsula—Florida.
Pine Tree—Maine.
Prairie—Illinois.
Sage-Brush—Nevada.
Silver—Nevada.
Sucker—Illinois.
Sunflower—Kansas.
Tree Planter—Nebraska.
Turpentine—North Carolina.
Web-Foot—Oregon.
Wolverine—Michigan.
Wooden Nutmeg—Connecticut.

Give a little more than you promise—but don't promise too much.

If the world cannot beat you down it will turn around and hail you as a genius.

A gentleman is a man who is just as well bred when he is at home as when he is out calling.

The average of human life is 33 years. The American life in different occupations follows: Storekeepers, 41.8 years; teamsters, 43.6; laborers, 44.6 years; seamen, 46.1 years; mechanics, 47.3 years; merchants, 48.4 years; lawyers, 52.6 years; and farmers, 64.2 years.

The fleur-de-lis is the national emblem of France; the violet is the national flower of Athens; the sugar maple is the national emblem of Canada; the shamrock is emblematic of the Emerald Isle; the linden is the national emblem of Prussia; England's national flower is the rose; Italy's emblem is the lily; Germany's emblem is the cornflower; the leek is the emblem of Wales, and the thistle of Scotland; the United States has no national flower.

THE LORDLY BUCKEYE RED

No variety of poultry in the last 5 years has created such a sensation as the Buckeye Red. It was originated by Mrs Frank Metcalf of Ohio, and has attracted wide attention because it breeds so true to type. It combines the good features of its parents, Indian Game Cochins and Plymouth Rocks. As a farm fowl it is noted for laying and for good size, and as a utility variety it is likely to be a strong rival of the other American varieties. The specimen shown was raised by W. G. Judson of Pennsylvania.

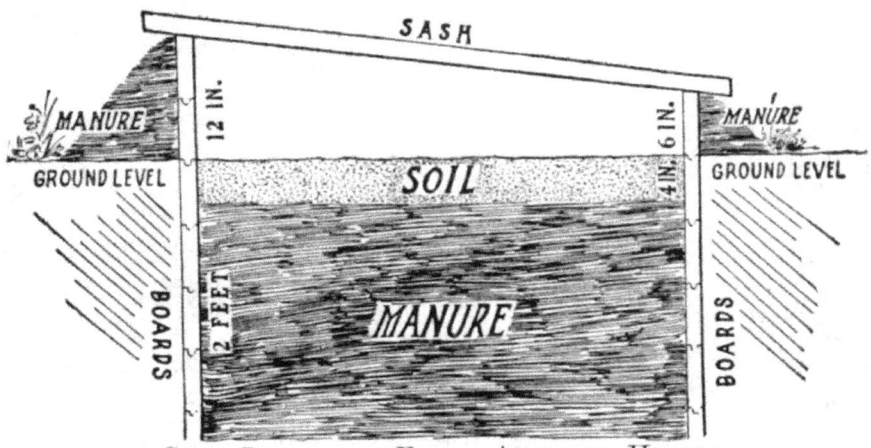

CROSS-SECTION OF EASILY ARRANGED HOTBED

INDEX

To the Man With the Reading Habit

If doing a thing better than the other fellow does it is worth your while—then it is worth your while to read high-class books, particularly those which will benefit you in your chosen field.

The successful farmer who really makes progress and gets ahead is the one who reads, studies and thinks for himself, and does not wait for others to do it for him. It is most essential that you form the habit of reading if you desire to keep abreast of the times.

Herewith we present a partial list of the newest and best books treating on every phase of rural life. All of these have been written by the most eminent authorities on their respective subjects, in plain language, and are quite easily understood; in many cases they have been especially prepared for self-instruction.

GENERAL

Farmers' Cyclopedia of Agriculture, E. V. Wilcox and C. B. Smith	$3.50—$4.50
Gardening for Profit, by Peter Henderson	1.50
Gardening for Pleasure, by Peter Henderson	1.50
Southern Gardeners' Practical Manual, by J. S. Newman	1.00
Money in the Garden, by P. T. Quinn	1.00
Market Gardening and Farm Notes, by Burnett Landreth	1.00
Gardening for Young and Old, by Joseph Harris	1.00
Play and Profit in My Garden, by E. P. Roe	1.00
Brill's Farm Gardening and Seed Growing, by Francis Brill	1.00
Truck Farming in the South, by A. Oemler	1.00
American Farm Book, by Richard L. Allen	2.00
Manual of Agriculture, by George B. Emerson and Charles L. Flint	1.00
Annals of Horticulture, L. H. Bailey	1.00
A B C of Agriculture	.50

FARM SCIENCE

How Crops Grow, by Prof. Samuel W. Johnson	$1.50
How Crops Feed, by Prof. Samuel W. Johnson	1.50
Chemistry of the Farm, by R. Warrington	1.00
Farm Machinery and Farm Motors, J. B. Davidson and L. W. Chase	net 2.00
First Lessons in Dairying, by Hubert E. Van Norman	net .50
First Principles of Soil Fertility, by Alfred Vivian	net 1.00
Soils, by Charles William Burkett	1.25
Rural School Agriculture, by Charles M. Davis	1.00
Practical Dairy Bacteriology, by Dr. H. W. Conn	1.25
Physical Properties of the Soils, by A. G. McCall	net .50
The Study of Breeds, by Thomas Shaw	1.50
Animal Breeding, by Thomas Shaw	1.50
Feeding Farm Animals, by Prof. Thomas Shaw	2.00
Plant Life on the Farm, by M. T. Masters	1.00
Soil Physics Laboratory Guide, by W. G. Stevenson and I. O. Schaub	.50
Agriculture Through the Laboratory and School Garden, by C. R. Jackson and L. S. Daugherty	net 1.50
Systematic Pomology, by F. A. Waugh	1.00
Manual of Corn Judging, by A. D. Shamel	.50
The Propagation of Plants, by Andrew S. Fuller	1.50
Soiling Crops and the Silo, by Thomas Shaw	1.50
Talks on Manures, by Joseph Harris	1.50
The Soil of the Farm, by John Scott and J. C. Morton	1.00
Bommer's Methods of Making Manures, by George Bommer	.25
How to Co-operate, by Herbert Myrick	1.00
Pedder's Land-measurer for Farmers	.50
Bookkeeping for Farmers, by T. Clark Atkeson	.25

IRRIGATION AND DRAINAGE

Draining for Profit and Draining for Health, by George E. Waring, Jr.	$1.00
Irrigation Farming, by Lucius M. Wilcox	2.00
Irrigation for the Farm, Garden and Orchard	1.00
Land Draining, by Manly Miles	1.00
Farm Drainage, by Judge French	1.00

HORSES AND THEIR CARE

Dadd's American Reformed Horse Book, by George H. Dadd	$2.00
Dadd's Modern Horse Doctor, by George H. Dadd	1.00
Herbert's Hints to Horse Keepers, by Henry William Herbert	1.50
Handbook of the Turf, by Samuel L. Boardman	1.00
Youatt and Spooner on the Horse, by Hon. Henry S. Randall	1.00
Farmers' Cyclopedia of Live Stock, by E. V. Wilcox and C. B. Smith	4.50—5.50
The Saddle Horse	1.00

The Horse: How to Buy and Sell, by Peter Howden...$ 1.00
The Family Horse, by George A. Martin...1.00
The Bridle Bits, by Col. J. C. Bettersby..1.00
How to Handle and Educate Vicious Horses, by Oscar R. Gleason..................................50
The Percheron Horse, by Col. M. C. Weld..50

CATTLE AND DAIRYING

The Business of Dairying, by C. B. Lane...net $1.75
Profitable Dairying, by C. L. Peck..75
Farmers' Cyclopedia of Live Stock, by E. V. Wilcox and C. B. Smith.........................4.50—5.50
Science and Practice of Cheese-Making, by Van Slyke and Publow.............................net 1.75
Clean Milk, by S. D. Belcher...1.00
Modern Methods of Testing Milk and Milk Products, by L. L. Van Slyke............................75
American Cattle Doctor, by George H. Dadd..2.00
The Dairyman's Manual, by Henry Stewart..1.50
Guenon's Treatise on Milch Cows, by Thomas J. Hand..1.00
Keeping One Cow...1.00
Key to Profitable Stock Feeding, by Herbert Myrick..25

SHEEP AND SWINE

Swine in America, by F. D. Coburn...net $2.50
Stewart's Shepherd's Manual, by Henry Stewart..1.00
The American Merino, by Stephen Powers...1.50
Coburn's Swine Husbandry, by F. D. Coburn..1.50
Harris on the Pig, by Joseph Harris..1.00
Farmer's Cyclopedia of Live Stock, by E. V. Wilcox and C. B. Smith.........................4.50—5.50
Farm Live Stock of Great Britain, by R. Wallace...net 5.00
Diseases of Swine, by Dr. R. A. Craig...75
Home Pork Making, by A. W. Fulton...50

POULTRY, BEES AND PETS

Making Poultry Pay, by Edwin C. Powell... $1.00
Profits in Poultry...1.00
The New Egg Farm, by H. H. Stoddard..1.00
Poultry Feeding and Fattening, by G. B. Fiske..50
Squabs for Profit, by William E. Rice and William E. Cox...50
Turkeys and How to Grow Them, by Herbert Myrick...1.00
Quimby's New Bee-keeping, by L. C. Root..1.00
American Bird Fancier, by D. J. Browne and Dr. Fuller Walker......................................50
Canary Birds..50

FARM CROPS

The Cereals in America, by Thomas F. Hunt.. $1.75
The Forage and Fiber Crops in America, by Thomas F. Hunt...1.75
The Book of Alfalfa, by F. D. Coburn...2.00
The Book of Wheat, by P. T. Dondlinger...net 2.00
The Book of Corn, by Herbert Myrick..1.50
Farm Grasses of the United States of America, by William Jasper Spillman........................1.00
Forage Crops Other Than Grasses, by Thomas Shaw...1.00
Soil and Crops of the Farm, by George E. Morrow and Thomas F. Hunt..............................1.00
Clovers and How to Grow Them, by Thomas Shaw...net 1.00
Tobacco Leaf, by J. N. Killebrew and Herbert Myrick...2.00
Tobacco Culture...25
The Hop, by Herbert Myrick...1.50
American Sugar Industry, by Herbert Myrick...1.50
Alfalfa, by F. D. Coburn..50
Silos, Ensilage and Silage, by Manly Miles...50
Hemp, by S. S. Boyce..50
Wheat Culture, by D. S. Curtis..50
Flax Culture..30
The Peanut Plant: Its Cultivation and Uses, by J. W. Jones.......................................50
Broom Corn and Brooms...50

FRUIT CROPS

American Fruit Culturist, by John J. Thomas...net $2.50
American Apple Orchard, by F. A. Waugh...net 1.00
Successful Fruit Culture, by Samuel T. Maynard..1.00
Plums and Plum Culture, by Prof. F. A. Waugh..1.50
Fruit Harvesting, Storing, Marketing, by F. A. Waugh..1.00
Dwarf Fruit Trees, by F. A. Waugh...50
The Fruit Garden, by P. Barry...1.50
Citrus Fruits and Their Culture, by H. Harold Hume...net 2.50
The Nut Culturist, by Andrew S. Fuller..1.50
American Grape Growing and Wine Making, by George Hussman.......................................1.50
Grape Culturist, by A. S. Fuller..1.50
Pear Culture for Profit, by P. T. Quinn...1.00
Quince Culture, by W. W. Meech..1.00
Peach Culture, by Hon. J. Alexander Fulton..1.00
Cranberry Culture, by Joseph J. White...1.00
Small Fruit Culturist, by Andrew S. Fuller..1.00
Field Notes on Apple Culture, by Prof. L. H. Bailey, Jr..75
Chorlton's Grape Grower's Guide, by William Chorlton...75
The Practical Fruit Grower, by S. T. Maynard...50
Cape Cod Cranberries, by James Webb..40
Strawberry Culturist, by Andrew S. Fuller..25

VEGETABLE CROPS

The Potato, by Samuel Fraser.. $0.75
Tomato Culture, by Will W. Tracy..50
Bean Culture, by Glenn C. Sevey...50
Celery Culture, by W. R. Beattie...50

The New Onion Culture, by T. Greiner.. $0.50
Onions; How to Raise Them Profitably.. .20
Cabbage, Cauliflower, Etc., by C. L. Allen.. .50
Asparagus, by F. M. Hexamer... .50
Sweet Potato Culture, by James Fitz.. .50
Mushrooms; How to Grow Them, by William Falconer...................................... 1.00
The New Rhubarb Culture, by J. E. Morse... .50

FLORICULTURE

Greenhouse Construction, by L. R. Taft.. $1.50
Greenhouse Management, by L. R. Taft.. 1.50
Practical Floriculture, by Peter Henderson... 1.50
Home Floriculture, by Eben E. Rexford.. 1.00
The Window Flower Garden, by Julius J. Heinrich...................................... .50
Your Plants, by James Sheehan.. .40
Bulbs and Tuberous-rooted Plants, by C. L. Allen..................................... 1.50
Parsons on the Rose, by Samuel B. Parsons.. 1.00
The Chrysanthemum, by A. Herrington.. .50

LANDSCAPE GARDENING AND FORESTRY

Landscape Gardening, by F. A. Waugh.. $0.50
Ornamental Gardening for Americans, by Elias A. Long................................. 1.50
Beautifying Country Homes, by Weidenmann... 10.00
Hedges, Windbreaks, Shelters and Live Fences, by E. P. Powell....................... .50
Forest Planting, by H. Nicholas Jarchow... 1.50
Practical Forestry, by Andrew S. Fuller.. 1.50

INSECTS AND PLANT DISEASES

Insects Injurious to Vegetables, by T. H. Chittenden................................ $1.50
Spraying Crops, by C. M. Weed.. .50
Fungi and Fungicides, by Prof. Clarence M. Weed..................................... 1.00
Insects and Insecticides, by Clarence M. Weed....................................... 1.50
Treat's Injurious Insects of the Farm and Garden, by Mrs. Mary Treat................ 1.50
Fumigation Methods, by Willis G. Johnson.. 1.00

BUILDINGS AND CONVENIENCES

Homes for Home Builders, by W. D. King.. $1.00
Modern House Plans for Everybody, by S. B. Reed..................................... 1.00
Cottage House, by S. B. Reed... 1.00
Barn Plans and Outbuildings.. 1.00
Poultry Architecture, by G. B. Fiske.. .50
Fences, Gates and Bridges.. .50
Farm Conveniences.. 1.00
Farm Appliances.. .50
Poultry Appliances and Handicraft, by G. B. Fiske................................... .50

SPECIAL SUBJECTS

The Ice Crop, by Theron L. Hiles.. $1.00
Cider Makers' Handbook, by J. M. Trowbridge... 1.00
Ginseng, by M. G. Kains.. .50
Silk Culture, by Mrs. C. E. Bamford... .30
Our Farm of Four Acres, and the Money We Made by It................................. .30
Ten Acres Enough, by I. P. Roberts.. 1.00
The Dogs of Great Britain, America and Other Countries.............................. 1.50
Murphy's American Game Bird Shooting, by John Mortimer Murphy....................... 1.50
Home Fishing and Home Waters, by Seth Green... .50
Hunter and Trapper, by Halsay Thrasher.. .50
Practical Taxidermy and Home Decoration, by Joseph G. Batty........................ 1.00
Hints on Rifle Practice with Military Arms, by C. E. Prescott...................... .25

MISCELLANEOUS

Cache la Poudre, by Herbert Myrick...$6.00—$1.50
The Promise of Life, by Herbert Myrick.. .50
A Swim for Life.. 1.00
The Bride's Primer... 1.50
The End of the World, by Edward Eggleston... 1.50
The Mystery of Metropolisville, by Edward Eggleston................................. 1.50
Left-Overs Made Palatable, by Isabel Gordon Curtis................................. 1.00
New Methods in Education, by J. Liberty Tadd.....................................3.00—2.00
The Hoosier Schoolmaster, by Edward Eggleston....................................... 1.25

Free to All

Our brief descriptive catalog (16 pages) will be sent free of charge to all applying for the same. Our New Large Illustrated Descriptive Catalog containing 116 pages, 6x9 inches, thoroughly indexed by titles and authors and containing detailed descriptions on rural and home topics, sent for 6 cents in stamps, which only covers mailing cost. It gives a line on the whole literature of agriculture.

We are prepared to supply any book published, at the retail price, and are always happy to correspond with our readers in reference to books, and cordially invite you to address us on any subject concerning any publication.

ORANGE JUDD COMPANY
Publishers and Importers

439-441 Lafayette St.   -   -   -   New York, N. Y.

Why Farmers Ought to Advertise—and How

The Farmer Finds It Pays to Advertise His Products Direct to the City Consumer and to His Brother Farmers

This is an age of advertising. It is the most powerful factor in modern business life, and while already some farmers have realized that they can make money by advertising, there are numbers of progressive farmers who have not advertised because they feel they don't know how to take the first step.

But it is really an easy matter to write an advertisement of your farm produce that will sell the goods for you.

How to Write an Advertisement

First of all, tell the truth. If the article you are advertising is not in first-class condition, don't say that it is, but rather say: " In fair condition."

Second, be brief, but to the point. Read over your adv and see if you could not say the same thing just as well in fewer words. Tell the essential things that a purchaser would be most likely to want to know. Tell about the *quality*.

Third, be sure to give names, dimensions and prices.

In other words, tell all the things *you* would want to know if you were an intending purchaser. If your article is of a well-known make or has a trade mark or a familiar name by which it is known, be sure to mention that.

Write Just as You Would Talk

to anyone you were trying to sell something to. Never mind the grammar, but say what you actually feel and your grammar will take care of itself, and when you have signed your full name and address you will be surprised to find what a readable adv you have written. Some people think careful grammar or big words necessary. Not at all. Just tell your story in a clear, straightforward way, so that everyone who reads it will know you mean exactly what you say, and be perfectly honest. If you misrepresent ever so slightly, your customer will feel he has been ill-used, and nine men out of ten would rather *give* you a dollar than feel they had been *done* out of ten cents. Honesty is not only the *best* policy—it's the *only* policy in advertising. Much better say of your advertised article, " It's in *fair* condition " or " It's been used a *little*," than say, " In

splendid condition," or "Never been used."

Display advertising is more profitable for many things than small classified advs. It does cost a little more, but your adv will stand right out and catch the eye the moment the page is turned. In display advertising you should have two or three words set in large type, so as to attract the attention. Usually the name of the article you are advertising should be the largest, and then you can add in a somewhat smaller type a short phrase —just five or six words—that *explains* a little more about your article or makes so strong an appeal to the reader that he will feel as though *he* must buy what *you* are advertising. There you have the foundation for a rattling good adv that will produce sales if the small text matter following gives, in the fewest possible words, *why* your article is different from others, *because* of the way it is made, the conditions under which it is produced, etc. Then give the dimensions, prices, etc.

You Can Tell the Space an Adv Will Occupy

by the following rules. For convenience, most newspapers have adopted what is termed the agate line standard, so they can readily tell the space occupied by any given adv, and the price is figured at so much per agate line.

This line is set in agate type

It contains seven words, and this is the average number of words contained in any agate line of a single newspaper column, although, of course, if the words are short ones, a line may contain nine or ten, while if the words are long, it may contain only five or six, but seven is the average.

It takes 14 of these lines, one below the other, to make an inch in depth, so if you want to write a 1-inch adv, you must limit yourself to 98 words, but of course you will want some of the words displayed in large type as headlines, so you will have to allow for the headlines by subtracting enough words to equal the space occupied by the display headlines. For instance, you want a quarter-inch headline. Now that is equal to four agate lines, so you will have to subtract 28 words from your 98 words, which will leave you but 70 words for your adv. Then if you want another line half as big, you will have to subtract 14 words more, which will leave 56 words, and then it is well to try and condense these 56 words into 40 words, so you can have a little white space, which will set off the display lines to better advantage.

You Can Figure the Cost of Display Advs

For display advs just decide how much space your adv is to occupy, seven lines to a half inch and 14 lines to a full inch. If you want your adv double column, just multiply by two, that is, 1-inch double column adv contains 2 inches of space, or 28 agate lines. Then multiply the number of agate lines in your adv by the price per line, which is 40 cents for the New England Homestead, 50 cents for the Orange Judd Farmer and 60 cents for the American Agriculturist. If you want an advertisement in all three of the papers, the combined rate is $1.25 an agate line.

We'll Help to Write Your Advs

if you will tell us just exactly what it is you want to sell, give us full particulars as to prices, etc.

Classified Advs are Money Makers, too

The small Want and For Sale advs in Orange Judd Company's classified adv departments, called the "Farmers' Exchange," "Our Real Estate Market" and "Our Help Bureau," are often as profitable as display advs—for advertising *some* things they are much *more* profitable.

If you want to reach your *brother farmers* and sell them surplus stock, some particular variety of seed, plants, vegetables or berry shrubs, that you have perfected; if you want to sell some of your property, implements or tools that you have outgrown.

Or, if you are in the market for live stock, seeds, plants, second-hand implements, etc; if you want to buy a few more acres, or a whole farm; or, if you are in want of farm help, or the good wife in want of a hired girl.

Everyone has Something to Advertise

at some time or other, if one stops to think. How many articles there are in woodsheds, lofts, lumber rooms and wagon sheds—implements, tools, vehicles, farm machinery and even furniture—that have not been used for months, perhaps, and which have been replaced by more modern and serviceable things, but whose days of usefulness are by no means over, and which would be eagerly bought by people who have to economize, but who perhaps have just the place for something that you can no longer find use for.

Readers Open Paper at Classified Advs

for the value of advertising by the farmer himself is more appreciated today than ever before, the up-to-date agriculturist naturally turning to the advertising columns of his farm paper when he wants live stock, seeds, implements or hired men. Then, when he has any surplus stock, produce, or tools that he has outgrown, but which some brother farmer might value, he inserts a small "FOR SALE" adv.

Low Rates for Classified Advs

To find the cost of a classified adv in the Farmers' Exchange, Help Bureau and Real Estate Market, you count the number of words, including your name and address, counting each initial and abbreviation as a word, and then multiply the number of words by 4 cents for the New England Homestead, 4 cents for the Orange Judd Farmer and 5 cents for the American Agriculturist.

Cash must accompany the order for both Display and Classified advs.

An Adv with Us Covers the United States

Of course, you know there are three editions of the American Agriculturist weeklies, the American Agriculturist proper, that covers the middle and southern Atlantic states, the Orange Judd Farmer, that covers all the states west of the Ohio and Mississippi rivers, and the New England Homestead, that covers the six New England states.

The maps on the next page show the number of paid subscribers each of these three journals has, while underneath the maps are the advertising rates for both the display columns and the classified advertising.

We Guarantee all Our Advertisers

so before we accept an adv from anyone, we require them to give us the full particulars regarding their proposition, which we hold as strictly confidential, and we reserve the right of rejecting any adv for reasons that may appear satisfactory to ourselves. We do not accept any liquor advs, or any objectionable advs of any kind, medical or otherwise. Write to us freely, ask all the questions you want. You will find an advertisement will "move the goods." Address your letters to the Advertising Manager at the office nearest you.

Orange Judd Company, *Publishers*

Headquarters
439 Lafayette Street, New York, N. Y.

Eastern Office
1-57 Worthington Street
Springfield, Mass.

Western Office
1448 Marquette Building
Chicago, Ill.